THE WORKS

OF

ARCHIMEDES.

London: C. J. CLAY AND SONS,
CAMBRIDGE UNIVERSITY PRESS WAREHOUSE.
AVE MARIA LANE.
Glasgow: 263, ARGYLE STREET.

Leipzig: F. A. BROCKHAUS.
New York: THE MACMILLAN COMPANY.

THE WORKS

of

ARCHIMEDES

.

EDITED IN MODERN NOTATION

WITH INTRODUCTORY CHAPTERS

BY

T. L. HEATH, Sc.D.,

SOMETIME FELLOW OF TRINITY COLLEGE, CAMBRIDGE.

CAMBRIDGE:
AT THE UNIVERSITY PRESS.
1897

PREFACE.

THIS book is intended to form a companion volume to my edition of the treatise of Apollonius on Conic Sections lately published. If it was worth while to attempt to make the work of "the great geometer" accessible to the mathematician of to-day who might not be able, in consequence of its length and of its form, either to read it in the original Greek or in a Latin translation, or, having read it, to master it and grasp the whole scheme of the treatise, I feel that I owe even less of an apology for offering to the public a reproduction, on the same lines, of the extant works of perhaps the greatest mathematical genius that the world has ever seen.

Michel Chasles has drawn an instructive distinction between the predominant features of the geometry of Archimedes and of the geometry which we find so highly developed in Apollonius. Their works may be regarded, says Chasles, as the origin and basis of two great inquiries which seem to share between them the domain of geometry. Apollonius is concerned with the *Geometry of Forms and Situations*, while in Archimedes we find the *Geometry of Measurements* dealing with the quadrature of curvilinear plane figures and with the quadrature and cubature of curved surfaces, investigations which "gave birth to the calculus of the infinite conceived and brought to perfection successively by Kepler, Cavalieri, Fermat, Leibniz, and Newton." But whether Archimedes is viewed as the man who, with the limited means at his disposal, nevertheless succeeded in performing what are really *integrations* for the purpose of finding the area of a parabolic segment and a

spiral, the surface and volume of a sphere and a segment
of a sphere, and the volume of any segments of the solids
of revolution of the second degree, whether he is seen finding
the centre of gravity of a parabolic segment, calculating
arithmetical approximations to the value of π, inventing a
system for expressing in words any number up to that which
we should write down with 1 followed by 80,000 billion
ciphers, or inventing the whole science of hydrostatics and at
the same time carrying it so far as to give a most complete
investigation of the positions of rest and stability of a right
segment of a paraboloid of revolution floating in a fluid, the
intelligent reader cannot fail to be struck by the remarkable
range of subjects and the mastery of treatment. And if these
are such as to create genuine enthusiasm in the student of
Archimedes, the style and method are no less irresistibly
attractive. One feature which will probably most impress the
mathematician accustomed to the rapidity and directness secured
by the generality of modern methods is the *deliberation* with
which Archimedes approaches the solution of any one of his
main problems. Yet this very characteristic, with its incidental
effects, is calculated to excite the more admiration because the
method suggests the tactics of some great strategist who
foresees everything, eliminates everything not immediately
conducive to the execution of his plan, masters every position
in its order, and then suddenly (when the very elaboration of
the scheme has almost obscured, in the mind of the spectator,
its ultimate object) strikes the final blow. Thus we read in
Archimedes proposition after proposition the bearing of which is
not immediately obvious but which we find infallibly used later
on; and we are led on by such easy stages that the difficulty of
the original problem, as presented at the outset, is scarcely
appreciated. As Plutarch says, "it is not possible to find in
geometry more difficult and troublesome questions, or more
simple and lucid explanations." But it is decidedly a rhetorical
exaggeration when Plutarch goes on to say that we are deceived

by the easiness of the successive steps into the belief that anyone could have discovered them for himself. On the contrary, the studied simplicity and the perfect finish of the treatises involve at the same time an element of mystery. Though each step depends upon the preceding ones, we are left in the dark as to how they were suggested to Archimedes. There is, in fact, much truth in a remark of Wallis to the effect that he seems "as it were of set purpose to have covered up the traces of his investigation as if he had grudged posterity the secret of his method of inquiry while he wished to extort from them assent to his results." Wallis adds with equal reason that not only Archimedes but nearly all the ancients so hid away from posterity their method of Analysis (though it is certain that they had one) that more modern mathematicians found it easier to invent a new Analysis than to seek out the old. This is no doubt the reason why Archimedes and other Greek geometers have received so little attention during the present century and why Archimedes is for the most part only vaguely remembered as the inventor of a screw, while even mathematicians scarcely know him except as the discoverer of the principle in hydrostatics which bears his name. It is only of recent years that we have had a satisfactory edition of the Greek text, that of Heiberg brought out in 1880–1, and I know of no complete translation since the German one of Nizze, published in 1824, which is now out of print and so rare that I had some difficulty in procuring a copy.

The plan of this work is then the same as that which I followed in editing the *Conics* of Apollonius. In this case, however, there has been less need as well as less opportunity for compression, and it has been possible to retain the numbering of the propositions and to enunciate them in a manner more nearly approaching the original without thereby making the enunciations obscure. Moreover, the subject matter is not so complicated as to necessitate absolute uniformity in the notation used (which is the only means whereby Apollonius can be made

even tolerably readable), though I have tried to secure as much
uniformity as was fairly possible. My main object has been to
present a perfectly faithful reproduction of the treatises as they
have come down to us, neither adding anything nor leaving out
anything essential or important. The notes are for the most
part intended to throw light on particular points in the text or
to supply proofs of propositions assumed by Archimedes as
known; sometimes I have thought it right to insert within
square brackets after certain propositions, and in the same type,
notes designed to bring out the exact significance of those
propositions, in cases where to place such notes in the Introduction or at the bottom of the page might lead to their being
overlooked.

Much of the Introduction is, as will be seen, historical; the
rest is devoted partly to giving a more general view of certain
methods employed by Archimedes and of their mathematical
significance than would be possible in notes to separate propositions, and partly to the discussion of certain questions arising
out of the subject matter upon which we have no positive
historical data to guide us. In these latter cases, where it is
necessary to put forward hypotheses for the purpose of explaining
obscure points, I have been careful to call attention to their
speculative character, though I have given the historical evidence
where such can be quoted in support of a particular hypothesis,
my object being to place side by side the authentic information
which we possess and the inferences which have been or may
be drawn from it, in order that the reader may be in a position
to judge for himself how far he can accept the latter as probable.
Perhaps I may be thought to owe an apology for the length of
one chapter on the so-called νεύσεις, or *inclinationes*, which goes
somewhat beyond what is necessary for the elucidation of
Archimedes; but the subject is interesting, and I thought it
well to make my account of it as complete as possible in
order to round off, as it were, my studies in Apollonius and
Archimedes.

I have had one disappointment in preparing this book for the press. I was particularly anxious to place on or opposite the title-page a portrait of Archimedes, and I was encouraged in this idea by the fact that the title-page of Torelli's edition bears a representation in medallion form on which are endorsed the words *Archimedis effigies marmorea in veteri anaglypho Romae asservato*. Caution was however suggested when I found two more portraits wholly unlike this but still claiming to represent Archimedes, one of them appearing at the beginning of Peyrard's French translation of 1807, and the other in Gronovius' *Thesaurus Graecarum Antiquitatum*; and I thought it well to inquire further into the matter. I am now informed by Dr A. S. Murray of the British Museum that there does not appear to be any authority for any one of the three, and that writers on iconography apparently do not recognise an Archimedes among existing portraits. I was, therefore, reluctantly obliged to give up my idea.

The proof sheets have, as on the former occasion, been read over by my brother, Dr R. S. Heath, Principal of Mason College, Birmingham; and I desire to take this opportunity of thanking him for undertaking what might well have seemed, to any one less genuinely interested in Greek geometry, a thankless task.

T. L. HEATH.

March, 1897.

LIST OF THE PRINCIPAL WORKS CONSULTED.

JOSEPH TORELLI, *Archimedis quae supersunt omnia cum Eutocii Ascalonitae commentariis.* (Oxford, 1792.)

ERNST NIZZE, *Archimedes von Syrakus vorhandene Werke aus dem griechischen übersetzt und mit erläuternden und kritischen Anmerkungen begleitet.* (Stralsund, 1824.)

J. L. HEIBERG, *Archimedis opera omnia cum commentariis Eutocii.* (Leipzig, 1880-1.)

J. L. HEIBERG, *Quaestiones Archimedeae.* (Copenhagen, 1879.)

F. HULTSCH, Article *Archimedes* in Pauly-Wissowa's *Real-Encyclopädie der classischen Altertumswissenschaften.* (Edition of 1895, II. 1, pp. 507-539.)

C. A. BRETSCHNEIDER, *Die Geometrie und die Geometer vor Euklides.* (Leipzig, 1870.)

M. CANTOR, *Vorlesungen über Geschichte der Mathematik*, Band I, zweite Auflage. (Leipzig, 1894.)

G. FRIEDLEIN, *Procli Diadochi in primum Euclidis elementorum librum commentarii.* (Leipzig, 1873.)

JAMES GOW, *A short history of Greek Mathematics.* (Cambridge, 1884.)

SIEGMUND GÜNTHER, *Abriss der Geschichte der Mathematik und der Naturwissenschaften im Altertum* in Iwan von Müller's *Handbuch der klassischen Altertumswissenschaft*, V. 1.

HERMANN HANKEL, *Zur Geschichte der Mathematik in Alterthum und Mittelalter.* (Leipzig, 1874.)

J. L. HEIBERG, *Litterargeschichtliche Studien über Euklid.* (Leipzig, 1882.)

J. L. HEIBERG, *Euclidis elementa.* (Leipzig, 1883-8.)

F. HULTSCH, Article *Arithmetica* in Pauly-Wissowa's *Real-Encyclopädie*, II. 1, pp. 1066-1116.

F. Hultsch, *Heronis Alexandrini geometricorum et stereometricorum reliquiae.* (Berlin, 1864.)

F. Hultsch, *Pappi Alexandrini collectionis quae supersunt.* (Berlin, 1876–8.)

Gino Loria, *Il periodo aureo della geometria greca.* (Modena, 1895.)

Maximilien Marie, *Histoire des sciences mathématiques et physiques,* Tome I. (Paris, 1883.)

J. H. T. Müller, *Beiträge zur Terminologie der griechischen Mathematiker.* (Leipzig, 1860.)

G. H. F. Nesselmann, *Die Algebra der Griechen.* (Berlin, 1842.)

F. Susemihl, *Geschichte der griechischen Litteratur in der Alexandrinerzeit,* Band I. (Leipzig, 1891.)

P. Tannery, *La Géométrie grecque,* Première partie, *Histoire générale de la Géométrie élémentaire.* (Paris, 1887.)

H. G. Zeuthen, *Die Lehre von den Kegelschnitten im Altertum.* (Copenhagen, 1886.)

H. G. Zeuthen, *Geschichte der Mathematik im Altertum und Mittelalter.* (Copenhagen, 1896.)

CONTENTS.

INTRODUCTION.

THE WORKS OF ARCHIMEDES.

INTRODUCTION.

CHAPTER I.

ARCHIMEDES.

A LIFE of Archimedes was written by one Heracleides*, but this biography has not survived, and such particulars as are known have to be collected from many various sources†. According to Tzetzes‡ he died at the age of 75, and, as he perished in the sack of Syracuse (B.C. 212), it follows that he was probably born about 287 B.C. He was the son of Pheidias the astronomer§, and was on intimate terms with, if not related to, king Hieron and his

---

* Eutocius mentions this work in his commentary on Archimedes' *Measurement of the circle*, ὡς φησιν Ἡρακλείδης ἐν τῷ Ἀρχιμήδους βίῳ. He alludes to it again in his commentary on Apollonius' *Conics* (ed. Heiberg, Vol. II. p. 168), where, however, the name is wrongly given as Ἡράκλειος. This Heracleides is perhaps the same as the Heracleides mentioned by Archimedes himself in the preface to his book *On Spirals*.

† An exhaustive collection of the materials is given in Heiberg's *Quaestiones Archimedeae* (1879). The preface to Torelli's edition also gives the main points, and the same work (pp. 363—370) quotes at length most of the original references to the mechanical inventions of Archimedes. Further, the article *Archimedes* (by Hultsch) in Pauly-Wissowa's *Real-Encyclopädie der classischen Altertumswissenschaften* gives an entirely admirable summary of all the available information. See also Susemihl's *Geschichte der griechischen Litteratur in der Alexandrinerzeit*, I. pp. 723—733.

‡ Tzetzes, *Chiliad.*, II. 35, 105.

§ Pheidias is mentioned in the *Sand-reckoner* of Archimedes, τῶν προτέρων ἀστρολόγων Εὐδόξου...Φειδία δὲ τοῦ ἁμοῦ πατρὸς (the last words being the correction of Blass for τοῦ Ἀκούπατρος, the reading of the text). Cf. Schol. Clark. in Gregor. Nazianz. Or. 34, p. 355 a Morel. Φειδίας τὸ μὲν γένος ἦν Συρακόσιος ἀστρολόγος ὁ Ἀρχιμήδους πατήρ.

son Gelon. It appears from a passage of Diodorus* that he spent a considerable time at Alexandria, where it may be inferred that he studied with the successors of Euclid. It may have been at Alexandria that he made the acquaintance of Conon of Samos (for whom he had the highest regard both as a mathematician and as a personal friend) and of Eratosthenes. To the former he was in the habit of communicating his discoveries before their publication, and it is to the latter that the famous Cattle-problem purports to have been sent. Another friend, to whom he dedicated several of his works, was Dositheus of Pelusium, a pupil of Conon, presumably at Alexandria though at a date subsequent to Archimedes' sojourn there.

After his return to Syracuse he lived a life entirely devoted to mathematical research. Incidentally he made himself famous by a variety of ingenious mechanical inventions. These things were however merely the "diversions of geometry at play †," and he attached no importance to them. In the words of Plutarch, "he possessed so high a spirit, so profound a soul, and such treasures of scientific knowledge that, though these inventions had obtained for him the renown of more than human sagacity, he yet would not deign to leave behind him any written work on such subjects, but, regarding as ignoble and sordid the business of mechanics and every sort of art which is directed to use and profit, he placed his whole ambition in those speculations in whose beauty and subtlety there is no admixture of the common needs of life ‡." In fact he wrote only one such mechanical book, *On Sphere-making*§, to which allusion will be made later.

Some of his mechanical inventions were used with great effect against the Romans during the siege of Syracuse. Thus he contrived

* Diodorus v. 37, 3, οὒς [τοὺς κοχλίας] Ἀρχιμήδης ὁ Συρακόσιος εὖρεν, ὅτε παρέβαλεν εἰς Αἴγυπτον.

† Plutarch, *Marcellus*, 14.

‡ *ibid.* 17.

§ Pappus VIII. p. 1026 (ed. Hultsch). Κάρπος δέ πού φησιν ὁ Ἀντιοχεὺς Ἀρχιμήδη τὸν Συρακόσιον ἕν μόνον βιβλίον συντεταχέναι μηχανικὸν τὸ κατὰ τὴν σφαιροποιίαν, τῶν δὲ ἄλλων οὐδὲν ἠξιωκέναι συντάξαι. καίτοι παρὰ τοῖς πολλοῖς ἐπὶ μηχανικῇ δοξασθεὶς καὶ μεγαλοφυής τις γενόμενος ὁ θαυμαστὸς ἐκεῖνος, ὥστε διαμεῖναι παρὰ πᾶσιν ἀνθρώποις ὑπερβαλλόντως ὑμνούμενος, τῶν τε προηγουμένων γεωμετρικῆς καὶ ἀριθμητικῆς ἐχομένων θεωρίας τὰ βραχύτατα δοκοῦντα εἶναι σπουδαίως συνέγραφεν· ὃς φαίνεται τὰς εἰρημένας ἐπιστήμας οὕτως ἀγαπήσας ὡς μηδὲν ἔξωθεν ὑπομένειν αὐταῖς ἐπεισάγειν.

catapults so ingeniously constructed as to be equally serviceable at long or short ranges, machines for discharging showers of missiles through holes made in the walls, and others consisting of long moveable poles projecting beyond the walls which either dropped heavy weights upon the enemy's ships, or grappled the prows by means of an iron hand or a beak like that of a crane, then lifted them into the air and let them fall again*. Marcellus is said to have derided his own engineers and artificers with the words, "Shall we not make an end of fighting against this geometrical Briareus who, sitting at ease by the sea, plays pitch and toss with our ships to our confusion, and by the multitude of missiles that he hurls at us outdoes the hundred-handed giants of mythology?†"; but the exhortation had no effect, the Romans being in such abject terror that "if they did but see a piece of rope or wood projecting above the wall, they would cry 'there it is again,' declaring that Archimedes was setting some engine in motion against them, and would turn their backs and run away, insomuch that Marcellus desisted from all conflicts and assaults, putting all his hope in a long siege‡."

If we are rightly informed, Archimedes died, as he had lived, absorbed in mathematical contemplation. The accounts of the exact circumstances of his death differ in some details. Thus Livy says simply that, amid the scenes of confusion that followed the capture of Syracuse, he was found intent on some figures which he had drawn in the dust, and was killed by a soldier who did not know who he was§. Plutarch gives more than one version in the following passage. "Marcellus was most of all afflicted at the death of Archimedes; for, as fate would have it, he was intent on working out some problem with a diagram and, having fixed his mind and his eyes alike on his investigation, he never noticed the incursion of the Romans nor the capture of the city. And when a soldier came up to him suddenly and bade him follow to

* Polybius, *Hist.* viii. 7—8 ; Livy xxiv. 34; Plutarch, *Marcellus*, 15—17.

† Plutarch, *Marcellus*, 17.

‡ *ibid.*

§ Livy xxv. 31. Cum multa irae, multa anaritiae foeda exempla ederentur, Archimedem memoriae proditum est in tanto tumultu, quantum pauor captae urbis in discursu diripientium militum ciere poterat, intentum formis, quas in puluere descripserat, ab ignaro milite quis esset interfectum ; aegre id Marcellum tulisse sepulturaeque curam habitam, et propinquis etiam inquisitis honori praesidioque nomen ac memoriam eius fuisse.

Marcellus, he refused to do so until he had worked out his problem to a demonstration; whereat the soldier was so enraged that he drew his sword and slew him. Others say that the Roman ran up to him with a drawn sword offering to kill him; and, when Archimedes saw him, he begged him earnestly to wait a short time in order that he might not leave his problem incomplete and unsolved, but the other took no notice and killed him. Again there is a third account to the effect that, as he was carrying to Marcellus some of his mathematical instruments, sundials, spheres, and angles adjusted to the apparent size of the sun to the sight, some soldiers met him and, being under the impression that he carried gold in the vessel, slew him*." The most picturesque version of the story is perhaps that which represents him as saying to a Roman soldier who came too close, "Stand away, fellow, from my diagram," whereat the man was so enraged that he killed him†. The addition made to this story by Zonaras, representing him as saying παρὰ κεφαλὰν καὶ μὴ παρὰ γραμμάν, while it no doubt recalls the second version given by Plutarch, is perhaps the most far-fetched of the touches put to the picture by later hands.

Archimedes is said to have requested his friends and relatives to place upon his tomb a representation of a cylinder circumscribing a sphere within it, together with an inscription giving the ratio which the cylinder bears to the sphere‡; from which we may infer that he himself regarded the discovery of this ratio [*On the Sphere and Cylinder*, I. 33, 34] as his greatest achievement. Cicero, when quaestor in Sicily, found the tomb in a neglected state and restored it§.

Beyond the above particulars of the life of Archimedes, we have nothing left except a number of stories, which, though perhaps not literally accurate, yet help us to a conception of the personality of the most original mathematician of antiquity which we would not willingly have altered. Thus, in illustration of his entire preoccupation by his abstract studies, we are told that he would forget all about his food and such necessities of life, and would be drawing geometrical figures in the ashes of the fire, or, when

---

* Plutarch, *Marcellus*, 19.
† Tzetzes, *Chil.* II. 35, 135; Zonaras IX. 5.
‡ Plutarch, *Marcellus*, 17 *ad fin.*
§ Cicero, *Tusc.* v. 64 sq.

anointing himself, in the oil on his body*. Of the same kind is
the well-known story that, when he discovered in a bath the
solution of the question referred to him by Hieron as to whether
a certain crown supposed to have been made of gold did not in
reality contain a certain proportion of silver, he ran naked through
the street to his home shouting εὕρηκα, εὕρηκα†.

According to Pappus‡ it was in connexion with his discovery
of the solution of the problem *To move a given weight by a given
force* that Archimedes uttered the famous saying, "Give me a
place to stand on, and I can move the earth (δός μοι ποῦ στῶ καὶ
κινῶ τὴν γῆν)." Plutarch represents him as declaring to Hieron
that any given weight could be moved by a given force, and
boasting, in reliance on the cogency of his demonstration, that, if
he were given another earth, he would cross over to it and move
this one. "And when Hieron was struck with amazement and asked
him to reduce the problem to practice and to give an illustration
of some great weight moved by a small force, he fixed upon a ship
of burden with three masts from the king's arsenal which had
only been drawn up with great labour and many men; and loading
her with many passengers and a full freight, sitting himself the
while far off, with no great endeavour but only holding the end
of a compound pulley (πολύσπαστος) quietly in his hand and pulling
at it, he drew the ship along smoothly and safely as if she were
moving through the sea§." According to Proclus the ship was one
which Hieron had had made to send to king Ptolemy, and, when all
the Syracusans with their combined strength were unable to launch
it, Archimedes contrived a mechanical device which enabled Hieron
to move it by himself, insomuch that the latter declared that
"from that day forth Archimedes was to be believed in everything that he might say‖." While however it is thus established
that Archimedes invented some mechanical contrivance for moving
a large ship and thus gave a practical illustration of his thesis,
it is not certain whether the machine used was simply a compound

---

* Plutarch, *Marcellus*, 17.
† Vitruvius, *Architect.* ix. 3. For an explanation of the manner in which
Archimedes probably solved this problem, see the note following *On floating
bodies*, i. 7 (p. 259 sq.).
‡ Pappus viii. p. 1060.
§ Plutarch, *Marcellus*, 14.
‖ Proclus, *Comm. on Eucl.* i., p. 63 (ed. Friedlein).

pulley (πολύσπαστος) as stated by Plutarch; for Athenaeus*, in describing the same incident, says that a *helix* was used. This term must be supposed to refer to a machine similar to the κοχλίας described by Pappus, in which a cog-wheel with oblique teeth moves on a cylindrical helix turned by a handle†. Pappus, however, describes it in connexion with the βαρουλκός of Heron, and, while he distinctly refers to Heron as his authority, he gives no hint that Archimedes invented either the βαρουλκός or the particular κοχλίας; on the other hand, the πολύσπαστος is mentioned by Galen ‡, and the τρίσπαστος (triple pulley) by Oribasius §, as one of the inventions of Archimedes, the τρίσπαστος being so called either from its having three wheels (Vitruvius) or three ropes (Oribasius). Nevertheless, it may well be that though the ship could easily be kept in motion, when once started, by the τρίσπαστος or πολύσπαστος, Archimedes was obliged to use an appliance similar to the κοχλίας to give the first impulse.

The name of yet another instrument appears in connexion with the phrase about moving the earth. Tzetzes' version is, "Give me a place to stand on (πᾶ βῶ), and I will move the whole earth with a χαριστίων ‖"; but, as in another passage¶ he uses the word τρίσπαστος, it may be assumed that the two words represented one and the same thing**.

It will be convenient to mention in this place the other mechanical inventions of Archimedes. The best known is the

---

* Athenaeus v. 207 a–b, κατασκευάσας γὰρ ἕλικα τὸ τηλικοῦτον σκάφος εἰς τὴν θάλασσαν κατήγαγε· πρῶτος δ' Ἀρχιμήδης εὗρε τὴν τῆς ἕλικος κατασκευήν. To the same effect is the statement of Eustathius *ad Il.* III. p. 114 (ed. Stallb.) λέγεται δὲ ἕλιξ καί τι μηχανῆς εἶδος, ὃ πρῶτος εὑρὼν ὁ Ἀρχιμήδης εὐδοκίμησέ, φασι, δι' αὐτοῦ.

† Pappus VIII. pp. 1066, 1108 sq.

‡ Galen, *in Hippocr. De artic.*, IV. 47 (=XVIII. p. 747, ed. Kühn).

§ Oribasius, *Coll. med.*, XLIX. 22 (IV. p. 407, ed. Bussemaker), Ἀπελλίδους ἢ Ἀρχιμήδους τρίσπαστον, described in the same passage as having been invented πρὸς τὰς τῶν πλοίων καθολκάς.

‖ Tzetzes, *Chil.* II. 130.

¶ *Ibid.*, III. 61, ὁ γῆν ἀνασπῶν μηχανῇ τῇ τρισπάστῳ βοῶν· ὅπα βῶ καὶ σαλεύσω τὴν χθόνα.

** Heiberg compares Simplicius, *Comm. in Aristot. Phys.* (ed. Diels, p. 1110, l. 2), ταύτῃ δὲ τῇ ἀναλογίᾳ τοῦ κινοῦντος καὶ τοῦ κινουμένου καὶ τοῦ διαστήματος τὸ σταθμιστικὸν ὄργανον τὸν καλούμενον χαριστίωνα συστήσας ὁ Ἀρχιμήδης ὡς μέχρι παντὸς τῆς ἀναλογίας προχωρούσης ἐκόμπασεν ἐκεῖνο τὸ πᾶ βῶ καὶ κινῶ τὰν γᾶν.

water-screw* (also called κοχλίας) which was apparently invented by him in Egypt, for the purpose of irrigating fields. It was also used for pumping water out of mines or from the hold of ships.

Another invention was that of a sphere constructed so as to imitate the motions of the sun, the moon, and the five planets in the heavens. Cicero actually saw this contrivance and gives a description of it†, stating that it represented the periods of the moon and the apparent motion of the sun with such accuracy that it would even (over a short period) show the eclipses of the sun and moon. Hultsch conjectures that it was moved by water‡. We know, as above stated, from Pappus that Archimedes wrote a book on the construction of such a sphere (περὶ σφαιροποιίας), and Pappus speaks in one place of "those who understand the making of spheres and produce a model of the heavens by means of the regular circular motion of water." In any case it is certain that Archimedes was much occupied with astronomy. Livy calls him "unicus spectator caeli siderumque." Hipparchus says§, "From these observations it is clear that the differences in the years are altogether small, but, as to the solstices, I almost think (οὐκ ἀπελπίζω) that both I and Archimedes have erred to the extent of a quarter of a day both in the observation and in the deduction therefrom." It appears therefore that Archimedes had considered the question of the length of the year, as Ammianus also states‖. Macrobius says that he discovered the distances of the planets¶. Archimedes himself describes in the *Sand-reckoner* the apparatus by which he measured the apparent diameter of the sun, or the angle subtended by it at the eye.

The story that he set the Roman ships on fire by an arrangement of burning-glasses or concave mirrors is not found in any

* Diodorus I. 34, v. 37; Vitruvius x. 16 (11); Philo III. p. 330 (ed. Pfeiffer); Strabo XVII. p. 807; Athenaeus v. 208 f.

† Cicero, *De rep.*, I. 21–22; *Tusc.*, I. 63; *De nat. deor.*, II. 88. Cf. Ovid, *Fasti*, VI. 277; Lactantius, *Instit.*, II. 5, 18; Martianus Capella, II. 212, VI. 583 sq.; Claudian, *Epigr.* 18; Sextus Empiricus, p. 416 (ed. Bekker).

‡ *Zeitschrift f. Math. u. Physik* (hist. litt. Abth.), XXII. (1877), 106 sq.

§ Ptolemy, σύνταξις, I. p. 153.

‖ Ammianus Marcell., XXVI. i. 8.

¶ Macrobius, *in Somn. Scip.*, II. 3.

authority earlier than Lucian*; and the so-called *loculus Archi-*
medius, which was a sort of puzzle made of 14 pieces of ivory of
different shapes cut out of a square, cannot be supposed to be his
invention, the explanation of the name being perhaps that it was
only a method of expressing that the puzzle was cleverly made,
in the same way as the πρόβλημα Ἀρχιμήδειον came to be simply
a proverbial expression for something very difficult†.

---

* The same story is told of Proclus in Zonaras xiv. 3. For the other
references on the subject see Heiberg's *Quaestiones Archimedeae*, pp. 39–41.

† Cf. also Tzetzes, *Chil.* xii. 270, τῶν Ἀρχιμήδους μηχανῶν χρείαν ἔχω.

CHAPTER II.

THE sources of the text and versions are very fully described
by Heiberg in the Prolegomena to Vol. III. of his edition of Archimedes, where the editor supplements and to some extent amends
what he had previously written on the same subject in his dis-
sertation entitled *Quaestiones Archimedeae* (1879). It will therefore suffice here to state briefly the main points of the discussion.

The MSS. of the best class all had a common origin in a MS.
which, so far as is known, is no longer extant. It is described
in one of the copies made from it (to be mentioned later and dating
from some time between A.D. 1499 and 1531) as 'most ancient'
(παλαιοτάτου), and all the evidence goes to show that it was written
as early as the 9th or 10th century. At one time it was in the
possession of George Valla, who taught at Venice between the
years 1486 and 1499; and many important inferences with regard
to its readings can be drawn from some translations of parts of
Archimedes and Eutocius made by Valla himself and published
in his book entitled *de expetendis et fugiendis rebus* (Venice, 1501).
It appears to have been carefully copied from an original belonging
to some one well versed in mathematics, and it contained figures
drawn for the most part with great care and accuracy, but there
was considerable confusion between the letters in the figures and
those in the text. This MS., after the death of Valla in 1499,
became the property of Albertus Pius Carpensis (Alberto Pio,
prince of Carpi). Part of his library passed through various hands
and ultimately reached the Vatican; but the fate of the Valla
MS. appears to have been different, for we hear of its being in
the possession of Cardinal Rodolphus Pius (Rodolfo Pio), a nephew
of Albertus, in 1544, after which it seems to have disappeared.

The three most important MSS. extant are:

F (= Codex Florentinus bibliothecae Laurentianae Mediceae plutei XXVIII. 4to.).

B (= Codex Parisinus 2360, olim Mediceus).

C (= Codex Parisinus 2361, Fonteblandensis).

Of these it is certain that B was copied from the Valla MS. This is proved by a note on the copy itself, which states that the archetype formerly belonged to George Valla and afterwards to Albertus Pius. From this it may also be inferred that B was written before the death of Albertus in 1531; for, if at the date of B the Valla MS. had passed to Rodolphus Pius, the name of the latter would presumably have been mentioned. The note referred to also gives a list of peculiar abbreviations used in the archetype, which list is of importance for the purpose of comparison with F and other MSS.

From a note on C it appears that that MS. was written by one Christophorus Auverus at Rome in 1544, at the expense of Georgius Armagniacus (Georges d'Armagnac), Bishop of Rodez, then on a mission from King Francis I. to Pope Paul III. Further, a certain Guilelmus Philander, in a letter to Francis I. published in an edition of Vitruvius (1552), mentions that he was allowed, by the kindness of Cardinal Rodolphus Pius, acting at the instance of Georgius Armagniacus, to see and make extracts from a volume of Archimedes which was destined to adorn the library founded by Francis at Fontainebleau. He adds that the volume had been the property of George Valla. We can therefore hardly doubt that C was the copy which Georgius Armagniacus had made in order to present it to the library at Fontainebleau.

Now F, B and C all contain the same works of Archimedes and Eutocius, and in the same order, viz. (1) two Books *de sphaera et cylindro*, (2) *de dimensione circuli*, (3) *de conoidibus*, (4) *de lineis spiralibus*, (5) *de planis aeque ponderantibus*, (6) *arenarius*, (7) *quadratura parabolae*, and the commentaries of Eutocius on (1) (2) and (5). At the end of the *quadratura parabolae* both F and B give the following lines:

εὐτυχοίης λέον γεώμετρα

πολλοὺς εἰς λυκάβαντας ἴοις πολὺ φίλτατε μούσαις.

F and C also contain *mensurae* from Heron and two fragments περὶ σταθμῶν and περὶ μέτρων, the order being the same in both and the contents only differing in the one respect that the last fragment περὶ μέτρων is slightly longer in F than in C.

A short preface to C states that the first page of the archetype was so rubbed and worn with age that not even the name of Archimedes could be read upon it, while there was no copy at Rome by means of which the defect could be made good, and further that the last page of Heron's *de mensuris* was similarly obliterated. Now in F the first page was apparently left blank at first and afterwards written in by a different hand with many gaps, while in B there are similar deficiencies and a note attached by the copyist is to the effect that the first page of the archetype was indistinct. In another place (p. 4 of Vol. III., ed. Heiberg) all three MSS. have the same lacuna, and the scribe of B notes that one whole page or even two are missing.

Now C could not have been copied from F because the last page of the fragment περὶ μέτρων is perfectly distinct in F; and, on the other hand, the archetype of F must have been illegible at the end because there is no word τέλος at the end of F, nor any other of the signs by which copyists usually marked the completion of their task. Again, Valla's translations show that his MS. had certain readings corresponding to correct readings in B and C instead of incorrect readings given by F. Hence F cannot have been Valla's MS. itself.

The positive evidence about F is as follows. Valla's translations, with the exception of the few readings just referred to, agree completely with the text of F. From a letter written at Venice in 1491 by Angelus Politianus (Angelo Poliziano) to Laurentius Mediceus (Lorenzo de' Medici), it appears that the former had found a MS. at Venice containing works by Archimedes and Heron and proposed to have it copied. As G. Valla then lived at Venice, the MS. can hardly have been any other but his, and no doubt F was actually copied from it in 1491 or soon after. Confirmatory evidence for this origin of F is found in the fact that the form of most of the letters in it is older than the 15th century, and the abbreviations etc., while they all savour of an ancient archetype, agree marvellously with the description which the note to B above referred to gives of the abbreviations used in Valla's MS. Further, it is remarkable that the corrupt passage corresponding to the illegible first page of the archetype just takes up one page of F, no more and no less.

The natural inference from all the evidence is that F, B and C all had their origin in the Valla MS.; and of the three F is the most trustworthy. For (1) the extreme care with which the copyist of F kept to the original is illustrated by a number of mistakes in it which correspond to Valla's readings but are corrected in B and C, and (2) there is no doubt that the writer of B was somewhat of an expert and made many alterations on his own authority, not always with success.

Passing to other MSS., we know that Pope Nicholas V. had a MS. of Archimedes which he caused to be translated into Latin. The translation was made by Jacobus Cremonensis (Jacopo Cassiani*), and one copy of this was written out by Joannes Regiomontanus (Johann Müller of Königsberg, near Hassfurt, in Franconia), about 1461, who not only noted in the margin a number of corrections of the Latin but added also in many places Greek readings from another MS. This copy by Regiomontanus is preserved at Nürnberg and was the source of the Latin translation given in the *editio princeps* of Thomas Gechauff Venatorius (Basel, 1544); it is called N^b by Heiberg. (Another copy of the same translation is alluded to by Regiomontanus, and this is doubtless the Latin MS. 327 of 15th c. still extant at Venice.) From the fact that the translation of Jacobus Cremonensis has the same lacuna as that in F, B and C above referred to (Vol. III., ed. Heiberg, p. 4), it seems clear that the translator had before him either the Valla MS. itself or (more likely) a copy of it, though the order of the books in the translation differs in one respect from that in our MSS., viz. that the *arenarius* comes after instead of before the *quadratura parabolae*.

It is probable that the Greek MS. used by Regiomontanus was V (= Codex Venetus Marcianus cccv. of the 15th c.), which is still extant and contains the same books of Archimedes and Eutocius with the same fragment of Heron as F has, and in the same order. If the above conclusion that F dates from 1491 or thereabouts is correct, then, as V belonged to Cardinal Bessarione who died in 1472, it cannot have been copied from F, and the simplest way of accounting for its similarity to F is to suppose that it too was derived from Valla's MS.

---

* Tiraboschi, *Storia della Letteratura Italiana*, Vol. VI. Pt. 1 (p. 358 of the edition of 1807). Cantor (*Vorlesungen üb. Gesch. d. Math.*, II. p. 192) gives the full name and title as Jacopo da S. Cassiano Cremonese canonico regolare.

Regiomontanus mentions, in a note inserted later than the rest and in different ink, two other Greek MSS., one of which he calls "exemplar vetus apud magistrum Paulum." Probably the monk Paulus (Albertini) of Venice is here meant, whose date was 1430 to 1475; and it is possible that the "exemplar vetus" is the MS. of Valla.

The two other inferior MSS., viz. A (= Codex Parisinus 2359, olim Mediceus) and D (= Cod. Parisinus 2362, Fonteblandensis), owe their origin to V.

It is next necessary to consider the probabilities as to the MSS. used by Nicolas Tartaglia for his Latin translation of certain of the works of Archimedes. The portion of this translation published at Venice in 1543 contained the books *de centris gravium vel de aequerepentibus I–II, tetragonismus [parabolae], dimensio circuli* and *de insidentibus aquae I*; the rest, consisting of Book II *de insidentibus aquae*, was published with Book I of the same treatise, after Tartaglia's death in 1557, by Troianus Curtius (Venice, 1565). Now the last-named treatise is not extant in any Greek MS. and, as Tartaglia adds it, without any hint of a separate origin, to the rest of the books which he says he took from a mutilated and almost illegible Greek MS., it might easily be inferred that the Greek MS. contained that treatise also. But it is established, by a letter written by Tartaglia himself eight years later (1551) that he then had no Greek text of the Books *de insidentibus aquae*, and it would be strange if it had disappeared in so short a time without leaving any trace. Further, Commandinus in the preface to his edition of the same treatise (Bologna, 1565) shows that he had never heard of a Greek text of it. Hence it is most natural to suppose that it reached Tartaglia from some other source and in the Latin translation only*.

The fact that Tartaglia speaks of the old MS. which he used as "fracti et qui vix legi poterant libri," at practically the same time as the writer of the preface to C was giving a similar description of Valla's MS., makes it probable that the two were

---

* The Greek fragment of Book I., περὶ τῶν ὕδατι ἐφισταμένων ἢ περὶ τῶν ὀχουμένων, edited by A. Mai from two Vatican MSS. (*Classici auct.* I. p. 426–30; Vol. II. of Heiberg's edition, pp. 356–8), seems to be of doubtful authenticity. Except for the first proposition, it contains enunciations only and no proofs. Heiberg is inclined to think that it represents an attempt at retranslation into Greek made by some mediaeval scholar, and he compares the similar attempt made by Rivault.

identical; and this probability is confirmed by a considerable agreement between the mistakes in Tartaglia and in Valla's versions.

But in the case of the *quadratura parabolae* and the *dimensio circuli* Tartaglia adopted bodily, without alluding in any way to the source of it, another Latin translation published by Lucas Gauricus "Iuphanensis ex regno Neapolitano" (Luca Gaurico of Gifuni) in 1503, and he copied it so faithfully as to reproduce most obvious errors and perverse punctuation, only filling up a few gaps and changing some figures and letters. This translation by Gauricus is seen, by means of a comparison with Valla's readings and with the translation of Jacobus Cremonensis, to have been made from the same MS. as the latter, viz. that of Pope Nicolas V.

Even where Tartaglia used the Valla MS. he does not seem to have taken very great pains to decipher it when it was not easily legible—it may be that he was unused to deciphering MSS.—and in such cases he did not hesitate to draw from other sources. In one place (*de planor. equilib.* II. 9) he actually gives as the Archimedean proof a paraphrase of Eutocius somewhat retouched and abridged, and in many other instances he has inserted corrections and interpolations from another Greek MS. which he once names. This MS. appears to have been a copy made from F, with interpolations due to some one not unskilled in the subject-matter; and this interpolated copy of F was apparently also the source of the Nürnberg MS. now to be mentioned.

N^a (= Codex Norimbergensis) was written in the 16th century and brought from Rome to Nürnberg by Wilibald Pirckheymer. It contains the same works of Archimedes and Eutocius, and in the same order, as F, but was evidently not copied from F direct, while, on the other hand, it agrees so closely with Tartaglia's version as to suggest a common origin. N^a was used by Venatorius in preparing the *editio princeps*, and Venatorius corrected many mistakes in it with his own hand by notes in the margin or on slips attached thereto; he also made many alterations in the body of it, erasing the original, and sometimes wrote on it directions to the printer, so that it was probably actually used to print from. The character of the MS. shows it to belong to the same class as the others; it agrees with them in the more important errors and in having a similar lacuna at the beginning. Some mistakes common to it and F alone show that its source was F, though at second hand, as above indicated.

It remains to enumerate the principal editions of the Greek text and the published Latin versions which are based, wholly or partially, upon direct collation of the MSS. These are as follows, in addition to Gaurico's and Tartaglia's translations.

1. The *editio princeps* published at Basel in 1544 by Thomas Gechauff Venatorius under the title *Archimedis opera quae quidem exstant omnia nunc primum graece et latine in lucem edita. Adiecta quoque sunt Eutocii Ascalonitae commentaria item graece et latine nunquam antea excusa.* The Greek text and the Latin version in this edition were taken from different sources, that of the Greek text being N^a, while the translation was Joannes Regiomontanus' revised copy (N^b) of the Latin version made by Jacobus Cremonensis from the MS. of Pope Nicolas V. The revision by Regiomontanus was effected by the aid of (1) another copy of the same translation still extant, (2) other Greek MSS., one of which was probably V, while another may have been Valla's MS. itself.

2. A translation by F. Commandinus (containing the following works, *circuli dimensio, de lineis spiralibus, quadratura parabolae, de conoidibus et sphaeroidibus, de arenae numero*) appeared at Venice in 1558 under the title *Archimedis opera nonnulla in latinum conversa et commentariis illustrata.* For this translation several MSS. were used, among which was V, but none preferable to those which we now possess.

3. D. Rivault's edition, *Archimedis opera quae exstant graece et latine novis demonstr. et comment. illustr.* (Paris, 1615), gives only the propositions in Greek, while the proofs are in Latin and somewhat retouched. Rivault followed the Basel *editio princeps* with the assistance of B.

4. Torelli's edition (Oxford, 1792) entitled Ἀρχιμήδους τὰ σωζόμενα μετὰ τῶν Εὐτοκίου Ἀσκαλωνίτου ὑπομνημάτων, *Archimedis quae supersunt omnia cum Eutocii Ascalonitae commentariis ex recensione J. Torelli Veronensis cum nova versione latina. Accedunt lectiones variantes ex codd. Mediceo et Parisiensibus.* Torelli followed the Basel *editio princeps* in the main, but also collated V. The book was brought out after Torelli's death by Abram Robertson, who added the collation of five more MSS., F, A, B, C, D, with the Basel edition. The collation however was not well done, and the edition was not properly corrected when in the press.

5. Last of all comes the definitive edition of Heiberg (*Archimedis opera omnia cum commentariis Eutocii. E codice Florentino recensuit, Latine uertit notisque illustrauit J. L. Heiberg*. Leipzig, 1880—1).

The relation of all the MSS. and the above editions and translations is well shown by Heiberg in the following scheme (with the omission, however, of his own edition):

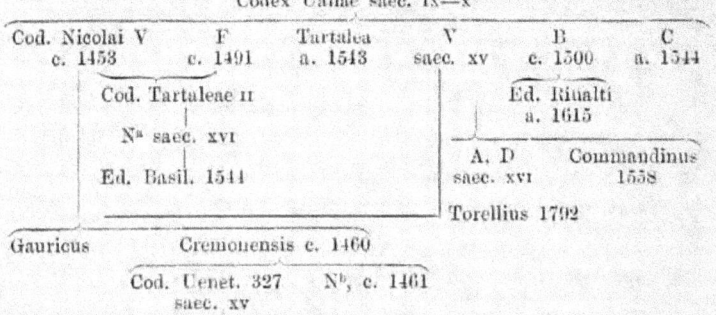

The remaining editions which give portions of Archimedes in Greek, and the rest of the translations of the complete works or parts of them which appeared before Heiberg's edition, were not based upon any fresh collation of the original sources, though some excellent corrections of the text were made by some of the editors, notably Wallis and Nizze. The following books may be mentioned.

Joh. Chr. Sturm, *Des unvergleichlichen Archimedis Kunstbücher, übersetzt und erläutert* (Nürnberg, 1670). This translation embraced all the works extant in Greek and followed three years after the same author's separate translation of the *Sand-reckoner*. It appears from Sturm's preface that he principally used the edition of Rivault.

Is. Barrow, *Opera Archimedis, Apollonii Pergaei conicorum libri, Theodosii sphaerica methodo novo illustrata et demonstrata* (London, 1675).

Wallis, *Archimedis arenarius et dimensio circuli, Eutocii in hanc commentarii cum versione et notis* (Oxford, 1678), also given in Wallis' *Opera*, Vol. III. pp. 509—546.

Karl Friedr. Hauber, *Archimeds zwei Bücher über Kugel und Cylinder. Ebendesselben Kreismessung. Uebersetzt mit Anmerkungen u. s. w. begleitet* (Tübingen, 1798).

F. Peyrard, *Œuvres d'Archimède, traduites littéralement, avec un commentaire, suivies d'un mémoire du traducteur, sur un nouveau miroir ardent, et d'un autre mémoire de M. Delambre, sur l'arithmétique des Grecs.* (Second edition, Paris, 1808.)

Ernst Nizze, *Archimedes von Syrakus vorhandene Werke, aus dem Griechischen übersetzt und mit erläuternden und kritischen Anmerkungen begleitet* (Stralsund, 1824).

The MSS. give the several treatises in the following order.

1. περὶ σφαίρας καὶ κυλίνδρου α΄ β΄, two Books *On the Sphere and Cylinder.*

2. κύκλου μέτρησις*, *Measurement of a Circle.*

3. περὶ κωνοειδέων καὶ σφαιροειδέων, *On Conoids and Spheroids.*

4. περὶ ἑλίκων, *On Spirals.*

5. ἐπιπέδων ἰσορροπιῶν α΄ β΄†, two Books *On the Equilibrium of Planes.*

6. ψαμμίτης, *The Sand-reckoner.*

7. τετραγωνισμὸς παραβολῆς (a name substituted later for that given to the treatise by Archimedes himself, which must undoubtedly have been τετραγωνισμὸς τῆς τοῦ ὀρθογωνίου κώνου τομῆς‡), *Quadrature of the Parabola.*

To these should be added

8. περὶ ὀχουμένων§, the Greek title of the treatise *On floating bodies*, only preserved in a Latin translation.

---

* Pappus alludes (I. p. 312, ed. Hultsch) to the κύκλου μέτρησις in the words ἐν τῷ περὶ τῆς τοῦ κύκλου περιφερείας.

† Archimedes himself twice alludes to properties proved in Book I. as demonstrated ἐν τοῖς μηχανικοῖς (*Quadrature of the Parabola*, Props. 6, 10). Pappus (VIII. p. 1034) quotes τὰ Ἀρχιμήδους περὶ ἰσορροπιῶν. The beginning of Book I. is also cited by Proclus in his *Commentary on Eucl.* I., p. 181, where the reading should be τοῦ ā ἰσορροπιῶν, and not τῶν ἀνισορροπιῶν (Hultsch).

‡ The name 'parabola' was first applied to the curve by Apollonius. Archimedes always used the old term 'section of a right-angled cone.' Cf. Eutocius (Heiberg, vol. III., p. 342) δέδεικται ἐν τῷ περὶ τῆς τοῦ ὀρθογωνίου κώνου τομῆς.

§ This title corresponds to the references to the book in Strabo I. p. 54 (Ἀρχιμήδης ἐν τοῖς περὶ τῶν ὀχουμένων) and Pappus VIII. p. 1024 (ὡς Ἀρχιμήδης ὀχουμένοις). The fragment edited by Mai has a longer title, περὶ τῶν ὕδατι ἐφισταμένων ἢ περὶ τῶν ὀχουμένων, where the first part corresponds to Tartaglia's version, *de insidentibus aquae,* and to that of Commandinus, *de iis quae vehuntur in aqua.* But Archimedes intentionally used the more general word ὑγρόν (fluid) instead of ὕδωρ; and hence the shorter title περὶ ὀχουμένων, *de iis quae in humido vehuntur* (Torelli and Heiberg), seems the better.

The books were not, however, written in the above order; and Archimedes himself, partly through his prefatory letters and partly by the use in later works of properties proved in earlier treatises, gives indications sufficient to enable the chronological sequence to be stated approximately as follows:

1. *On the equilibrium of planes*, I.
2. *Quadrature of the Parabola*.
3. *On the equilibrium of planes*, II.
4. *On the Sphere and Cylinder*, I, II.
5. *On Spirals*.
6. *On Conoids and Spheroids*.
7. *On floating bodies*, I, II.
8. *Measurement of a circle*.
9. *The Sand-reckoner*.

It should however be observed that, with regard to (7), no more is certain than that it was written after (6), and with regard to (8) no more than that it was later than (4) and before (9).

In addition to the above we have a collection of Lemmas (*Liber Assumptorum*) which has reached us through the Arabic. The collection was first edited by S. Foster, *Miscellanea* (London, 1659), and next by Borelli in a book published at Florence, 1661, in which the title is given as *Liber assumptorum Archimedis interprete Thebit ben Kora et exponente doctore Almochtasso Abilhasan*. The Lemmas cannot, however, have been written by Archimedes in their present form, because his name is quoted in them more than once. The probability is that they were propositions collected by some Greek writer* of a later date for the purpose of elucidating some ancient work, though it is quite likely that some of the propositions were of Archimedean origin, e.g. those concerning the geometrical figures called respectively ἄρβηλος† (literally

---

* It would seem that the compiler of the *Liber Assumptorum* must have drawn, to a considerable extent, from the same sources as Pappus. The number of propositions appearing substantially in the same form in both collections is, I think, even greater than has yet been noticed. Tannery (*La Géométrie grecque*, p. 162) mentions, as instances, Lemmas 1, 4, 5, 6; but it will be seen from the notes in this work that there are several other coincidences.

† Pappus gives (p. 208) what he calls an 'ancient proposition' (ἀρχαία πρότασις) about the same figure, which he describes as χωρίον, ὅ δὴ καλοῦσιν ἄρβηλον. Cf. the note to Prop. 6 (p. 308). The meaning of the word is gathered

'shoemaker's knife') and σάλινον (probably a 'salt-cellar'*), and Prop. 8 which bears on the problem of trisecting an angle.

from the Scholia to Nicander, *Theriaca*, 423 : ἄρβηλοι λέγονται τὰ κυκλοτερῆ σιδήρια, οἷς οἱ σκυτοτόμοι τέμνουσι καὶ ξύουσι τὰ δέρματα. Cf. Hesychius, ἀνάρβηλα, τὰ μὴ ἐξεσμένα δέρματα· ἄρβηλοι γὰρ τὰ σμιλία.

* The best authorities appear to hold that in any case the name σάλινον was not applied to the figure in question by Archimedes himself but by some later writer. Subject to this remark, I believe σάλινον to be simply a Graecised form of the Latin word *salinum*. We know that a salt-cellar was an essential part of the domestic apparatus in Italy from the early days of the Roman Republic. "All who were raised above poverty had one of silver which descended from father to son (Hor., *Carm.* II. 16, 13, Liv. XXVI. 36), and was accompanied by a silver *patella* which was used together with the salt-cellar in the domestic sacrifices (Pers. III. 24, 25). These two articles of silver were alone compatible with the simplicity of Roman manners in the early times of the Republic (Plin., *H. N.* XXXIII. § 153, Val. Max. IV. 4, § 3). ...In shape the *salinum* was probably in most cases a round shallow bowl" [*Dict. of Greek and Roman Antiquities*, article *salinum*]. Further we have in the early chapters of Mommsen's *History of Rome* abundant evidence of similar transferences of Latin words to the Sicilian dialect of Greek. Thus (Book I., ch. xiii.) it is shown that, in consequence of Latino-Sicilian commerce, certain words denoting measures of weight, *libra, triens, quadrans, sextans, uncia,* found their way into the common speech of Sicily in the third century of the city under the forms λίτρα, τριᾶς, τετρᾶς, ἑξᾶς, οὐγκία. Similarly Latin law-terms (ch. xi.) were transferred ; thus *mutuum* (a form of loan) became μοῖτον, *carcer* (a prison) κάρκαρον. Lastly, the Latin word for lard, *arvina*, became in Sicilian Greek ἀρβίνη, and *patina* (a dish) πατάνη. The last word is as close a parallel for the supposed transfer of *salinum* as could be wished. Moreover the explanation of σάλινον as *salinum* has two obvious advantages in that (1) it does not require any alteration in the word, and

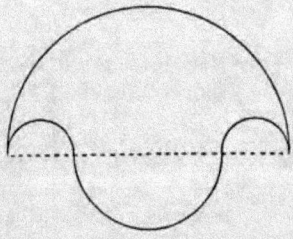

(2) the resemblance of the lower curve to an ordinary type of salt-cellar is evident. I should add, as confirmation of my hypothesis, that Dr A. S. Murray, of the British Museum, expresses the opinion that we cannot be far wrong in accepting as a *salinum* one of the small silver bowls in the Roman ministerium

Archimedes is further credited with the authorship of the *Cattle-problem* enunciated in the epigram edited by Lessing in 1773. According to the heading prefixed to the epigram it was communicated by Archimedes to the mathematicians at Alexandria in a letter to Eratosthenes*. There is also in the Scholia to Plato's *Charmides* 165 E a reference to the problem "called by Archimedes the Cattle-problem" (τὸ κληθὲν ὑπ᾽ Ἀρχιμήδους βοεικὸν πρόβλημα). The question whether Archimedes really propounded the problem, or whether his name was only prefixed to it in order to mark the extraordinary difficulty of it, has been much debated. A complete account of the arguments for and against is given in an article by Krumbiegel in the *Zeitschrift für Mathematik und Physik* (*Hist. litt. Abtheilung*) xxv. (1880), p. 121 sq., to which Amthor added (*ibid.* p. 153 sq.) a discussion of the problem itself. The general result of Krumbiegel's investigation is to show (1) that

at the Museum which was found at Chaourse (Aisne) in France and is of a section sufficiently like the curve in the Salinon.

The other explanations of σάλινον which have been suggested are as follows.

(1) Cantor connects it with σάλος, "das Schwanken des hohen Meeres," and would presumably translate it as *wave-line*. But the resemblance is not altogether satisfactory, and the termination -*ινον* would need explanation.

(2) Heiberg says the word is "sine dubio ab Arabibus deprauatum," and suggests that it should be σέλινον, *parsley* ("ex similitudine frondis apii"). But, whatever may be thought of the resemblance, the theory that the word is corrupted is certainly not supported by the analogy of ἄρβηλος which is correctly reproduced by the Arabs, as we know from the passage of Pappus referred to in the last note.

(3) Dr Gow suggests that σάλινον may be a 'sieve,' comparing σάλαξ. But this guess is not supported by any evidence.

* The heading is, Πρόβλημα ὅπερ Ἀρχιμήδης ἐν ἐπιγράμμασιν εὑρὼν τοῖς ἐν Ἀλεξανδρείᾳ περὶ ταῦτα πραγματευομένοις ζητεῖν ἀπέστειλεν ἐν τῇ πρὸς Ἐρατοσθένην τὸν Κυρηναῖον ἐπιστολῇ. Heiberg translates this as "the problem which Archimedes discovered and sent in an epigram...in a letter to Eratosthenes." He admits however that the order of words is against this, as is also the use of the plural ἐπιγράμμασιν. It is clear that to take the two expressions ἐν ἐπιγράμμασιν and ἐν ἐπιστολῇ as both following ἀπέστειλεν is very awkward. In fact there seems to be no alternative but to translate, as Krumbiegel does, in accordance with the order of the words, "a problem which Archimedes found among (some) epigrams and sent...in his letter to Eratosthenes"; and this sense is certainly unsatisfactory. Hultsch remarks that, though the mistake πραγματουμένοις for πραγματευομένοις and the composition of the heading as a whole betray the hand of a writer who lived some centuries after Archimedes, yet he must have had an earlier source of information, because he could hardly have invented the story of the letter to Eratosthenes.

the epigram can hardly have been written by Archimedes in its present form, but (2) that it is possible, nay probable, that the problem was in substance originated by Archimedes. Hultsch* has an ingenious suggestion as to the occasion of it. It is known that Apollonius in his ὠκυτόκιον had calculated a closer approximation to the value of π than that of Archimedes, and he must therefore have worked out more difficult multiplications than those contained in the *Measurement of a circle*. Also the other work of Apollonius on the multiplication of large numbers, which is partly preserved in Pappus, was inspired by the *Sand-reckoner* of Archimedes; and, though we need not exactly regard the treatise of Apollonius as polemical, yet it did in fact constitute a criticism of the earlier book. Accordingly, that Archimedes should then reply with a problem which involved such a manipulation of immense numbers as would be difficult even for Apollonius is not altogether outside the bounds of possibility. And there is an unmistakable vein of satire in the opening words of the epigram "Compute the number of the oxen of the Sun, giving thy mind thereto, if thou hast a share of wisdom," in the transition from the first part to the second where it is said that ability to solve the first part would entitle one to be regarded as "not unknowing nor unskilled in numbers, but still not yet to be numbered among the wise," and again in the last lines. Hultsch concludes that in any case the problem is not much later than the time of Archimedes and dates from the beginning of the 2nd century B.C. at the latest.

Of the extant books it is certain that in the 6th century A.D. only three were generally known, viz. *On the Sphere and Cylinder*, the *Measurement of a circle*, and *On the equilibrium of planes*. Thus Eutocius of Ascalon who wrote commentaries on these works only knew the *Quadrature of the Parabola* by name and had never seen it nor the book *On Spirals*. Where passages might have been elucidated by references to the former book, Eutocius gives explanations derived from Apollonius and other sources, and he speaks vaguely of the discovery of a straight line equal to the circumference of a given circle "by means of certain spirals," whereas, if he had known the treatise *On Spirals*, he would have quoted Prop. 18. There is reason to suppose that only the three treatises on which Eutocius commented were contained in the

---

* Pauly-Wissowa's *Real-Encyclopädie*, II. 1, pp. 534, 5.

ordinary editions of the time such as that of Isidorus of Miletus, the teacher of Eutocius, to which the latter several times alludes.

In these circumstances the wonder is that so many more books have survived to the present day. As it is, they have lost to a considerable extent their original form. Archimedes wrote in the Doric dialect*, but in the best known books (*On the Sphere and Cylinder* and the *Measurement of a circle*) practically all traces of that dialect have disappeared, while a partial loss of Doric forms has taken place in other books, of which however the *Sand-reckoner* has suffered least. Moreover in all the books, except the *Sand-reckoner*, alterations and additions were first of all made by an interpolator who was acquainted with the Doric dialect, and then, at a date subsequent to that of Eutocius, the book *On the Sphere and Cylinder* and the *Measurement of a circle* were completely recast.

Of the lost works of Archimedes the following can be identified.

1. Investigations relating to *polyhedra* are referred to by Pappus who, after alluding (v. p. 352) to the five regular polyhedra, gives a description of thirteen others discovered by Archimedes which are semi-regular, being contained by polygons equilateral and equiangular but not similar.

2. A book of arithmetical content, entitled ἀρχαί *Principles* and dedicated to Zeuxippus. We learn from Archimedes himself that the book dealt with the *naming of numbers* (κατονόμαξις τῶν ἀριθμῶν)† and expounded a system of expressing numbers higher

---

* Thus Eutocius in his commentary on Prop. 4 of Book II. *On the Sphere and Cylinder* speaks of the fragment, which he found in an old book and which appeared to him to be the missing supplement to the proposition referred to, as "preserving in part Archimedes' favourite Doric dialect" (ἐν μέρει δὲ τὴν Ἀρχιμήδει φίλην Δωρίδα γλῶσσαν ἀπέσωξον). From the use of the expression ἐν μέρει Heiberg concludes that the Doric forms had by the time of Eutocius begun to disappear in the books which have come down to us no less than in the fragment referred to.

† Observing that in all the references to this work in the *Sand-reckoner* Archimedes speaks of the *naming of numbers* or of *numbers which are named or have their names* (ἀριθμοὶ κατωνομασμένοι, τὰ ὀνόματα ἔχοντες, τὰν κατονομαξίαν ἔχοντες), Hultsch (Pauly-Wissowa's *Real-Encyclopädie*, II. 1, p. 511) speaks of κατονόμαξις τῶν ἀριθμῶν as the name of the work; and he explains the words τινὰς τῶν ἐν ἀρχαῖς <ἀριθμῶν> τῶν κατονομαξίαν ἐχόντων as meaning "some of the numbers mentioned *at the beginning* which have a special name," where "at the beginning" refers to the passage in which Archimedes first mentions τῶν

than those which could be expressed in the ordinary Greek notation. This system embraced all numbers up to the enormous figure which we should now represent by a 1 followed by 80,000 billion ciphers; and, in setting out the same system in the *Sandreckoner*, Archimedes explains that he does so for the benefit of those who had not had the opportunity of seeing the earlier work addressed to Zeuxippus.

3. περὶ ζυγῶν, *On balances* or *levers*, in which Pappus says (VIII. p. 1068) that Archimedes proved that "greater circles overpower (κατακρατοῦσι) lesser circles when they revolve about the same centre." It was doubtless in this book that Archimedes proved the theorem assumed by him in the *Quadrature of the Parabola*, Prop. 6, viz. that, if a body hangs at rest from a point, the centre of gravity of the body and the point of suspension are in the same vertical line.

4. κεντροβαρικά, *On centres of gravity*. This work is mentioned by Simplicius on Aristot. *de caelo* II. (*Scholia in Arist.* 508 a 30). Archimedes may be referring to it when he says (*On the equilibrium of planes* I. 4) that it has before been proved that the centre of gravity of two bodies taken together lies on the line joining the centres of gravity of the separate bodies. In the treatise *On floating bodies* Archimedes assumes that the centre of gravity of a segment of a paraboloid of revolution is on the axis of the segment at a distance from the vertex equal to ⅔rds of its length. This may perhaps have been proved in the κεντροβαρικά, if it was not made the subject of a separate work.

Doubtless both the περὶ ζυγῶν and the κεντροβαρικά preceded the extant treatise *On the equilibrium of planes*.

5. κατοπτρικά, an optical work, from which Theon (on Ptolemy, *Synt.* I. p. 29, ed. Halma) quotes a remark about refraction. Cf. Olympiodorus *in Aristot. Meteor.*, II. p. 94, ed. Ideler.

ὑφ' ἁμῶν κατωνομασμένων ἀριθμῶν καὶ ἐνδεδομένων ἐν τοῖς ποτὶ Ζεύξιππον γεγραμμένοις. But ἐν ἀρχαῖς seems a less natural expression for "at the beginning" than ἐν ἀρχῇ or κατ' ἀρχάς would have been. Moreover, there being no participial expression except κατονομαξίαν ἐχόντων to be taken with ἐν ἀρχαῖς in this sense, the meaning would be unsatisfactory; for the numbers are not *named* at the beginning, but only *referred to*, and therefore some word like εἰρημένων should have been used. For these reasons I think that Heiberg, Cantor and Susemihl are right in taking ἀρχαί to be the name of the treatise.

6. περὶ σφαιροποιίας, *On sphere-making*, a mechanical work on the construction of a sphere representing the motions of the heavenly bodies as already mentioned (p. xxi).

7. ἐφόδιον, a *Method*, noticed by Suidas, who says that Theodosius wrote a commentary on it, but gives no further information about it.

8. According to Hipparchus Archimedes must have written on the *Calendar* or the length of the year (cf. p. xxi).

Some Arabian writers attribute to Archimedes works (1) On a heptagon in a circle, (2) On circles touching one another, (3) On parallel lines, (4) On triangles, (5) On the properties of right-angled triangles, (6) a book of *Data*; but there is no confirmatory evidence of his having written such works. A book translated into Latin from the Arabic by Gongava (Louvain, 1548) and entitled *antiqui scriptoris de speculo comburente concavitatis parabolae* cannot be the work of Archimedes, since it quotes Apollonius.

CHAPTER III.

An extraordinarily large proportion of the subject matter of the writings of Archimedes represents entirely new discoveries of his own. Though his range of subjects was almost encyclopaedic, embracing geometry (plane and solid), arithmetic, mechanics, hydrostatics and astronomy, he was no compiler, no writer of textbooks; and in this respect he differs even from his great successor Apollonius, whose work, like that of Euclid before him, largely consisted of systematising and generalising the methods used, and the results obtained, in the isolated efforts of earlier geometers. There is in Archimedes no mere working-up of existing materials; his objective is always some new thing, some definite addition to the sum of knowledge, and his complete originality cannot fail to strike any one who reads his works intelligently, without any corroborative evidence such as is found in the introductory letters prefixed to most of them. These introductions, however, are eminently characteristic of the man and of his work; their directness and simplicity, the complete absence of egoism and of any effort to magnify his own achievements by comparison with those of others or by emphasising their failures where he himself succeeded: all these things intensify the same impression. Thus his manner is to state simply what particular discoveries made by his predecessors had suggested to him the possibility of extending them in new directions; e.g. he says that, in connexion with the efforts of earlier geometers to square the circle and other figures, it occurred to him that no one had endeavoured to square a parabola, and he accordingly attempted the problem and finally solved it. In like manner, he speaks, in the preface of his treatise *On the*

Sphere and Cylinder, of his discoveries with reference to those solids as supplementing the theorems about the pyramid, the cone and the cylinder proved by Eudoxus. He does not hesitate to say that certain problems baffled him for a long time, and that the solution of some took him many years to effect; and in one place (in the preface to the book *On Spirals*) he positively insists, for the sake of pointing a moral, on specifying two propositions which he had enunciated and which proved on further investigation to be wrong. The same preface contains a generous eulogy of Conon, declaring that, but for his untimely death, Conon would have solved certain problems before him and would have enriched geometry by many other discoveries in the meantime.

In some of his subjects Archimedes had no fore-runners, e.g. in hydrostatics, where he invented the whole science, and (so far as mathematical demonstration was concerned) in his mechanical investigations. In these cases therefore he had, in laying the foundations of the subject, to adopt a form more closely resembling that of an elementary textbook, but in the later parts he at once applied himself to specialised investigations.

Thus the historian of mathematics, in dealing with Archimedes' obligations to his predecessors, has a comparatively easy task before him. But it is necessary, first, to give some description of the use which Archimedes made of the general methods which had found acceptance with the earlier geometers, and, secondly, to refer to some particular results which he mentions as having been previously discovered and as lying at the root of his own investigations, or which he tacitly assumes as known.

§ 1. Use of traditional geometrical methods.

In my edition of the *Conics* of Apollonius*, I endeavoured, following the lead given in Zeuthen's work, *Die Lehre von den Kegelschnitten im Altertum*, to give some account of what has been fitly called the *geometrical algebra* which played such an important part in the works of the Greek geometers. The two main methods included under the term were (1) the use of the *theory of proportions*, and (2) the method of *application of areas*, and it was shown that, while both methods are fully expounded in the *Elements* of Euclid, the second was much the older of the two, being attributed by the pupils of Eudemus (quoted by Proclus) to the

---

* *Apollonius of Perga*, pp. ci sqq.

Pythagoreans. It was pointed out that the *application of areas*, as set forth in the second Book of Euclid and extended in the sixth, was made by Apollonius the means of expressing what he takes as the fundamental properties of the conic sections, namely the properties which we express by the Cartesian equations

$$y^2 = px,$$
$$y^2 = px \mp \frac{p}{d} x^2,$$

referred to any diameter and the tangent at its extremity as axes; and the latter equation was compared with the results obtained in the 27th, 28th and 29th Props. of Euclid's Book VI, which are equivalent to the solution, by geometrical means, of the quadratic equations

$$ax \pm \frac{b}{c} x^2 = D.$$

It was also shown that Archimedes does not, as a rule, connect his description of the central conics with the method of application of areas, as Apollonius does, but that Archimedes generally expresses the fundamental property in the form of a proportion

$$\frac{y^2}{x \cdot x_1} = \frac{y'^2}{x' \cdot x_1'},$$

and, in the case of the ellipse,

$$\frac{y^2}{x \cdot x_1} = \frac{b^2}{a^2},$$

where x, x_1 are the abscissae measured from the ends of the diameter of reference.

It results from this that the application of areas is of much less frequent occurrence in Archimedes than in Apollonius. It is however used by the former in all but the most general form. The simplest form of "applying a rectangle" to a given straight line which shall be equal to a given area occurs e.g. in the proposition *On the equilibrium of Planes* II. 1; and the same mode of expression is used (as in Apollonius) for the property $y^2 = px$ in the parabola, px being described in Archimedes' phrase as the rectangle "applied to" (παραπίπτον παρά) a line equal to p and "having at its width" (πλάτος ἔχον) the abscissa (x). Then in Props. 2, 25, 26, 29 of the book *On Conoids and Spheroids* we have the complete expression which is the equivalent of solving the equation

$$ax + x^2 = b^2,$$

"let a rectangle be applied (to a certain straight line) exceeding by a square figure ($\pi\alpha\rho\alpha\pi\epsilon\pi\tau\omega\kappa\acute{\epsilon}\tau\omega$ $\chi\omega\rho\acute{\iota}o\nu$ $\acute{\upsilon}\pi\epsilon\rho\beta\acute{\alpha}\lambda\lambda o\nu$ $\epsilon\breve{\iota}\delta\epsilon\iota$ $\tau\epsilon\tau\rho\alpha\gamma\acute{\omega}\nu\dot{\omega}$) and equal to (a certain rectangle)." Thus a rectangle of this sort has to be made (in Prop. 25) equal to what we have above called $x \cdot x_1$ in the case of the hyperbola, which is the same thing as $x(a + x)$ or $ax + x^2$, where a is the length of the transverse axis. But, curiously enough, we do not find in Archimedes the application of a rectangle "*falling short* by a square figure," which we should obtain in the case of the ellipse if we substituted $x(a - x)$ for $x \cdot x_1$. In the case of the ellipse the area $x \cdot x_1$ is represented (*On Conoids and Spheroids*, Prop. 29) as a gnomon which is the difference between the rectangle $h \cdot h_1$ (where h, h_1 are the abscissae of the ordinate bounding a segment of an ellipse) and a rectangle applied to $h_1 - h$ and exceeding by a square figure whose side is $h - x$; and the rectangle $h \cdot h_1$ is simply constructed from the sides h, h_1. Thus Archimedes avoids* the application of a rectangle *falling short* by a square, using for $x \cdot x_1$ the rather complicated form

$$h \cdot h_1 - \{(h_1 - h)(h - x) + (h - x)^2\}.$$

It is easy to see that this last expression is equal to $x \cdot x_1$, for it reduces to

$$h \cdot h_1 - \{h_1(h - x) - x(h - x)\}$$
$$= x(h_1 + h) - x^2,$$
$$= ax - x^2, \text{ since } h_1 + h = a,$$
$$= x \cdot x_1.$$

It will readily be understood that the transformation of rectangles and squares in accordance with the methods of Euclid, Book II, is just as important to Archimedes as to other geometers, and there is no need to enlarge on that form of geometrical algebra.

The theory of *proportions*, as expounded in the fifth and sixth Books of Euclid, including the transformation of ratios (denoted by the terms *componendo*, *dividendo*, etc.) and the composition or multiplication of ratios, made it possible for the ancient geometers to deal with magnitudes in general and to work out relations between them with an effectiveness not much inferior to that of modern algebra. Thus the addition and subtraction of ratios could be effected by procedure equivalent to what we should in algebra

---

* The object of Archimedes was no doubt to make the Lemma in Prop. 2 (dealing with the summation of a series of terms of the form $a \cdot rx + (rx)^2$, where r successively takes the values 1, 2, 3, ...) serve for the hyperboloid of revolution and the spheroid as well.

call bringing to a common denominator. Next, the composition or multiplication of ratios could be indefinitely extended, and hence the algebraical operations of multiplication and division found easy and convenient expression in the geometrical algebra. As a particular case, suppose that there is a series of magnitudes in continued proportion (i.e. in geometrical progression) as $a_0, a_1, a_2, \ldots a_n$, so that

$$\frac{a_0}{a_1} = \frac{a_1}{a_2} = \ldots = \frac{a_{n-1}}{a_n}.$$

We have then, by multiplication,

$$\frac{a_n}{a_0} = \left(\frac{a_1}{a_0}\right)^n, \text{ or } \frac{a_1}{a_0} = \sqrt[n]{\frac{a_n}{a_0}}.$$

It is easy to understand how powerful such a method as that of proportions would become in the hands of an Archimedes, and a few instances are here appended in order to illustrate the mastery with which he uses it.

1. A good example of a reduction in the order of a ratio after the manner just shown is furnished by *On the equilibrium of Planes* II. 10. Here Archimedes has a ratio which we will call a^3/b^3, where $a^2/b^2 = c/d$; and he reduces the ratio between cubes to a ratio between straight lines by taking two lines x, y such that

$$\frac{c}{x} = \frac{x}{d} = \frac{d}{y}.$$

It follows from this that

$$\left(\frac{c}{x}\right)^2 = \frac{c}{d} = \frac{a^2}{b^2},$$

or

$$\frac{a}{b} = \frac{c}{x};$$

and hence

$$\frac{a^3}{b^3} = \left(\frac{c}{x}\right)^3 = \frac{c}{x} \cdot \frac{x}{d} \cdot \frac{d}{y} = \frac{c}{y}.$$

2. In the last example we have an instance of the use of auxiliary fixed lines for the purpose of simplifying ratios and thereby, as it were, economising power in order to grapple the more successfully with a complicated problem. With the aid of such auxiliary lines or (what is the same thing) auxiliary fixed points in a figure, combined with the use of proportions, Archimedes is able to effect some remarkable *eliminations*.

Thus in the proposition *On the Sphere and Cylinder* II. 4 he obtains three relations connecting three as yet undetermined points, and proceeds at once to eliminate two of the points, so that the problem is then reduced to finding the remaining point by means of one equation. Expressed in an algebraical form, the three original relations amount to the three equations

$$\left.\begin{array}{c} \dfrac{3a-x}{2a-x} = \dfrac{y}{x} \\[2mm] \dfrac{a+x}{x} = \dfrac{z}{2a-x} \\[2mm] \dfrac{y}{z} = \dfrac{m}{n} \end{array}\right\},$$

and the result, after the elimination of y and z, is stated by Archimedes in a form equivalent to

$$\frac{m+n}{n} \cdot \frac{a+x}{a} = \frac{4a^2}{(2a-x)^2}.$$

Again the proposition *On the equilibrium of Planes* II. 9 proves by the same method of proportions that, if a, b, c, d, x, y, are straight lines satisfying the conditions

$$\left.\begin{array}{c} \dfrac{a}{b} = \dfrac{b}{c} = \dfrac{c}{d}, \quad (a > b > c > d) \\[2mm] \dfrac{d}{a-d} = \dfrac{x}{\frac{2}{3}(a-c)}, \\[2mm] \dfrac{2a+4b+6c+3d}{5a+10b+10c+5d} = \dfrac{y}{a-c}, \end{array}\right\}$$

and

then $\qquad\qquad x + y = \frac{2}{5}a.$

The proposition is merely brought in as a subsidiary lemma to the proposition following, and is not of any intrinsic importance; but a glance at the proof (which again introduces an auxiliary line) will show that it is a really extraordinary instance of the manipulation of proportions.

3. Yet another instance is worth giving here. It amounts to the proof that, if

$$\frac{x^2}{a^2} + \frac{y^2}{b^2} = 1,$$

then $\qquad \dfrac{2a+x}{a+x} \cdot y^2(a-x) + \dfrac{2a-x}{a-x} \cdot y^2(a+x) = 4ab^2.$

A, A' are the points of contact of two parallel tangent planes to a spheroid; the plane of the paper is the plane through AA' and the axis of the spheroid, and PP' is the intersection of this plane with another plane at right angles to it (and therefore parallel to the tangent planes), which latter plane divides the spheroid into two segments whose axes are AN, $A'N$. Another plane is drawn through

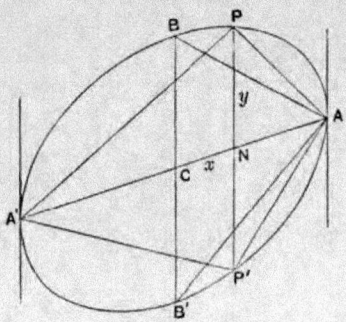

the centre and parallel to the tangent plane, cutting the spheroid into two halves. Lastly cones are drawn whose bases are the sections of the spheroid by the parallel planes as shown in the figure.

Archimedes' proposition takes the following form [*On Conoids and Spheroids*, Props. 31, 32].

APP' being the smaller segment of the two whose common base is the section through PP', and x, y being the coordinates of P, he has proved in preceding propositions that

$$\frac{\text{(volume of) segment } APP'}{\text{(volume of) cone } APP'} = \frac{2a+x}{a+x} \quad \dots\dots\dots\dots(a),$$

and

$$\frac{\text{half spheroid } ABB'}{\text{cone } ABB'} = 2; \quad \dots\dots\dots\dots(\beta),$$

and he seeks to prove that

$$\frac{\text{segment } A'PP'}{\text{cone } A'PP'} = \frac{2a-x}{a-x}.$$

The method is as follows.

We have

$$\frac{\text{cone } ABB'}{\text{cone } APP'} = \frac{a}{a-x} \cdot \frac{b^2}{y^2} = \frac{a}{a-x} \cdot \frac{a^2}{a^2-x^2}.$$

If we suppose

$$\frac{z}{a} = \frac{a}{a-x} \quad \dots\dots\dots\dots\dots\dots(\gamma),$$

the ratio of the cones becomes $\dfrac{za}{a^2-x^2}.$

Next, by hypothesis (a),

$$\frac{\text{cone } APP'}{\text{segmt. } APP'} = \frac{a+x}{2a+x}.$$

Therefore, *ex aequali*,

$$\frac{\text{cone } ABB'}{\text{segmt. } APP'} = \frac{za}{(a-x)(2a+x)}.$$

It follows from (β) that

$$\frac{\text{spheroid}}{\text{segmt. } APP'} = \frac{4za}{(a-x)(2a+x)},$$

whence

$$\frac{\text{segmt. } A'PP'}{\text{segmt. } APP'} = \frac{4za-(a-x)(2a+x)}{(a-x)(2a+x)}$$

$$= \frac{z(2a-x)+(2a+x)(z-a-x)}{(a-x)(2a+x)}.$$

Now we have to obtain the ratio of the segment $A'PP'$ to the *cone* $A'PP'$, and the comparison between the segment APP' and the cone $A'PP'$ is made by combining two ratios *ex aequali*. Thus

$$\frac{\text{segmt. } APP'}{\text{cone } APP'} = \frac{2a+x}{a+x}, \text{ by } (\alpha),$$

and

$$\frac{\text{cone } APP'}{\text{cone } A'PP'} = \frac{a-x}{a+x}.$$

Thus combining the last *three* proportions, *ex aequali*, we have

$$\frac{\text{segmt. } A'PP'}{\text{cone } A'PP'} = \frac{z(2a-x)+(2a+x)(z-a-x)}{a^2+2ax+x^2}$$

$$= \frac{z(2a-x)+(2a+x)(z-a-x)}{z(a-x)+(2a+x)x},$$

since $\qquad a^2 = z(a-x)$, by (γ).

[The object of the transformation of the numerator and denominator of the last fraction, by which $z(2a-x)$ and $z(a-x)$ are made the first terms, is now obvious, because $\dfrac{2a-x}{a-x}$ is the fraction which Archimedes wishes to arrive at, and, in order to prove that the required ratio is equal to this, it is only necessary to show that

$$\frac{2a-x}{a-x} = \frac{z-(a-x)}{x}.]$$

Now
$$\frac{2a-x}{a-x} = 1 + \frac{a}{a-x}$$

$$= 1 + \frac{z}{a}, \text{ by } (\gamma),$$

$$= \frac{a+z}{a}$$

$$= \frac{z-(a-x)}{x} \text{ } (dividendo),$$

so that
$$\frac{\text{segmt. } A'PP'}{\text{cone } A'PP'} = \frac{2a-x}{a-x}.$$

4. One use by Euclid of the method of proportions deserves mention because Archimedes *does not* use it in similar circumstances. Archimedes (*Quadrature of the Parabola*, Prop. 23) sums a particular geometric series

$$a + a\left(\tfrac{1}{4}\right) + a\left(\tfrac{1}{4}\right)^2 + \ldots + a\left(\tfrac{1}{4}\right)^{n-1}$$

in a manner somewhat similar to that of our text-books, whereas Euclid (IX. 35) sums any geometric series of any number of terms by means of proportions thus.

Suppose $a_1, a_2, \ldots a_n, a_{n+1}$ to be $(n+1)$ terms of a geometric series in which a_{n+1} is the greatest term. Then

$$\frac{a_{n+1}}{a_n} = \frac{a_n}{a_{n-1}} = \frac{a_{n-1}}{a_{n-2}} = \ldots = \frac{a_2}{a_1}.$$

Therefore
$$\frac{a_{n+1}-a_n}{a_n} = \frac{a_n - a_{n-1}}{a_{n-1}} = \ldots = \frac{a_2 - a_1}{a_1}.$$

Adding all the antecedents and all the consequents, we have

$$\frac{a_{n+1}-a_1}{a_1 + a_2 + a_3 + \ldots + a_n} = \frac{a_2 - a_1}{a_1},$$

which gives the sum of n terms of the series.

§ 2. Earlier discoveries affecting quadrature and cubature.

Archimedes quotes the theorem that *circles are to one another as the squares on their diameters* as having being proved by earlier geometers, and he also says that it was proved by means of a certain lemma which he states as follows: "Of unequal lines, unequal surfaces, or unequal solids, the greater exceeds the less by such a magnitude as is capable, if added [continually] to itself, of exceeding any given magnitude of those which are comparable with one another (τῶν πρὸς ἄλληλα λεγομένων)." We know that Hippocrates of Chios proved the theorem that circles are to one another as the squares on their diameters, but no clear conclusion can be established as to the method which he used. On the other hand, Eudoxus (who is mentioned in the preface to *The Sphere and Cylinder* as having proved two theorems in solid geometry to be mentioned presently) is generally credited with the invention of the *method of exhaustion* by which Euclid proves the proposition in question in XII. 2. The lemma stated by Archimedes to have been used in the original proof is not however found in that form in Euclid and is not used in the proof of XII. 2, where the lemma used is that proved by him in X. 1, viz. that "Given two unequal magnitudes, if from the greater [a part] be subtracted greater than the half, if from the remainder [a part] greater than the half be subtracted, and so on continually, there will be left some magnitude which will be less than the lesser given magnitude." This last lemma is frequently assumed by Archimedes, and the application of it to equilateral polygons inscribed in a circle or sector in the manner of XII. 2 is referred to as having been handed down in the *Elements**, by which it is clear that only Euclid's *Elements* can be meant. The apparent difficulty caused by the mention of *two* lemmas in connexion with the theorem in question can, however, I think, be explained by reference to the proof of X. 1 in Euclid. He there takes the lesser magnitude and says that it is possible, by multiplying it, to make it some time exceed the greater, and this statement he clearly bases on the 4th definition of Book V. to the effect that "magnitudes are said to bear a ratio to one another, which can, if multiplied, exceed one another." Since then the smaller magnitude in X. 1 may be regarded as the difference between some two unequal magnitudes, it is clear that the lemma first quoted by Archimedes is in substance used to prove the lemma in X. 1 which appears to play so much larger a part in the investigations in quadrature and cubature which have come down to us.

The two theorems which Archimedes attributes to Eudoxus by name† are

(1) that *any pyramid is one third part of the prism which has the same base as the pyramid and equal height*, and

---

* *On the Sphere and Cylinder*, I. 6.
† *ibid*. Preface.

(2) that *any cone is one third part of the cylinder which has the same base as the cone and equal height.*

The other theorems in solid geometry which Archimedes quotes as having been proved by earlier geometers are* :

(3) *Cones of equal height are in the ratio of their bases, and conversely.*

(4) *If a cylinder be divided by a plane parallel to the base, cylinder is to cylinder as axis to axis.*

(5) *Cones which have the same bases as cylinders and equal height with them are to one another as the cylinders.*

(6) *The bases of equal cones are reciprocally proportional to their heights, and conversely.*

(7) *Cones the diameters of whose bases have the same ratio as their axes are in the triplicate ratio of the diameters of their bases.*

In the preface to the *Quadrature of the Parabola* he says that earlier geometers had also proved that

(8) *Spheres have to one another the triplicate ratio of their diameters;* and he adds that this proposition and the first of those which he attributes to Eudoxus, numbered (1) above, were proved by means of the same lemma, viz. that the difference between any two unequal magnitudes can be so multiplied as to exceed any given magnitude, while (if the text of Heiberg is right) the second of the propositions of Eudoxus, numbered (2), was proved by means of "a lemma similar to that aforesaid." As a matter of fact, all the propositions (1) to (8) are given in Euclid's twelfth Book, except (5), which, however, is an easy deduction from (2); and (1), (2), (3), and (7) all depend upon the same lemma [x. 1] as that used in Eucl. XII. 2.

The proofs of the above seven propositions, excluding (5), as given by Euclid are too long to quote here, but the following sketch will show the line taken in the proofs and the order of the propositions. Suppose *ABCD* to be a pyramid with a triangular base, and suppose it to be cut by two planes, one bisecting *AB*, *AC*, *AD* in *F*, *G*, *E* respectively, and the other bisecting *BC*, *BD*, *BA* in *H*, *K*, *F* respectively. These planes are then each parallel to one face, and they cut off two pyramids each similar to the original

---

* Lemmas placed between Props. 16 and 17 of Book I. *On the Sphere and Cylinder.*

pyramid and equal to one another, while the remainder of the pyramid is proved to form two equal prisms which, taken together,

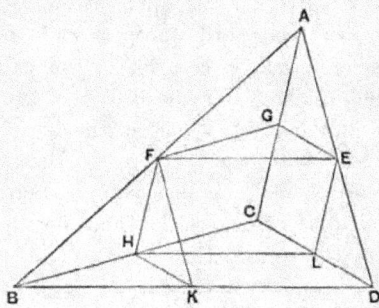

are greater than one half of the original pyramid [XII. 3]. It is next proved [XII. 4] that, if there are two pyramids with triangular bases and equal height, and if they are each divided in the manner shown into two equal pyramids each similar to the whole and two prisms, the sum of the prisms in one pyramid is to the sum of the prisms in the other in the ratio of the bases of the whole pyramids respectively. Thus, if we divide in the same manner the two pyramids which remain in each, then all the pyramids which remain, and so on continually, it follows on the one hand, by X. 1, that we shall ultimately have pyramids remaining which are together less than any assigned solid, while on the other hand the sums of all the prisms resulting from the successive subdivisions are in the ratio of the bases of the original pyramids. Accordingly Euclid is able to use the regular method of exhaustion exemplified in XII. 2, and to establish the proposition [XII. 5] that pyramids with the same height and with triangular bases are to one another as their bases. The proposition is then extended [XII. 6] to pyramids with the same height and with polygonal bases. Next [XII. 7] a prism with a triangular base is divided into three pyramids which are shown to be equal by means of XII. 5; and it follows, as a corollary, that any pyramid is one third part of the prism which has the same base and equal height. Again, two similar and similarly situated pyramids are taken and the solid parallelepipeds are completed, which are then seen to be six times as large as the pyramids respectively; and, since (by XI. 33) similar parallelepipeds are in the triplicate ratio of corresponding sides, it follows that the same is true of the pyramids [XII. 8]. A corollary gives the obvious extension to the case of similar pyramids with polygonal bases. The proposition [XII. 9] that, in equal pyramids with triangular bases, the bases are reciprocally proportional to the heights is proved by the same method of completing the parallelepipeds and using XI. 34 ; and similarly for the converse. It is next proved [XII. 10] that, if in the circle which is the base of a cylinder a square be described, and then polygons be successively described by bisecting the arcs remaining in each case, and so doubling the number of sides, and if prisms of the same height as the cylinder be erected on the square and the polygons as bases respectively, the prism with the square base will be greater than half the cylinder, the next prism will add to it more than half of the remainder, and so on. And each prism is triple of the pyramid with the same base and altitude. Thus the same method of exhaustion as that in XII. 2 proves that any cone is one third part of the cylinder with the same base and equal height. Exactly the same method is used to prove [XII. 11] that cones and cylinders which have the same height are to one another as their bases, and [XII. 12] that similar cones and cylinders are to one another in the triplicate ratio of the diameters of their bases (the latter proposition depending of course on the similar proposition XII. 8 for pyramids). The next three propositions are proved without fresh recourse to X. 1. Thus the criterion of equimultiples laid down in Def. 5 of Book v. is used to prove [XII. 13] that, if a cylinder be cut by a plane parallel to its bases, the resulting cylinders are to one another as their axes. It is an easy deduction [XII. 14] that cones and cylinders which have equal bases are proportional to their heights, and [XII. 15] that in equal cones and cylinders the bases are reciprocally proportional to the heights, and, conversely, that cones or cylinders having this property are equal. Lastly, to prove that spheres are to one another in the triplicate ratio of their diameters [XII. 18], a new procedure is adopted, involving two preliminary propositions. In the first of these [XII. 16] it is proved, by an application of the usual lemma X. 1, that, if two concentric circles are given (however nearly equal), an equilateral polygon can be inscribed in the outer circle whose sides do not touch the inner ; the second proposition [XII. 17] uses the result of the first to prove that, given two concentric spheres, it is possible to inscribe a certain polyhedron in the outer so that it does not anywhere touch the inner, and a corollary adds the proof that, if a similar polyhedron be inscribed in a second sphere, the volumes of the polyhedra are to one another in the triplicate ratio of the diameters of the respective spheres. This last property is then applied [XII. 18] to prove that spheres are in the triplicate ratio of their diameters.

§ 3. Conic Sections.

In my edition of the *Conics* of Apollonius there is a complete account of all the propositions in conics which are used by Archimedes, classified under three headings, (1) those propositions which he expressly attributes to earlier writers, (2) those which are assumed without any such reference, (3) those which appear to represent new developments of the theory of conics due to Archimedes himself.   As all these properties will appear in this volume in their proper places, it will suffice here to state only such propositions as come under the first heading and a few under the second which may safely be supposed to have been previously known.

Archimedes says that the following propositions "are proved in the elements of conics," i.e. in the earlier treatises of Euclid and Aristaeus.

1. In the *parabola*

(*a*) if PV be the diameter of a segment and QVq the chord parallel to the tangent at P, then $QV = Vq$;

(*b*) if the tangent at Q meet VP produced in T, then $PV = PT$;

(*c*) if two chords QVq, $Q'V'q'$ each parallel to the tangent at P meet the diameter PV in V, V' respectively,

$$PV : PV' = QV^2 : Q'V'^2.$$

2. If straight lines drawn from the same point touch *any conic section* whatever, and if two chords parallel to the respective tangents intersect one another, then the rectangles under the segments of the chords are to one another as the squares on the parallel tangents respectively.

3. The following proposition is quoted as proved "in the conics." If in a parabola p_a be the parameter of the principal ordinates,

QQ' any chord not perpendicular to the axis which is bisected in V by the diameter PV, p the parameter of the ordinates to PV, and if QD be drawn perpendicular to PV, then

$$QV^2 : QD^2 = p : p_a.$$

[*On Conoids and Spheroids*, Prop. 3, which see.]

The properties of a parabola, $PN^2 = p_a . AN$, and $QV^2 = p . PV$, were already well known before the time of Archimedes. In fact the former property was used by Menaechmus, the discoverer of conic sections, in his duplication of the cube.

It may be taken as certain that the following properties of the ellipse and hyperbola were proved in the *Conics* of Euclid.

1. For the ellipse

$$PN^2 : AN . A'N = P'N'^2 : AN' . A'N' = CB^2 : CA^2$$

and $$QV^2 : PV . P'V = Q'V'^2 : PV' . P'V' = CD^2 : CP^2.$$

(Either proposition could in fact be derived from the proposition about the rectangles under the segments of intersecting chords above referred to.)

2. For the hyperbola

$$PN^2 : AN . A'N = P'N'^2 : AN' . A'N'$$

and $$QV^2 : PV . P'V = Q'V'^2 : PV' . P'V',$$

though in this case the absence of the conception of the double hyperbola as one curve (first found in Apollonius) prevented Euclid, and Archimedes also, from equating the respective ratios to those of the squares on the parallel semidiameters.

3. In a hyperbola, if P be any point on the curve and PK, PL be each drawn parallel to one asymptote and meeting the other,

$$PK . PL = (\text{const.})$$

This property, in the particular case of the rectangular hyperbola, was known to Menaechmus.

It is probable also that the property of the subnormal of the parabola ($NG = \frac{1}{2} p_a$) was known to Archimedes' predecessors. It is tacitly assumed, *On floating bodies*, II. 4, etc.

From the assumption that, in the hyperbola, $AT < AN$ (where N is the foot of the ordinate from P, and T the point in which the tangent at P meets the transverse axis) we may perhaps infer that the harmonic property

$$TP : TP' = PV : P'V,$$

or at least the particular case of it,

$$TA : TA' = AN : A'N,$$

was known before Archimedes' time.

Lastly, with reference to the genesis of conic sections from cones and cylinders, Euclid had already stated in his *Phaenomena* that, "if a cone or cylinder be cut by a plane not parallel to the base, the resulting section is a section of an acute-angled cone [an ellipse] which is similar to a θυρεός." Though it is not probable that Euclid had in mind any other than a right cone, the statement should be compared with *On Conoids and Spheroids*, Props. 7, 8, 9.

§ 4. Surfaces of the second degree.

Prop. 11 of the treatise *On Conoids and Spheroids* states without proof the nature of certain plane sections of the conicoids of revolution. Besides the obvious facts (1) that sections perpendicular to the axis of revolution are circles, and (2) that sections through the axis are the same as the generating conic, Archimedes asserts the following.

1. In a paraboloid of revolution any plane section parallel to the axis is a parabola equal to the generating parabola.

2. In a hyperboloid of revolution any plane section parallel to the axis is a hyperbola similar to the generating hyperbola.

3. In a hyperboloid of revolution a plane section through the vertex of the enveloping cone is a hyperbola which is not similar to the generating hyperbola.

4. In any spheroid a plane section parallel to the axis is an ellipse similar to the generating ellipse.

Archimedes adds that "the proofs of all these propositions are manifest (φανεραί)." The proofs may in fact be supplied as follows.

1. *Section of a paraboloid of revolution by a plane parallel to the axis.*

Suppose that the plane of the paper represents the plane section through the axis AN which intersects the given plane section at right angles, and let $A'O$ be the line of intersection.

Let POP' be any double ordinate to AN in the section through the axis, meeting $A'O$ and AN at right angles in O, N respectively. Draw $A'M$ perpendicular to AN.

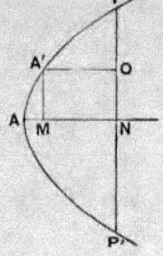

Suppose a perpendicular drawn from O to $A'O$ in the plane of the given section parallel to the axis, and let y be the length intercepted by the surface on this perpendicular.

Then, since the extremity of y is on the circular section whose diameter is PP',

$$y^2 = PO \cdot OP'.$$

If $A'O = x$, and if p is the principal parameter of the generating parabola, we have then

$$y^2 = PN^2 - ON^2$$
$$= PN^2 - A'M^2$$
$$= p\,(AN - AM)$$
$$= px,$$

so that the section is a parabola equal to the generating parabola.

2. *Section of a hyperboloid of revolution by a plane parallel to the axis.*

Take, as before, the plane section through the axis which intersects

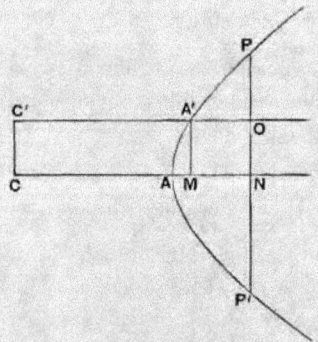

the given plane section at right angles in $A'O$. Let the hyperbola

PAP' in the plane of the paper represent the plane section through the axis, and let C be the centre (or the vertex of the enveloping cone). Draw CC' perpendicular to CA, and produce OA' to meet it in C'. Let the rest of the construction be as before.

Suppose that

$$CA = a, \quad C'A' = a', \quad C'O = x,$$

and let y have the same meaning as before.

Then $$y^2 = PO \cdot OP' = PN^2 - A'M^2.$$

And, by the property of the original hyperbola,

$$PN^2 : CN^2 - CA^2 = A'M^2 : CM^2 - CA^2 \text{ (which is constant).}$$

Thus $$A'M^2 : CM^2 - CA^2 = PN^2 : CN^2 - CA^2$$

$$= PN^2 - A'M^2 : CN^2 - CM^2$$

$$= y^2 : x^2 - a'^2,$$

whence it appears that the section is a hyperbola similar to the original one.

3. *Section of a hyperboloid of revolution by a plane passing through the centre (or the vertex of the enveloping cone).*

I think there can be no doubt that Archimedes would have proved his proposition about this section by means of the same general property of conics which he uses to prove Props. 3 and 12–14 of the same treatise, and which he enunciates at the beginning of Prop. 3 as a known theorem proved in the "elements of conics," viz. that the rectangles under the segments of intersecting chords are as the squares of the parallel tangents.

Let the plane of the paper represent the plane section through the axis which intersects the given plane passing through the centre at right angles. Let $CA'O$ be the line of intersection, C being the centre, and A' being the point where $CA'O$ meets the surface. Suppose $CAMN$ to be the axis of the hyperboloid, and POp, $P'O'p'$ two double ordinates to it in the plane section through the axis, meeting $CA'O$ in O, O' respectively; similarly let $A'M$ be the ordinate from A'. Draw the tangents at A and A' to the section through the axis meeting in T, and let QOq, $Q'O'q'$ be the two double ordinates in the same section which are parallel to the tangent at A' and pass through O, O' respectively.

Suppose, as before, that y, y' are the lengths cut off by the surface from the perpendiculars at O and O' to OC in the plane of the given section through $CA'O$, and that

$$CO = x, \quad CO' = x', \quad CA = a, \quad CA' = a'.$$

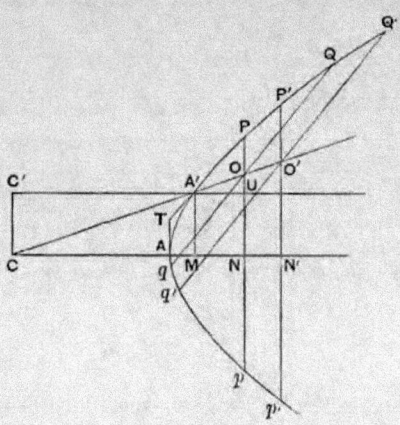

Then, by the property of the intersecting chords, we have, since $QO = Oq$,

$$PO \cdot Op : QO^2 = TA^2 : TA'^2$$
$$= P'O' \cdot O'p' : Q'O'^2.$$

Also $\qquad y^2 = PO \cdot Op, \quad y'^2 = P'O' \cdot O'p',$

and, by the property of the hyperbola,

$$QO^2 : x^2 - a'^2 = Q'O'^2 : x'^2 - a'^2.$$

It follows, *ex aequali*, that

$$y^2 : x^2 - a'^2 = y'^2 : x'^2 - a'^2 \quad\ldots\ldots\ldots\ldots\ldots(a),$$

and therefore that the section is a hyperbola.

To prove that this hyperbola is not similar to the generating hyperbola, we draw CC' perpendicular to CA, and $C'A'$ parallel to CA meeting CC' in C' and Pp in U.

If then the hyperbola (a) is similar to the original hyperbola, it must by the last proposition be similar to the hyperbolic section made by the plane through $C'A'U$ at right angles to the plane of the paper.

Now $\quad CO^2 - CA'^2 = (C'U^2 - C'A'^2) + (CC' + OU)^2 - CC'^2$
$$> C'U^2 - C'A'^2,$$

and $\qquad\qquad PO \cdot Op < PU \cdot Up.$

Therefore $PO . Op : CO^2 - CA'^2 < PU . Up : C'U^2 - C'A'^2,$

and it follows that the hyperbolas are not similar*.

4. *Section of a spheroid by a plane parallel to the axis.*

That this is an ellipse similar to the generating ellipse can of course be proved in exactly the same way as theorem (2) above for the hyperboloid.

---

* I think Archimedes is more likely to have used this proof than one on the lines suggested by Zeuthen (p. 421). The latter uses the equation of the hyperbola simply and proceeds thus. If y have the same meaning as above, and if the coordinates of P referred to CA, CC' as axes be z, x, while those of O referred to the same axes are z, x', we have, for the point P,

$$x^2 = \kappa (z^2 - a^2),$$

where κ is constant.

Also, since the angle $A'CA$ is given, $x' = az$, where a is constant.

Thus $\qquad y^2 = x^2 - x'^2 = (\kappa - a^2) z^2 - \kappa a^2.$

Now z is proportional to CO, being in fact equal to $\dfrac{CO}{\sqrt{1+a^2}}$, and the equation becomes

$$y^2 = \frac{\kappa - a^2}{1 + a^2} . CO^2 - \kappa a^2 \dots\dots\dots\dots\dots\dots\dots(1),$$

which is clearly a hyperbola, since $a^2 < \kappa$.

Now, though the Greeks could have worked out the proof in a geometrical form equivalent to the above, I think that it is alien from the manner in which Archimedes regarded the equations to central conics. These he always expressed in the form of a proportion

$$\frac{y^2}{x^2 \sim a^2} = \frac{y'^2}{x'^2 \sim a^2} \quad \left[= \frac{b^2}{a^2} \text{ in the case of the ellipse} \right],$$

and never in the form of an equation between areas like that used by Apollonius, viz.

$$y^2 = px \pm \frac{p}{d} x^2.$$

Moreover the occurrence of the two different constants and the necessity of expressing them geometrically as ratios between areas and lines respectively would have made the proof very long and complicated; and, as a matter of fact, Archimedes never does express the ratio $y^2/(x^2 - a^2)$ in the case of the *hyperbola* in the form of a ratio between constant areas like b^2/a^2. Lastly, when the equation of the given section through $CA'O$ was found in the form (1), assuming that the Greeks had actually found the geometrical equivalent, it would still have been held necessary, I think, to verify that

$$CA'^2 = \frac{\kappa (1 + a^2)}{\kappa - a^2} . a^2,$$

before it was finally pronounced that the hyperbola represented by the equation and the section made by the plane were one and the same thing.

We are now in a position to consider the meaning of Archimedes'
remark that "the proofs of all these properties are manifest."   In
the first place, it is not likely that "manifest" means "known" as
having been proved by earlier geometers ; for Archimedes' habit is
to be precise in stating the fact whenever he uses important
propositions due to his immediate predecessors, as witness his
references to Eudoxus, to the *Elements* [of Euclid], and to the
"elements of conics."   When we consider the remark with reference
to the cases of the sections parallel to the axes of the surfaces
respectively, a natural interpretation of it is to suppose that
Archimedes meant simply that the theorems are such as can easily
be deduced from the fundamental properties of the three conics now
expressed by their equations, coupled with the consideration that
the sections by planes perpendicular to the axes are circles.   But I
think that this particular explanation of the "manifest" character
of the proofs is not so applicable to the third of the theorems
stating that any plane section of a hyperboloid of revolution
through the vertex of the enveloping cone but not through the axis
is a hyperbola.   This fact is indeed no more "manifest" in the
ordinary sense of the term than is the like theorem about the
spheroid, viz. that any section through the centre but not through
the axis is an ellipse.   But this latter theorem is not given along
with the other in Prop. 11 as being "manifest"; the proof of it is
included in the more general proposition (14) that any section of a
spheroid not perpendicular to the axis is an ellipse, and that parallel
sections are similar.   Nor, seeing that the propositions are essentially similar in character, can I think it possible that Archimedes
wished it to be understood, as Zeuthen suggests, that the proposition
about the hyperboloid alone, and not the other, should be proved
directly by means of the geometrical equivalent of the Cartesian
equation of the conic, and not by means of the property of the
rectangles under the segments of intersecting chords, used earlier
[Prop. 3] with reference to the parabola and later for the case of
the spheroid and the elliptic sections of the conoids and spheroids
generally.   This is the more unlikely, I think, because the proof
by means of the equation of the conic alone would present much
more difficulty to the Greek, and therefore could hardly be called
"manifest."

It seems necessary therefore to seek for another explanation,
and I think it is the following.   The theorems, numbered 1, 2, and

4 above, about sections of conoids and spheroids parallel to the axis are used afterwards in Props. 15—17 relating to tangent planes; whereas the theorem (3) about the section of the hyperboloid by a plane through the centre but not through the axis is not used in connexion with tangent planes, but only for formally proving that a straight line drawn from any point on a hyperboloid parallel to any transverse diameter of the hyperboloid falls, on the convex side of the surface, without it, and on the concave side within it. Hence it does not seem so probable that the four theorems were collected in Prop. 11 on account of the use made of them later, as that they were inserted in the particular place with special reference to the three propositions (12—14) immediately following and treating of the elliptic sections of the three surfaces. The main object of the whole treatise was the determination of the volumes of segments of the three solids cut off by planes, and hence it was first necessary to determine all the sections which were ellipses or circles and therefore could form the bases of the segments. Thus in Props. 12—14 Archimedes addresses himself to finding the elliptic sections, but, before he does this, he gives the theorems grouped in Prop. 11 by way of clearing the ground, so as to enable the propositions about elliptic sections to be enunciated with the utmost precision. Prop. 11 contains, in fact, *explanations* directed to defining the scope of the three following propositions rather than theorems definitely enunciated for their own sake; Archimedes thinks it necessary to explain, before passing to elliptic sections, that sections perpendicular to the axis of each surface are not ellipses but circles, and that some sections of each of the two conoids are neither ellipses nor circles, but parabolas and hyperbolas respectively. It is as if he had said, "My object being to find the volumes of segments of the three solids cut off by circular or elliptic sections, I proceed to consider the various elliptic sections; but I should first explain that sections at right angles to the axis are not ellipses but circles, while sections of the conoids by planes drawn in a certain manner are neither ellipses nor circles, but parabolas and hyperbolas respectively. With these last sections I am not concerned in the next propositions, and I need not therefore cumber my book with the proofs; but, as some of them can be easily supplied by the help of the ordinary properties of conics, and others by means of the methods illustrated in the propositions now about to be given, I leave them as an exercise for the reader." This will, I think, completely explain the assumption

of all the theorems except that concerning the sections of a *spheroid* parallel to the axis; and I think this is mentioned along with the others for symmetry, and because it can be proved in the same way as the corresponding one for the hyperboloid, whereas, if mention of it had been postponed till Prop. 14 about the elliptic sections of a spheroid generally, it would still require a proposition for itself, since the axes of the sections dealt with in Prop. 14 make an angle with the axis of the spheroid and are not parallel to it.

At the same time the fact that Archimedes omits the proofs of the theorems about sections of conoids and spheroids parallel to the axis as "manifest" is in itself sufficient to raise the presumption that contemporary geometers were familiar with the idea of three dimensions and knew how to apply it in practice. This is no matter for surprise, seeing that we find Archytas, in his solution of the problem of the two mean proportionals, using the intersection of a certain cone with a curve of double curvature traced on a right circular cylinder*. But, when we look for other instances of early investigations in geometry of three dimensions, we find practically nothing except a few vague indications as to the contents of a lost treatise of Euclid's consisting of two Books entitled *Surface-loci* (τόποι πρὸς ἐπιφανείᾳ)†. This treatise is mentioned by Pappus among other works by Aristaeus, Euclid and Apollonius grouped as forming the so-called τόπος ἀναλυόμενος‡. As the other works in the list which were on plane subjects dealt only with straight lines, circles and conic sections, it is *a priori* likely that the *surface-loci* of

---

* Cf. Eutocius on Archimedes (Vol. III. pp. 98—102), or *Apollonius of Perga*, pp. xxii.—xxiii.

† By this term we conclude that the Greeks meant "loci which are surfaces" as distinct from loci which are lines. Cf. Proclus' definition of a locus as "a position of a line or a surface involving one and the same property" (γραμμῆς ἢ ἐπιφανείας θέσις ποιοῦσα ἓν καὶ ταὐτὸν σύμπτωμα), p. 394. Pappus (pp. 660—2) gives, quoting from the *Plane Loci* of Apollonius, a classification of loci according to their *order* in relation to that of which they are the loci. Thus, he says, loci are (1) ἐφεκτικοί, i.e. *fixed*, e.g. in this sense the locus of a point is a point, of a line a line, and so on; (2) διεξοδικοί or *moving along*, a line being in this sense the locus of a point, a surface of a line, and a solid of a surface; (3) ἀναστροφικοί, *turning backwards*, i.e., presumably, moving backwards and forwards, a surface being in this sense the locus of a point, and a solid of a line. Thus a *surface-locus* might apparently be either the locus of a point or the locus of a line moving in space.

‡ Pappus, pp. 634, 636.

Euclid included at least such loci as were cones, cylinders and spheres. Beyond this, all is conjecture based upon two lemmas given by Pappus in connexion with the treatise.

First lemma to the Surface-loci *of Euclid*.*

The text of this lemma and the attached figure are not satisfactory as they stand, but they have been explained by Tannery in a way which requires a change in the figure, but only the very slightest alteration in the text, as follows†.

"If AB be a straight line and CD be parallel to a straight line given in position, and if the ratio $AD \cdot DB : DC^2$ be [given], the

point C lies on a conic section. If now AB be no longer given in position and A, B be no longer given but lie on straight lines AE, EB given in position‡, the point C raised above [the plane containing AE, EB] is on a surface given in position. And this was proved."

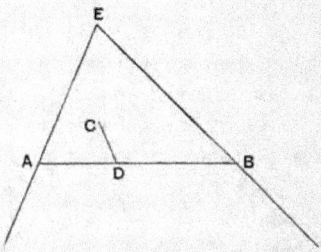

According to this interpretation, it is asserted that, if AB moves with one extremity on each of the lines AE, EB which are fixed, while DC is in a fixed direction and $AD \cdot DB : DC^2$ is constant, then C lies on a certain surface. So far as the first sentence is concerned, AB remains of constant length, but it is not made precisely clear whether, when AB is no longer given in position, its length may also vary§. If however AB remains of constant length for all positions which it assumes, the surface which is the locus of C would be a complicated one which we cannot suppose that Euclid could have profitably investigated. It may, therefore, be that Pappus purposely left the enunciation somewhat vague in order to make it appear to cover several surface-loci which, though belonging to the same type, were separately discussed by Euclid as involving

---

* Pappus, p. 1004.

† *Bulletin des sciences math.*, 2ᵉ Série, VI. 149.

‡ The words of the Greek text are γένηται δὲ πρὸς θέσει εὐθεῖα ταῖς AE, EB, and the above translation only requires εὐθείαις instead of εὐθεῖα. The figure in the text is so drawn that ADB, AEB are represented as two parallel lines, and CD is represented as perpendicular to ADB and meeting AEB in E.

§ The words are simply "if AB be deprived of its position (στερηθῇ τῆς θέσεως) and the points A, B be deprived of their [character of] being given" (στερηθῇ τοῦ δοθέντος εἶναι).

in each case somewhat different sets of conditions limiting the generality of the theorem.

It is at least open to conjecture, as Zeuthen has pointed out*, that two cases of the type were considered by Euclid, namely, (1) that in which AB remains of constant length while the two fixed straight lines on which A, B respectively move are parallel instead of meeting in a point, and (2) that in which the two fixed straight lines meet in a point while AB moves always parallel to itself and varies in length accordingly.

(1) In the first case, where the length of AB is constant and the two fixed lines parallel, we should have a surface described by a conic moving bodily†. This surface would be a cylindrical surface, though it would only have been called a "cylinder" by the ancients in the case where the moving conic was an *ellipse*, since the essence of a "cylinder" was that it could be bounded between two parallel circular sections. If then the moving conic was an ellipse, it would not be difficult to find the circular sections of the cylinder; this could be done by first taking a section at right angles to the axis, after which it could be proved, after the manner of Archimedes, *On Conoids and Spheroids*, Prop. 9, first that the section is an ellipse or a circle, and then, in the former case, that a section made by a plane drawn at a certain inclination to the ellipse and passing through, or parallel to, the major axis is a circle. There was nothing to prevent Euclid from investigating the surface similarly generated by a moving hyperbola or parabola; but there would be no circular sections, and hence the surfaces might perhaps not have been considered as of very great importance.

(2) In the second case, where AE, BE meet at a point and AB moves always parallel to itself, the surface generated is of course a cone. Some particular cases of this sort may easily have been discussed by Euclid, but he could hardly have dealt with the general case, where DC has any direction whatever, up to the point of showing that the surface was really a cone in the sense in which the Greeks understood the term, or (in other words) of finding the circular sections. To do this it would have been necessary to determine the principal planes, or to solve the dis-

---

* Zeuthen, *Die Lehre von den Kegelschnitten*, pp. 425 sqq.

† This would give a surface generated by a moving line, διεξοδικὸς γραμμῆς as Pappus has it.

criminating cubic, which we cannot suppose Euclid to have done.
Moreover, if Euclid had found the circular sections in the most
general case, Archimedes would simply have referred to the fact
instead of setting himself to do the same thing in the particular
case where the plane of symmetry is given. These remarks apply
to the case where the conic which is the locus of C is an ellipse;
there is still less ground for supposing that Euclid could have
proved the existence of circular sections where the conic was a
hyperbola, for there is no evidence that Euclid even knew that
hyperbolas and parabolas could be obtained by cutting an oblique
circular cone.

Second lemma to the Surface-loci.

In this Pappus states, and gives a complete proof of the proposition, that *the locus of a point whose distance from a given point
is in a given ratio to its distance from a fixed line is a conic
section, which is an ellipse, a parabola, or a hyperbola according
as the given ratio is less than, equal to, or greater than unity**.
Two conjectures are possible as to the application of this theorem
by Euclid in the treatise referred to.

(1) Consider a plane and a straight line meeting it at any angle.
Imagine any plane drawn at right angles to the straight line and
meeting the first plane in another straight line which we will call
X. If then the given straight line meets the plane at right angles
to it in the point S, a conic can be described in that plane with
S for focus and X for directrix; and, as the perpendicular on X
from any point on the conic is in a constant ratio to the perpendicular from the same point on the original plane, all points
on the conic have the property that their distances from S are in
a given ratio to their distances from the given plane respectively.
Similarly, by taking planes cutting the given straight line at right
angles in any number of other points besides S, we see that *the locus
of a point whose distance from a given straight line is in a given
ratio to its distance from a given plane is a cone whose vertex is
the point in which the given line meets the given plane, while the
plane of symmetry passes through the given line and is at right
angles to the given plane.* If the given ratio was such that the
guiding conic was an ellipse, the circular sections of the surface

---

* See Pappus, pp. 1006—1014, and Hultsch's Appendix, pp. 1270—1273; or
cf. *Apollonius of Perga*, pp. xxxvi.—xxxviii.

could, in that case at least, be found by the same method as that used by Archimedes (*On Conoids and Spheroids*, Prop. 8) in the rather more general case where the perpendicular from the vertex of the cone on the plane of the given elliptic section does not necessarily pass through the focus.

(2) Another natural conjecture would be to suppose that, by means of the proposition given by Pappus, Euclid found *the locus of a point whose distance from a given point is in a given ratio to its distance from a fixed plane.* This would have given surfaces identical with the conoids and spheroids discussed by Archimedes excluding the spheroid generated by the revolution of an ellipse about the *minor* axis. We are thus brought to the same point as Chasles who conjectured that the *Surface-loci* of Euclid dealt with surfaces of revolution of the second degree and sections of the same*. Recent writers have generally regarded this theory as improbable. Thus Heiberg says that the conoids and spheroids were without any doubt discovered by Archimedes himself; otherwise he would not have held it necessary to give exact definitions of them in his introductory letter to Dositheus; hence they could not have been the subject of Euclid's treatise†. I confess I think that the argument of Heiberg, so far from being conclusive against the probability of Chasles' conjecture, is not of any great weight. To suppose that Euclid found, by means of the theorem enunciated and proved by Pappus, the locus of a point whose distance from a given point is in a given ratio to its distance from a fixed plane does not oblige us to assume either that he gave a name to the loci or that he investigated them further than to show that sections through the perpendicular from the given point on the given plane were conics, while sections at right angles to the same perpendicular were circles; and of course these facts would readily suggest themselves. Seeing however that the object of Archimedes was to find the volumes of segments of each surface, it is not surprising that he should have preferred to give a definition of them which would indicate their form more directly than a description of them as loci would have done; and we have a parallel case in the distinction drawn between conics as such and conics regarded as loci, which is illustrated by the different titles of Euclid's *Conics* and the *Solid Loci* of Aristaeus, and also by the fact that Apollonius,

* *Aperçu historique*, pp. 273, 4.
† *Litterargeschichtliche Studien über Euklid*, p. 79.

though he speaks in his preface of some of the theorems in his *Conics* as useful for the synthesis of 'solid loci' and goes on to mention the 'locus with respect to three or four lines,' yet enunciates no proposition stating that the locus of such and such a point is a conic. There was a further special reason for defining the conoids and spheroids as surfaces described by the revolution of a conic about its axis, namely that this definition enabled Archimedes to include the spheroid which he calls 'flat' (ἐπιπλατὺ σφαιροειδές), i.e. the spheroid described by the revolution of an ellipse about its *minor* axis, which is not one of the loci which the hypothesis assumes Euclid to have discovered. Archimedes' new definition had the incidental effect of making the nature of the sections through and perpendicular to the axis of revolution even more obvious than it would be from Euclid's supposed way of treating the surfaces; and this would account for Archimedes' omission to state that the two classes of sections had been known before, for there would have been no point in attributing to Euclid the proof of propositions which, with the new definition of the surfaces, became self-evident. The further definitions given by Archimedes may be explained on the same principle. Thus the *axis*, as defined by him, has special reference to his definition of the surfaces, since it means the *axis of revolution*, whereas the axis of a *conic* is for Archimedes a *diameter*. The *enveloping cone* of the hyperboloid, which is generated by the revolution of the asymptotes about the axis, and the centre regarded as the point of intersection of the asymptotes were useful to Archimedes' discussion of the surfaces, but need not have been brought into Euclid's description of the surfaces as loci. Similarly with the *axis* and *vertex* of a *segment* of each surface. And, generally, it seems to me that all the definitions given by Archimedes can be explained in like manner without prejudice to the supposed discovery of three of the surfaces by Euclid.

I think, then, that we may still regard it as possible that Euclid's *Surface-loci* was concerned, not only with cones, cylinders and (probably) spheres, but also (to a limited extent) with three other surfaces of revolution of the second degree, viz. the paraboloid, the hyperboloid and the prolate spheroid. Unfortunately however we are confined to the statement of possibilities; and certainty can hardly be attained unless as the result of the discovery of fresh documents.

§ 5. Two mean proportionals in continued proportion.

Archimedes assumes the construction of two mean proportionals in two propositions (*On the Sphere and Cylinder* II. 1, 5). Perhaps he was content to use the constructions given by Archytas, Menaechmus*, and Eudoxus. It is worth noting, however, that Archimedes does not introduce the two geometric means where they are merely convenient but not necessary; thus, when (*On the Sphere and Cylinder* I. 34) he has to substitute for a ratio $\left(\dfrac{\beta}{\gamma}\right)^{\frac{3}{2}}$, where $\beta > \gamma$, a ratio between lines, and it is sufficient for his purpose that the required ratio cannot be *greater* than $\left(\dfrac{\beta}{\gamma}\right)^{\frac{3}{2}}$ but may be less, he takes two *arithmetic* means between β, γ, as δ, ϵ, and then assumes† as a known result that

$$\frac{\beta^3}{\delta^3} < \frac{\beta}{\gamma}.$$

* The constructions of Archytas and Menaechmus are given by Eutocius [*Archimedes*, Vol. III. pp. 92—102]; or see *Apollonius of Perga*, pp. xix—xxiii.

† The proposition is proved by Eutocius; see the note to *On the Sphere and Cylinder* I. 34 (p. 42).

CHAPTER IV.

ARITHMETIC IN ARCHIMEDES.

Two of the treatises, the *Measurement of a circle* and the *Sand-reckoner*, are mostly arithmetical in content. Of the *Sand-reckoner* nothing need be said here, because the system for expressing numbers of any magnitude which it unfolds and applies cannot be better described than in the book itself; in the *Measurement of a circle*, however, which involves a great deal of manipulation of numbers of considerable size though expressible by means of the ordinary Greek notation for numerals, Archimedes merely gives the results of the various arithmetical operations, multiplication, extraction of the square root, etc., without setting out any of the operations themselves. Various interesting questions are accordingly involved, and, for the convenience of the reader, I shall first give a short account of the Greek system of numerals and of the methods by which other Greek mathematicians usually performed the various operations included under the general term λογιστική (the art of *calculating*), in order to lead up to an explanation (1) of the way in which Archimedes worked out approximations to the square roots of large numbers, (2) of his method of arriving at the two approximate values of $\sqrt{3}$ which he simply sets down without any hint as to how they were obtained*.

* In writing this chapter I have been under particular obligations to Hultsch's articles *Arithmetica* and *Archimedes* in Pauly-Wissowa's *Real-Encyclopädie*, II. 1, as well as to the same scholar's articles (1) *Die Näherungswerthe irrationaler Quadratwurzeln bei Archimedes* in the *Nachrichten von der kgl. Gesellschaft der Wissenschaften zu Göttingen* (1893), pp. 367 sqq., and (2) *Zur Kreismessung des Archimedes* in the *Zeitschrift für Math. u. Physik (Hist. litt. Abtheilung)* xxxix. (1894), pp. 121 sqq. and 161 sqq. I have also made use, in the earlier part of the chapter, of Nesselmann's work *Die Algebra der Griechen* and the histories of Cantor and Gow.

§ 1. Greek numeral system.

It is well known that the Greeks expressed all numbers from 1 to 999 by means of the letters of the alphabet reinforced by the addition of three other signs, according to the following scheme, in which however the accent on each letter might be replaced by a short horizontal stroke above it, as $\bar{a}$.

a', β', γ', δ', ϵ', ς', ζ', η', θ' are 1, 2, 3, 4, 5, 6, 7, 8, 9 respectively.

ι', κ', λ', μ', ν', ξ', o', π', ς' „ 10, 20, 30, ... 90 „

ρ', σ', τ', υ', ϕ', χ', ψ', ω', λ'„ 100, 200, 300,......900 „

Intermediate numbers were expressed by simple juxtaposition (representing in this case addition), the largest number being placed on the left, the next largest following it, and so on in order. Thus the number 153 would be expressed by $\rho\nu\gamma'$ or $\overline{\rho\nu\gamma}$. There was no sign for zero, and therefore 780 was $\psi\pi'$, and 306 $\tau\varsigma'$ simply.

Thousands ($\chi\iota\lambda\iota\acute{a}\delta\epsilon\varsigma$) were taken as units of a higher order, and 1,000, 2,000, ... up to 9,000 (spoken of as $\chi\acute{\iota}\lambda\iota\omicron\iota$, $\delta\iota\sigma\chi\acute{\iota}\lambda\iota\omicron\iota$, $\kappa.\tau.\lambda.$) were represented by the same letters as the first nine natural numbers but with a small dash in front and below the line; thus e.g. $,\delta'$ was 4,000, and, on the same principle of juxtaposition as before, 1,823 was expressed by $,a\omega\kappa\gamma'$ or $,\overline{a\omega\kappa\gamma}$, 1,007 by $,a\zeta'$, and so on.

Above 9,999 came a *myriad* ($\mu\upsilon\rho\iota\acute{a}\varsigma$), and 10,000 and higher numbers were expressed by using the ordinary numerals with the substantive $\mu\upsilon\rho\iota\acute{a}\delta\epsilon\varsigma$ taken as a new denomination (though the words $\mu\acute{\upsilon}\rho\iota\omicron\iota$, $\delta\iota\sigma\mu\acute{\upsilon}\rho\iota\omicron\iota$, $\tau\rho\iota\sigma\mu\acute{\upsilon}\rho\iota\omicron\iota$, $\kappa.\tau.\lambda.$ are also found, following the analogy of $\chi\acute{\iota}\lambda\iota\omicron\iota$, $\delta\iota\sigma\chi\acute{\iota}\lambda\iota\omicron\iota$ and so on). Various abbreviations were used for the word $\mu\upsilon\rho\iota\acute{a}\varsigma$, the most common being M or Mυ; and, where this was used, the number of myriads, or the multiple of 10,000, was generally written over the abbreviation, though sometimes before it and even after it. Thus 349,450 was M$\overset{\lambda\delta}{\theta\upsilon\nu'}$*.

Fractions ($\lambda\epsilon\pi\tau\acute{a}$) were written in a variety of ways. The most usual was to express the denominator by the ordinary numeral with two accents affixed. When the numerator was unity, and it was therefore simply a question of a symbol for a single word such as

---

* Diophantus denoted myriads followed by thousands by the ordinary signs for numbers of units, only separating them by a dot from the thousands. Thus for 3,069,000 he writes $\tau\varsigma$. $\bar{\theta}$, and $\overline{\lambda\gamma}$. $_{,}a\psi_{oς}$ for 331,776. Sometimes myriads were represented by the ordinary letters with two dots above, as $\overset{..}{\rho}=100$ myriads (1,000,000), and myriads of myriads with two pairs of dots, as $\overset{::}{\iota}$ for 10 myriad-myriads (1,000,000,000).

τρίτον, $\frac{1}{3}$, there was no need to express the numerator, and the symbol was γ″; similarly $\varsigma''=\frac{1}{6}$, $\iota\epsilon''=\frac{1}{15}$, and so on. When the numerator was not unity and a certain number of fourths, fifths, etc., had to be expressed, the ordinary numeral was used for the numerator; thus $\theta'\ \iota\alpha''=\frac{9}{11}$, $\iota'\ oa''=\frac{10}{71}$. In Heron's *Geometry* the denominator was written twice in the latter class of fractions; thus $\frac{2}{5}$ (δύο πέμπτα) was β′ε″ε″, $\frac{23}{33}$ (λεπτὰ τριακοστότριτα κγ′ or εἰκοσιτρία τριακοστότριτα) was κγ′ λγ″ λγ″. The sign for $\frac{1}{2}$, ἥμισυ, is in Archimedes, Diophantus and Eutocius ∟″, in Heron ᑕ or a sign similar to a capital **S***.

A favourite way of expressing fractions with numerators greater than unity was to separate them into component fractions with numerator unity, when juxtaposition as usual meant addition. Thus $\frac{3}{4}$ was written ∟″δ″ $=\frac{1}{2}+\frac{1}{4}$; $\frac{15}{16}$ was ᑕδ″η″ιϛ″ $=\frac{1}{2}+\frac{1}{4}+\frac{1}{8}+\frac{1}{16}$; Eutocius writes ∟″ξδ″ or $\frac{1}{2}+\frac{1}{64}$ for $\frac{33}{64}$, and so on. Sometimes the same fraction was separated into several different sums; thus in Heron (p. 119, ed. Hultsch) $\frac{163}{224}$ is variously expressed as

(a) $\frac{1}{2}+\frac{1}{7}+\frac{1}{11}+\frac{1}{112}+\frac{1}{224}$,

(b) $\frac{1}{2}+\frac{1}{8}+\frac{1}{16}+\frac{1}{32}+\frac{1}{112}$,

and   (c) $\frac{1}{2}+\frac{1}{6}+\frac{1}{21}+\frac{1}{112}+\frac{1}{224}$.

Sexagesimal fractions. This system has to be mentioned because the only instances of the working out of some arithmetical operations which have been handed down to us are calculations expressed in terms of such fractions; and moreover they are of special interest as having much in common with the modern system of *decimal* fractions, with the difference of course that the submultiple is 60 instead of 10. The scheme of sexagesimal fractions was used by the Greeks in astronomical calculations and appears fully developed in the σύνταξις of Ptolemy. The circumference of a circle, and along with it the four right angles subtended by it at the centre, are divided into 360 parts (τμήματα or μοῖραι) or as we should say *degrees*, each μοῖρα into 60 parts called (*first*) *sixtieths*, (πρῶτα) ἑξηκοστά, or *minutes* (λεπτά), each of these again into δεύτερα ἑξηκοστά (*seconds*), and so on. A similar division of the radius of the circle into 60

---

* Diophantus has a general method of expressing fractions which is the exact reverse of modern practice; the denominator is written *above* the numerator, thus $\frac{\gamma}{\epsilon}=5/3$, $\frac{\kappa\epsilon}{\kappa\alpha}=21/25$, and $\frac{\alpha.\omega\iota\varsigma}{\rho\kappa\varsigma'.\phi\xi\eta}=1{,}270{,}568/10{,}816$. Sometimes he writes down the numerator and then introduces the denominator with ἐν μορίῳ or μορίου, e.g. $\overline{\tau\varsigma}.\overline{\theta}$ μορ. $\overline{\lambda\gamma}$, ͵αψοϛ $=3{,}069{,}000/331{,}776$.

parts (τμήματα) was also made, and these were each subdivided into sixtieths, and so on. Thus a convenient fractional system was available for general arithmetical calculations, expressed in units of any magnitude or character, so many of the fractions which we should represent by $\frac{1}{60}$, so many of those which we should write $(\frac{1}{60})^2$, $(\frac{1}{60})^3$, and so on to any extent. It is therefore not surprising that Ptolemy should say in one place "In general we shall use the method of numbers according to the sexagesimal manner because of the inconvenience of the [ordinary] fractions." For it is clear that the successive submultiples by 60 formed a sort of frame with fixed compartments into which any fractions whatever could be located, and it is easy to see that e.g. in additions and subtractions the sexagesimal fractions were almost as easy to work with as decimals are now, 60 units of one denomination being equal to one unit of the next higher denomination, and "carrying" and "borrowing" being no less simple than it is when the number of units of one denomination necessary to make one of the next higher is 10 instead of 60. In expressing the units of the circumference, *degrees*, μοῖραι or the symbol $μ̊$ was generally used along with the ordinary numeral which had a stroke above it; *minutes, seconds*, etc. were expressed by one, two, etc. accents affixed to the numerals. Thus $μ̊\ \bar{β} = 2°$, μοιρῶν $\bar{μζ}\ μβ'\ μ'' = 47°\ 42'\ 40''$. Also where there was *no* unit in any particular denomination O was used, signifying οὐδεμία μοῖρα, οὐδὲν ἑξηκοστόν and the like; thus $\bar{Ο}\ α'\ β''\ O''' = 0°\ 1'\ 2''\ 0'''$. Similarly, for the units representing the divisions of the radius the word τμήματα or some equivalent was used, and the fractions were represented as before; thus τμημάτων $\overline{ξζ}\ δ'\ νε'' = 67$ (units) 4' 55".

§ 2. Addition and Subtraction.

There is no doubt that, in writing down numbers for these purposes, the several powers of 10 were kept separate in a manner corresponding practically to our system of numerals, and the hundreds, thousands, etc., were written in separate vertical rows. The following would therefore be a typical form of a sum in addition;

$$
\begin{array}{ll}
_{,}αν\,κ\,δ' = & 1424 \\
ρ\ \ γ' & 103 \\
\overset{a}{\mathrm{M}}\,βσπ\,α' & 12281 \\
\overset{γ}{\mathrm{M}}\ \ λ' & 30030 \\
\hline
\overset{δ}{\mathrm{M}}\,γωλ\,η' & 43838
\end{array}
$$

and the mental part of the work would be the same for the Greek as
for us.

Similarly a subtraction would be represented as follows:

$$\overset{\theta}{\text{M}} \gamma \chi \lambda \varsigma' = 93636$$

$$\overset{\beta}{\text{M}} \gamma \upsilon \ \theta' \quad 23409$$

$$\overset{\varsigma}{\text{M}} \ \sigma \kappa \zeta \quad 70227$$

§ 3. Multiplication.

A number of instances are given in Eutocius' commentary on
the *Measurement of a circle*, and the similarity to our procedure is
just as marked as in the above cases of addition and subtraction.
The multiplicand is written first, and below it the multiplier preceded
by ἐπί (= "into"). Then the highest power of 10 in the multiplier
is taken and multiplied into the terms containing the separate
multiples of the successive powers of 10, beginning with the highest
and descending to the lowest; after which the next highest power
of 10 in the multiplier is multiplied into the various denominations
in the multiplicand in the same order. The same procedure is
followed where either or both of the numbers to be multiplied
contain fractions. Two instances from Eutocius are appended from
which the whole procedure will be understood.

(1)

$\psi \pi'$			780	
ἐπὶ $\psi \pi'$			× 780	
$\overset{\mu\theta \ \epsilon}{\text{MM}\varsigma'}$			490000	56000
$\overset{\epsilon}{\text{M}\varsigma,\varsigma\upsilon'}$			56000	6400
ὁμοῦ $\overset{\varsigma}{\text{M}}\eta\upsilon'$			*sum* 608400	

(2)

$\gamma\iota\gamma'$ $\llcorner''\delta''$		$3013\frac{1}{2}\frac{1}{4}$ $[=3013\frac{3}{4}]$			
ἐπὶ $\gamma\iota\gamma'$ $\llcorner''\delta''$		× $3013\frac{1}{2}\frac{1}{4}$			
$\overset{\lambda \ \gamma}{\text{MM}}\theta,\alpha\phi\psi\nu'$	9,000,000	30,000	9,000	1500	750
$\overset{\gamma}{\text{M}}\rho\lambda\epsilon'\beta'$ $\llcorner''$	30,000	100	30	5	$2\frac{1}{2}$
$\theta\lambda\theta'\alpha'$ $\llcorner''$ $\llcorner''\delta''$	9,000	30	9	$1\frac{1}{2}$ $\frac{1}{2}+\frac{1}{4}$	
$\alpha\phi'\epsilon\alpha'$ $\llcorner''\delta''\eta''$	1,500	5	$1\frac{1}{2}$	$\frac{1}{4}$	$\frac{1}{8}$
$\psi\nu'\beta'$ $\llcorner''$ $\llcorner''\delta''\eta''\iota\varsigma''$	750	$2\frac{1}{2}$	$\frac{1}{2}+\frac{1}{4}$	$\frac{1}{8}$	$\frac{1}{16}$

$[\text{ὁμοῦ}]\ \overset{\lambda\eta}{\text{M}}\beta\chi\pi\theta'\iota\varsigma''$ $[9,041,250 + 30,137\frac{1}{2} + 9,041\frac{1}{4} + 1506 + \frac{1}{2} + \frac{1}{4} + \frac{1}{8}$
$$+ 753 + \frac{1}{4} + \frac{1}{8} + \frac{1}{16}]$$

$$= 9,082,689\tfrac{1}{16}.$$

One instance of a similar multiplication of numbers involving fractions may be given from Heron (pp. 80, 81). It is only one of many, and, for brevity, the Greek notation will be omitted. Heron has to find the product of $4\frac{33}{64}$ and $7\frac{62}{64}$, and proceeds as follows:

$$4 \cdot 7 = 28,$$
$$4 \cdot \frac{62}{64} = \frac{248}{64},$$
$$\frac{33}{64} \cdot 7 = \frac{231}{64},$$
$$\frac{33}{64} \cdot \frac{62}{64} = \frac{2046}{64} \cdot \frac{1}{64} = \frac{31}{64} + \frac{62}{64} \cdot \frac{1}{64}.$$

The result is accordingly

$$28 + \frac{510}{64} + \frac{62}{64} \cdot \frac{1}{64} = 28 + 7 + \frac{62}{64} + \frac{62}{64} \cdot \frac{1}{64}$$
$$= 35 + \frac{62}{64} + \frac{62}{64} \cdot \frac{1}{64}.$$

The multiplication of $37°$ $4'$ $55''$ (in the sexagesimal system) by itself is performed by Theon of Alexandria in his commentary on Ptolemy's σύνταξις in an exactly similar manner.

§ 4. Division.

The operation of dividing by a number of one digit only was easy for the Greeks as for us, and what we call "long division" was with them performed, *mutatis mutandis*, in the same way as now with the help of multiplication and subtraction. Suppose, for instance, that the operation in the first case of multiplication given above had to be reversed and that $\overset{\xi}{\mathrm{M}}\eta\nu'$ (608,400) had to be divided by $\psi\pi'$ (780). The terms involving the different powers of 10 would be mentally kept separate as in addition and subtraction, and the first question would be, how many times will 7 hundreds go into 60 myriads, due allowance being made for the fact that the 7 hundreds have 80 behind them and that 780 is not far short of 8 hundreds? The answer is 7 hundreds or ψ', and this multiplied by the divisor $\psi\pi'$ (780) would give $\overset{\nu\delta}{\mathrm{M}}\varsigma'$ (546,000) which, subtracted from $\overset{\xi}{\mathrm{M}}\eta\nu'$ (608,400), leaves the remainder $\overset{\varsigma}{\mathrm{M}}\beta\upsilon'$ (62,400). This remainder has then to be divided by 780 or a number approaching 8 hundreds, and 8 tens or π' would have to be tried. In the particular case the result would then be complete, the quotient being $\psi\pi'$ (780), and there being no remainder, since π' (80) multiplied by $\psi\pi'$ (780) gives the exact figure $\overset{\varsigma}{\mathrm{M}}\beta\upsilon'$ (62,400).

An actual case of long division where the dividend and divisor contain sexagesimal fractions is described by Theon. The problem is to divide 1515 20′ 15″ by 25 12′ 10″, and Theon's account of the process comes to this.

Divisor	Dividend			Quotient
25 12′ 10″	1515	20′	15″	First term 60
	25 . 60 = 1500			
	Remainder 15 = 900′			
	Sum 920′			
	12′ . 60 = 720′			
	Remainder 200′			
	10″ . 60 = 10′			
	Remainder 190′			Second term 7′
	25 . 7′ = 175′			
	15′ = 900″			
	Sum 915″			
	12′ . 7′ 84″			
	Remainder 831″			
	10″ . 7′ 1″ 10‴			
	Remainder 829″ 50‴			Third term 33″
	25 . 33″ 825″			
	Remainder 4″ 50‴ = 290‴			
	12′ . 33″ 396‴			
	(too great by) 106‴			

Thus the quotient is something less than 60 7′ 33″. It will be observed that the difference between this operation of Theon's and that followed in dividing $\overset{\xi}{\text{M}}\gamma\nu'$ (608,400) by $\psi\pi'$ (780) as above is that Theon makes *three* subtractions for one term of the quotient, whereas the remainder was arrived at in the other case after *one* subtraction. The result is that, though Theon's method is quite clear, it is longer, and moreover makes it less easy to foresee what will be the proper figure to try in the quotient, so that more time would be apt to be lost in making unsuccessful trials.

§ 5. Extraction of the square root.

We are now in a position to see how the operation of extracting the square root would be likely to be attacked. First, as in the case of division, the given whole number whose square root is required would be separated, so to speak, into compartments each containing

such and such a number of units and of the separate powers of 10.
Thus there would be so many units, so many tens, so many hundreds,
etc., and it would have to be borne in mind that the squares of
numbers from 1 to 9 would lie between 1 and 99, the squares of
numbers from 10 to 90 between 100 and 9900, and so on. Then the
first term of the square root would be some number of tens or
hundreds or thousands, and so on, and would have to be found in
much the same way as the first term of a quotient in a "long
division," by trial if necessary. If A is the number whose square
root is required, while a represents the first term or denomination of
the square root and x the next term or denomination still to be
found, it would be necessary to use the identity $(a + x)^2 = a^2 + 2ax + x^2$
and to find x so that $2ax + x^2$ might be somewhat less than the
remainder $A - a^2$. Thus by trial the highest possible value of x
satisfying the condition would be easily found. If that value were
b, the further quantity $2ab + b^2$ would have to be subtracted from
the first remainder $A - a^2$, and from the second remainder thus left
a third term or denomination of the square root would have to be
derived, and so on. That this was the actual procedure adopted is
clear from a simple case given by Theon in his commentary on the
σύνταξις. Here the square root of 144 is in question, and it is
obtained by means of Eucl. II. 4. The highest possible denomination (i.e. power of 10) in the square root is 10 ; 10^2 subtracted from
144 leaves 44, and this must contain not only twice the product of
10 and the next term of the square root but also the square of that
next term itself. Now, since $2 . 10$ itself produces 20, the division
of 44 by 20 suggests 2 as the next term of the square root; and
this turns out to be the exact figure required, since

$$2 . 20 + 2^2 = 44.$$

The same procedure is illustrated by Theon's explanation of
Ptolemy's method of extracting square roots according to the
sexagesimal system of fractions. The problem is to find approximately the square root of 4500 μοῖραι or *degrees*, and a geometrical
figure is used which makes clear the essentially Euclidean basis of
the whole method. Nesselmann gives a complete reproduction of
the passage of Theon, but the following purely arithmetical representation of its purport will probably be found clearer, when looked at
side by side with the figure.

Ptolemy has first found the integral part of $\sqrt{4500}$ to be 67.

Now $67^2 = 4489$, so that the remainder is 11. Suppose now that the rest of the square root is expressed by means of the usual sexagesimal fractions, and that we may therefore put

$$\sqrt{4500} = \sqrt{67^2 + 11} = 67 + \frac{x}{60} + \frac{y}{60^2},$$

where x, y are yet to be found. Thus x must be such that $\dfrac{2.67x}{60}$ is somewhat less than 11, or x must be somewhat less than $\dfrac{11.60}{2.67}$ or $\dfrac{330}{67}$, which is at the same time greater than 4. On trial, it turns out that 4 will satisfy the conditions of the problem, namely that $\left(67 + \dfrac{4}{60}\right)^2$ must be less than 4500, so that a remainder will be left by means of which y may be found.

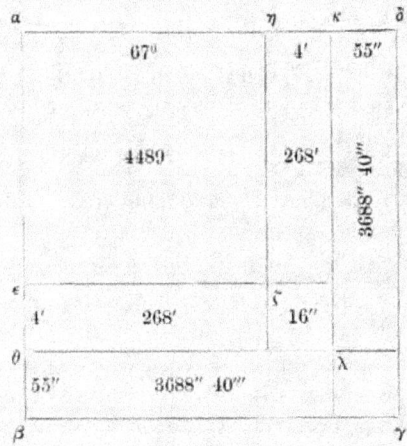

Now $11 - \dfrac{2.67.4}{60} - \left(\dfrac{4}{60}\right)^2$ is the remainder, and this is equal to

$$\frac{11.60^2 - 2.67.4.60 - 16}{60^2} = \frac{7424}{60^2}.$$

Thus we must suppose that $2\left(67 + \dfrac{4}{60}\right)\dfrac{y}{60^2}$ approximates to $\dfrac{7424}{60^2}$, or that $8048y$ is approximately equal to 7424.60.

Therefore y is approximately equal to 55. We have then to subtract

$$2\left(67 + \frac{4}{60}\right)\frac{55}{60^2} + \left(\frac{55}{60^2}\right)^2, \text{ or } \frac{442640}{60^3} + \frac{3025}{60^4},$$

from the remainder $\frac{7424}{60^2}$ above found.

The subtraction of $\frac{442640}{60^3}$ from $\frac{7424}{60^2}$ gives $\frac{2800}{60^3}$, or $\frac{46}{60^2} + \frac{40}{60^3}$; but Theon does not go further and subtract the remaining $\frac{3025}{60^4}$, instead of which he merely remarks that the square of $\frac{55}{60^2}$ approximates to $\frac{46}{60^2} + \frac{40}{60^3}$. As a matter of fact, if we deduct the $\frac{3025}{60^4}$ from $\frac{2800}{60^3}$, so as to obtain the correct remainder, it is found to be $\frac{164975}{60^4}$.

To show the power of this method of extracting square roots by means of sexagesimal fractions, it is only necessary to mention that Ptolemy gives $\frac{103}{60} + \frac{55}{60^2} + \frac{23}{60^3}$ as an approximation to $\sqrt{3}$, which approximation is equivalent to $1 \cdot 7320509$ in the ordinary decimal notation and is therefore correct to 6 places.

But it is now time to pass to the question how Archimedes obtained the two approximations to the value of $\sqrt{3}$ which he assumes in the *Measurement of a circle*. In dealing with this subject I shall follow the historical method of explanation adopted by Hultsch, in preference to any of the mostly *a priori* theories which the ingenuity of a multitude of writers has devised at different times.

§ 6. Early investigations of surds or incommensurables.

From a passage in Proclus' commentary on Eucl. I.* we learn that it was Pythagoras who discovered the *theory of irrationals* (ἡ τῶν ἀλόγων πραγματεία). Further Plato says (*Theaetetus* 147 D), "On square roots this Theodorus [of Cyrene] wrote a work in

---

* p. 65 (ed. Friedlein).

which he proved to us, with reference to those of 3 or 5 [square] feet that they are incommensurable in length with the side of one square foot, and proceeded similarly to select, one by one, each [of the other incommensurable roots] as far as the root of 17 square feet, beyond which for some reason he did not go." The reason why $\sqrt{2}$ is not mentioned as an incommensurable square root must be, as Cantor says, that it was before known to be such. We may therefore conclude that it was the square root of 2 which was geometrically constructed by Pythagoras and proved to be incommensurable with the side of a square in which it represented the diagonal. A clue to the method by which Pythagoras investigated the value of $\sqrt{2}$ is found by Cantor and Hultsch in the famous passage of Plato (*Rep.* VIII. 546 B, C) about the 'geometrical' or 'nuptial' number. Thus, when Plato contrasts the ῥητή and ἄρρητος διάμετρος τῆς πεμπάδος, he is referring to the diagonal of a square whose side contains five units of length; the ἄρρητος διάμετρος, or the irrational diagonal, is then $\sqrt{50}$ itself, and the nearest rational number is $\sqrt{50}-1$, which is the ῥητή διάμετρος. We have herein the explanation of the way in which Pythagoras must have made the first and most readily comprehensible approximation to $\sqrt{2}$; he must have taken, instead of 2, an improper fraction equal to it but such that the denominator was a square in any case, while the numerator was as near as possible to a complete square. Thus Pythagoras chose $\frac{50}{25}$, and the first approximation to $\sqrt{2}$ was accordingly $\frac{7}{5}$, it being moreover obvious that $\sqrt{2} > \frac{7}{5}$. Again, Pythagoras cannot have been unaware of the truth of the proposition, proved in Eucl. II. 4, that $(a+b)^2 = a^2 + 2ab + b^2$, where a, b are any two straight lines, for this proposition depends solely upon propositions in Book I. which precede the Pythagorean proposition I. 47 and which, as the basis of I. 47, must necessarily have been in substance known to its author. A slightly different geometrical proof would give the formula $(a-b)^2 = a^2 - 2ab + b^2$, which must have been equally well known to Pythagoras. It could not therefore have escaped the discoverer of the first approximation $\sqrt{50}-1$ for $\sqrt{50}$ that the use of the formula with the *positive* sign would give a much nearer approximation, viz. $7 + \frac{1}{14}$, which is only greater than $\sqrt{50}$ to the extent of $\left(\dfrac{1}{14}\right)^2$. Thus we may properly assign to Pythagoras the discovery of the fact represented by

$$7\frac{1}{14} > \sqrt{50} > 7.$$

The consequential result that $\sqrt{2} > \dfrac{1}{5}\sqrt{50-1}$ is used by Aristarchus of Samos in the 7th proposition of his work *On the size and distances of the sun and moon**.

With reference to the investigations of the values of $\sqrt{3}$, $\sqrt{5}$, $\sqrt{6},\ldots\ldots\sqrt{17}$ by Theodorus, it is pretty certain that $\sqrt{3}$ was geometrically represented by him, in the same way as it appears

---

* Part of the proof of this proposition was a sort of foretaste of the first part of Prop. 3 of Archimedes' *Measurement of a circle*, and the substance of it is accordingly appended as reproduced by Hultsch.

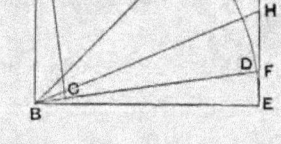

$ABEK$ is a square, KB a diagonal, $\angle HBE = \frac{1}{2}\angle KBE$, $\angle FBE = 3°$, and AC is perpendicular to BF so that the triangles ACB, BEF are similar.

Aristarchus seeks to prove that

$$AB : BC > 18 : 1.$$

If R denote a right angle, the angles KBE, HBE, FBE are respectively $\frac{30}{60}R$, $\frac{15}{60}R$, $\frac{2}{60}R$.

Then $HE : FE : \angle HBE : \angle FBE$.

[This is assumed as a known lemma by Aristarchus as well as Archimedes.]

Therefore $\qquad HE : FE > 15 : 2$..(a).

Now, by construction, $\qquad BK^2 = 2BE^2$.

Also [Eucl. VI. 3] $\qquad BK : BE = KH : HE$;

whence $\qquad KH = \sqrt{2}\, HE.$

And, since $\qquad\qquad \sqrt{2} > \sqrt{\dfrac{50-1}{25}},$

$$KH : HE > 7 : 5,$$

so that $\qquad KE : EH > 12 : 5$ ....(β).

From (a) and (β), *ex aequali*,

$$KE : FE > 18 : 1.$$

Therefore, since $\qquad\qquad BF > BE\ (\text{or } KE),$

$$BF : FE > 18 : 1,$$

so that, by similar triangles,

$$AB : BC > 18 : 1.$$

afterwards in Archimedes, as the perpendicular from an angular point of an equilateral triangle on the opposite side. It would thus be readily comparable with the side of the "1 square foot" mentioned by Plato. The fact also that it is the side of three square *feet* (τρίπους δύναμις) which was proved to be incommensurable suggests that there was some special reason in Theodorus' proof for specifying *feet*, instead of units of length simply; and the explanation is probably that Theodorus subdivided the sides of his triangles in the same way as the Greek foot was divided into halves, fourths, eighths and sixteenths. Presumably therefore, exactly as Pythagoras had approximated to $\sqrt{2}$ by putting $\frac{50}{25}$ for 2, Theodorus started from the identity $3 = \frac{48}{16}$. It would then be clear that

$$\sqrt{3} < \sqrt{\frac{48+1}{16}}, \text{ i.e. } \frac{7}{4},$$

To investigate $\sqrt{48}$ further, Theodorus would put it in the form $\sqrt{49-1}$, as Pythagoras put $\sqrt{50}$ into the form $\sqrt{49+1}$, and the result would be

$$\sqrt{48}\,(=\sqrt{49-1}) < 7 - \frac{1}{14}.$$

We know of no further investigations into incommensurable square roots until we come to Archimedes.

§ 7. Archimedes' approximations to $\sqrt{3}$.

Seeing that Aristarchus of Samos was still content to use the first and very rough approximation to $\sqrt{2}$ discovered by Pythagoras, it is all the more astounding that Aristarchus' younger contemporary Archimedes should all at once, without a word of explanation, give out that

$$\frac{1351}{780} > \sqrt{3} > \frac{265}{153},$$

as he does in the *Measurement of a circle*.

In order to lead up to the explanation of the probable steps by which Archimedes obtained these approximations, Hultsch adopts the same method of *analysis* as was used by the Greek geometers in solving problems, the method, that is, of supposing the problem solved and following out the necessary consequences. To compare the two fractions $\dfrac{265}{153}$ and $\dfrac{1351}{780}$, we first divide both denominators into their smallest factors, and we obtain

$$780 = 2 \cdot 2 \cdot 3 \cdot 5 \cdot 13,$$
$$153 = 3 \cdot 3 \cdot 17.$$

We observe also that $2 \cdot 2 \cdot 13 = 52$, while $3 \cdot 17 = 51$, and we may therefore show the relations between the numbers thus,

$$780 = 3 \cdot 5 \cdot 52,$$
$$153 = 3 \cdot 51.$$

For convenience of comparison we multiply the numerator and denominator of $\dfrac{265}{153}$ by 5; the two original fractions are then

$$\frac{1351}{15 \cdot 52} \text{ and } \frac{1325}{15 \cdot 51},$$

so that we can put Archimedes' assumption in the form

$$\frac{1351}{52} > 15\sqrt{3} > \frac{1325}{51},$$

and this is seen to be equivalent to

$$26 - \frac{1}{52} > 15\sqrt{3} > 26 - \frac{1}{51}.$$

Now $26 - \dfrac{1}{52} = \sqrt{26^2 - 1 + \left(\dfrac{1}{52}\right)^2}$, and the latter expression is an approximation to $\sqrt{26^2 - 1}$.

We have then $\qquad 26 - \dfrac{1}{52} > \sqrt{26^2 - 1}.$

As $26 - \dfrac{1}{52}$ was compared with $15\sqrt{3}$, and we want an approximation to $\sqrt{3}$ itself, we divide by 15 and so obtain

$$\frac{1}{15}\left(26 - \frac{1}{52}\right) > \frac{1}{15}\sqrt{26^2 - 1}.$$

But $\dfrac{1}{15}\sqrt{26^2 - 1} = \sqrt{\dfrac{676 - 1}{225}} = \sqrt{\dfrac{675}{225}} = \sqrt{3}$, and it follows that

$$\frac{1}{15}\left(26 - \frac{1}{52}\right) > \sqrt{3}.$$

The lower limit for $\sqrt{3}$ was given by

$$\sqrt{3} > \frac{1}{15}\left(26 - \frac{1}{51}\right),$$

f

and a glance at this suggests that it may have been arrived at by simply substituting $(52-1)$ for 52.

Now as a matter of fact the following proposition is true. *If $a^2 \pm b$ is a whole number which is not a square, while a^2 is the nearest square number (above or below the first number, as the case may be), then*

$$a \pm \frac{b}{2a} > \sqrt{a^2 \pm b} > a \pm \frac{b}{2a \pm 1}.$$

Hultsch proves this pair of inequalities in a series of propositions formulated after the Greek manner, and there can be little doubt that Archimedes had discovered and proved the same results in substance, if not in the same form. The following circumstances confirm the probability of this assumption.

(1) Certain approximations given by Heron show that he knew and frequently used the formula

$$\sqrt{a^2 \pm b} \backsim a \pm \frac{b}{2a},$$

(where the sign $\backsim$ denotes "is approximately equal to").

Thus he gives $\qquad \sqrt{50} \backsim 7 + \dfrac{1}{14},$

$$\sqrt{63} \backsim 8 - \frac{1}{16},$$

$$\sqrt{75} \backsim 8 + \frac{11}{16}.$$

(2) The formula $\sqrt{a^2 + b} \backsim a + \dfrac{b}{2a+1}$ is used by the Arabian Alkarkhī (11th century) who drew from Greek sources (Cantor, p. 719 sq.).

It can therefore hardly be accidental that the formula

$$a \pm \frac{b}{2a} > \sqrt{a^2 \pm b} > a \pm \frac{b}{2a+1}$$

gives us what we want in order to obtain the two Archimedean approximations to $\sqrt{3}$, and that in direct connexion with one another*.

* Most of the *a priori* theories as to the origin of the approximations are open to the serious objection that, as a rule, they give series of approximate values in which the two now in question do not follow consecutively, but are separated by others which do not appear in Archimedes. Hultsch's explanation is much preferable as being free from this objection. But it is fair to say that the actual formula used by Hultsch appears in Hunrath's solution of the puzzle

We are now in a position to work out the synthesis as follows. From the geometrical representation of $\sqrt{3}$ as the perpendicular from an angle of an equilateral triangle on the opposite side we obtain $\sqrt{2^2-1} = \sqrt{3}$ and, as a first approximation,

$$2-\frac{1}{4} > \sqrt{3}.$$

Using our formula we can transform this at once into

$$\sqrt{3} > 2-\frac{1}{4-1}, \text{ or } 2-\frac{1}{3}.$$

Archimedes would then square $\left(2-\frac{1}{3}\right)$, or $\frac{5}{3}$, and would obtain $\frac{25}{9}$, which he would compare with 3, or $\frac{27}{9}$; i.e. he would put $\sqrt{3} = \sqrt{\frac{25+2}{9}}$ and would obtain

$$\frac{1}{3}\left(5+\frac{1}{5}\right) > \sqrt{3}, \text{ i.e. } \frac{26}{15} > \sqrt{3}.$$

To obtain a still nearer approximation, he would proceed in the same manner and compare $\left(\frac{26}{15}\right)^2$, or $\frac{676}{225}$, with 3, or $\frac{675}{225}$, whence it would appear that

$$\sqrt{3} = \sqrt{\frac{26^2-1}{225}},$$

and therefore that

$$\frac{1}{15}\left(26 - \frac{1}{52}\right) > \sqrt{3},$$

that is,

$$\frac{1351}{780} > \sqrt{3}.$$

The application of the formula would then give the result

$$\sqrt{3} > \frac{1}{15}\left(26 - \frac{1}{52-1}\right),$$

that is,

$$\sqrt{3} > \frac{1326-1}{15.51}, \text{ or } \frac{265}{153}.$$

The complete result would therefore be

$$\frac{1351}{780} > \sqrt{3} > \frac{265}{153}.$$

(*Die Berechnung irrationaler Quadratwurzeln vor der Herrschaft der Decimalbrüche*, Kiel, 1884, p. 21; cf. *Ueber das Ausziehen der Quadratwurzel bei Griechen und Indern*, Hadersleben, 1883), and the same formula is implicitly used in one of the solutions suggested by Tannery (*Sur la mesure du cercle d'Archimède* in *Mémoires de la société des sciences physiques et naturelles de Bordeaux*, 2ᵉ série, IV. (1882), p. 313–337).

$f\,2$

Thus Archimedes probably passed from the first approximation $\frac{7}{4}$ to $\frac{5}{3}$, from $\frac{5}{3}$ to $\frac{26}{15}$, and from $\frac{26}{15}$ directly to $\frac{1351}{780}$, the closest approximation of all, from which again he derived the less close approximation $\frac{265}{153}$. The reason why he did not proceed to a still nearer approximation than $\frac{1351}{780}$ is probably that the squaring of this fraction would have brought in numbers much too large to be conveniently used in the rest of his calculations. A similar reason will account for his having started from $\frac{5}{3}$ instead of $\frac{7}{4}$; if he had used the latter, he would first have obtained, by the same method,

$$\sqrt{3} = \sqrt{\frac{49-1}{16}}, \text{ and thence } \frac{7-\frac{1}{14}}{4} > \sqrt{3}, \text{ or } \frac{97}{56} > \sqrt{3}; \text{ the squaring}$$

of $\frac{97}{56}$ would have given $\sqrt{3} = \frac{\sqrt{97^2-1}}{56}$, and the corresponding approximation would have given $\frac{18817}{56 \cdot 194}$, where again the numbers are inconveniently large for his purpose.

§ 8. Approximations to the square roots of large numbers.

Archimedes gives in the *Measurement of a circle* the following approximate values:

(1) $\qquad 3013\frac{3}{4} > \sqrt{9082321}$,

(2) $\qquad 1838\frac{9}{11} > \sqrt{3380929}$,

(3) $\qquad 1009\frac{1}{6} > \sqrt{1018405}$,

(4) $\qquad 2017\frac{1}{4} > \sqrt{4069284\frac{1}{36}}$,

(5) $\qquad 591\frac{1}{8} < \sqrt{349450}$,

(6) $\qquad 1172\frac{1}{8} < \sqrt{1373943\frac{33}{64}}$,

(7) $\qquad 2339\frac{1}{4} < \sqrt{5472132\frac{1}{16}}$.

There is no doubt that in obtaining the integral portion of the square root of these numbers Archimedes used the method based on the Euclidean theorem $(a+b)^2 = a^2 + 2ab + b^2$ which has already been exemplified in the instance given above from Theon, where an approximation to $\sqrt{4500}$ is found in sexagesimal fractions. The method does not substantially differ from that now followed; but whereas, to take the first case, $\sqrt{9082321}$, we can at once see what will be the number of digits in the square root by marking off pairs of digits in the given number, beginning from the end, the absence of a sign for 0 in Greek made the number of digits in the square root less easy to ascertain because, as written in Greek, the number $\overset{\overset{\lambda\eta}{}}{\text{M}}\beta\tau\kappa\alpha'$ only contains six signs representing digits instead of seven. Even in the Greek notation however it would not be difficult to see that, of the denominations, units, tens, hundreds, etc. in the square root, the units would correspond to $\kappa\alpha'$ in the original number, the tens to $\beta\tau$, the hundreds to $\overset{\eta}{\text{M}}$, and the thousands to $\overset{\lambda}{\text{M}}$. Thus it would be clear that the square root of 9082321 must be of the form

$$1000x + 100y + 10z + w,$$

where x, y, z, w can only have one or other of the values $0, 1, 2, \ldots 9$. Supposing then that x is found, the remainder $N - (1000x)^2$, where N is the given number, must next contain $2 . 1000x . 100y$ and $(100y)^2$, then $2 (1000x + 100y) . 10z$ and $(10z)^2$, after which the remainder must contain two more numbers similarly formed.

In the particular case (1) clearly $x = 3$. The subtraction of $(3000)^2$ leaves 82321, which must contain $2 . 3000 . 100y$. But, even if y is as small as 1, this product would be 600,000, which is greater than 82321. Hence there is no digit representing *hundreds* in the square root. To find z, we know that 82321 must contain

$$2 . 3000 . 10z + (10z)^2,$$

and z has to be obtained by dividing 82321 by 60,000. Therefore $z = 1$. Again, to find w, we know that the remainder

$$(82321 - 2 . 3000 . 10 - 10^2),$$

or 22221, must contain $2 . 3010w + w^2$, and dividing 22221 by $2 . 3010$ we see that $w = 3$. Thus 3013 is the integral portion of the square root, and the remainder is $22221 - (2 . 3010 . 3 + 3^2)$, or 4152.

The conditions of the proposition now require that the approximate value to be taken for the square root must not be less than the real value, and therefore the fractional part to be added to 3013 must be if anything too great. Now it is easy to see that the fraction to be added is greater than $\frac{1}{2}$ because $2 \cdot 3013 \cdot \frac{1}{2} + \left(\frac{1}{2}\right)^2$ is less than the remainder 4152. Suppose then that the number required (which is nearer to 3014 than to 3013) is $3014 - \frac{p}{q}$, and $\frac{p}{q}$ has to be if anything too small.

Now $(3014)^2 = (3013)^2 + 2 \cdot 3013 + 1 = (3013)^2 + 6027$

$$= 9082321 - 4152 + 6027,$$

whence $9082321 = (3014)^2 - 1875.$

By applying Archimedes' formula $\sqrt{a^2 \pm b} < a \pm \frac{b}{2a}$, we obtain

$$3014 - \frac{1875}{2 \cdot 3014} > \sqrt{9082321}.$$

The required value $\frac{p}{q}$ has therefore to be not greater than $\frac{1875}{6028}$.

It remains to be explained why Archimedes put for $\frac{p}{q}$ the value $\frac{1}{4}$ which is equal to $\frac{1507}{6028}$. In the first place, he evidently preferred fractions with unity for numerator and some power of 2 for denominator because they contributed to ease in working, e.g. when two such fractions, being equal to each other, had to be added. (The exceptions, the fractions $\frac{9}{11}$ and $\frac{1}{6}$, are to be explained by exceptional circumstances presently to be mentioned.) Further, in the particular case, it must be remembered that in the subsequent work 2911 had to be added to $3014 - \frac{p}{q}$ and the sum divided by 780, or $2 \cdot 2 \cdot 3 \cdot 5 \cdot 13$. It would obviously lead to simplification if a factor could be divided out, e.g. the best for the purpose, 13. Now, dividing $2911 + 3014$, or 5925, by 13, we obtain the quotient 455, and a remainder 10, so that $10 - \frac{p}{q}$ remains to be divided by 13. Therefore $\frac{p}{q}$ has to be so chosen that $10q - p$ is divisible by 13, while $\frac{p}{q}$ approximates to, but is not greater than, $\frac{1875}{6028}$. The solution $p = 1, q = 4$ would therefore be natural and easy.

(2) $\sqrt{3380929}$.

The usual process for extraction of the square root gave as the integral part of it 1838, and as the remainder 2685. As before, it was easy to see that the exact root was nearer to 1839 than to 1838, and that

$$\sqrt{3380929} = 1838^2 + 2685 = 1839^2 - 2 \cdot 1838 - 1 + 2685$$
$$= 1839^2 - 992.$$

The Archimedean formula then gave

$$1839 - \frac{992}{2 \cdot 1839} > \sqrt{3380929}.$$

It could not have escaped Archimedes that $\frac{1}{4}$ was a near approximation to $\frac{992}{3678}$ or $\frac{1984}{7356}$, since $\frac{1}{4} = \frac{1839}{7356}$; and $\frac{1}{4}$ would have satisfied the necessary condition that the fraction to be taken must be less than the real value. Thus it is clear that, in taking $\frac{2}{11}$ as the approximate value of the fraction, Archimedes had in view the simplification of the subsequent work by the elimination of a factor. If the fraction be denoted by $\frac{p}{q}$, the sum of $1839 - \frac{p}{q}$ and 1823, or $3662 - \frac{p}{q}$, had to be divided by 240, i.e. by $6 \cdot 40$. Division of 3662 by 40 gave 22 as remainder, and then p, q had to be so chosen that $22 - \frac{p}{q}$ was conveniently divisible by 40, while $\frac{p}{q}$ was less than but approximately equal to $\frac{992}{3678}$. The solution $p = 2$, $q = 11$ was easily seen to satisfy the conditions.

(3) $\sqrt{1018405}$.

The usual procedure gave $1018405 = 1009^2 + 324$ and the approximation

$$1009 \frac{324}{2018} > \sqrt{1018405}.$$

It was here necessary that the fraction to replace $\frac{324}{2018}$ should be greater but approximately equal to it, and $\frac{1}{6}$ satisfied the conditions, while the subsequent work did not require any change in it.

(4) $\sqrt{4069284\frac{1}{36}}$.

The usual process gave $4069284\frac{1}{36} = 2017^2 + 995\frac{1}{36}$; it followed that

$$2017 + \frac{36 \cdot 995 + 1}{36 \cdot 2 \cdot 2017} > \sqrt{4069284\frac{1}{36}},$$

and $2017\frac{1}{4}$ was an obvious value to take as an approximation somewhat greater than the left side of the inequality.

(5) $\sqrt{349450}$.

In the case of this and the two following roots an approximation had to be obtained which was *less*, instead of greater, than the true value. Thus Archimedes had to use the second part of the formula

$$a \pm \frac{b}{2a} > \sqrt{a^2 \pm b} > a \pm \frac{b}{2a \pm 1}.$$

In the particular case of $\sqrt{349450}$ the integral part of the root is 591, and the remainder is 169. This gave the result

$$591 + \frac{169}{2 \cdot 591} > \sqrt{349450} > 591 + \frac{169}{2 \cdot 591 + 1},$$

and since $169 = 13^2$, while $2 \cdot 591 + 1 = 7 \cdot 13^2$, it resulted without further calculation that

$$\sqrt{349450} > 591\frac{1}{7}.$$

Why then did Archimedes take, instead of this approximation, another which was not so close, viz. $591\frac{1}{8}$? The answer which the subsequent working and the other approximations in the first part of the proof suggest is that he preferred, for convenience of calculation, to use for his approximations fractions of the form $\frac{1}{2^n}$ only. But he could not have failed to see that to take the nearest fraction of this form, $\frac{1}{8}$, instead of $\frac{1}{7}$ might conceivably affect his final result and make it less near the truth than it need be. As a matter of fact, as Hultsch shows, it does not affect the result to take $591\frac{1}{4}$ and to work onwards from that figure. Hence we must suppose that Archimedes had satisfied himself, by taking $591\frac{1}{4}$ and proceeding on that basis for some distance, that he would not be introducing any appreciable error in taking the more convenient though less accurate approximation $591\frac{1}{8}$.

(6) $\sqrt{1373943\frac{33}{64}}$.

In this case the integral portion of the root is 1172, and the remainder $359\frac{33}{64}$. Thus, if R denote the root,

$$R > 1172 + \frac{359\frac{33}{64}}{2.1172+1}$$

$$> 1172 + \frac{359}{2.1172+1}, \ a \ fortiori.$$

Now $2.1172+1 = 2345$; the fraction accordingly becomes $\dfrac{359}{2345}$, and $\dfrac{1}{7}\left(=\dfrac{359}{2513}\right)$ satisfies the necessary conditions, viz. that it must be approximately equal to, but not greater than, the given fraction. Here again Archimedes would have taken $1172\frac{1}{7}$ as the approximate value but that, for the same reason as in the last case, $1172\frac{1}{8}$ was more convenient.

(7) $\sqrt{5472132\frac{1}{16}}$.

The integral portion of the root is here 2339, and the remainder $1211\frac{1}{16}$, so that, if R is the exact root,

$$R > 2339 + \frac{1211\frac{1}{16}}{2.2339+1}$$

$$> 2339\frac{1}{4}, \ a \ fortiori.$$

A few words may be added concerning Archimedes' ultimate reduction of the inequalities

$$3 + \frac{667\frac{1}{2}}{4673\frac{1}{2}} > \pi > 3 + \frac{284\frac{1}{4}}{2017\frac{1}{4}}$$

to the simpler result $\qquad 3\dfrac{1}{7} > \pi > 3\dfrac{10}{71}$.

As a matter of fact $\dfrac{1}{7} = \dfrac{667\frac{1}{2}}{4672\frac{1}{2}}$, so that in the first fraction it was only necessary to make the small change of diminishing the denominator by 1 in order to obtain the simple $3\dfrac{1}{7}$.

As regards the *lower* limit for π, we see that $\dfrac{284\frac{1}{4}}{2017\frac{1}{4}} = \dfrac{1137}{8069}$; and Hultsch ingeniously suggests the method of trying the effect of increasing the denominator of the latter fraction by 1. This produces $\dfrac{1137}{8070}$ or $\dfrac{379}{2690}$; and, if we divide 2690 by 379, the quotient is between 7 and 8, so that

$$\frac{1}{7} > \frac{379}{2690} > \frac{1}{8}.$$

Now it is a known proposition (proved in Pappus VII. p. 689) that, if $\dfrac{a}{b} > \dfrac{c}{d}$, then

$$\frac{a}{b} > \frac{a+c}{b+d}.$$

Similarly it may be proved that

$$\frac{a+c}{b+d} > \frac{c}{d}.$$

It follows in the above case that

$$\frac{379}{2690} > \frac{379+1}{2690+8} > \frac{1}{8},$$

which exactly gives

$$\frac{10}{71} > \frac{1}{8},$$

and $\dfrac{10}{71}$ is very much nearer to $\dfrac{379}{2690}$ than $\dfrac{1}{8}$ is.

Note on alternative hypotheses with regard to the
approximations to $\sqrt{3}$.

For a description and examination of all the various theories put forward, up to the year 1882, for the purpose of explaining Archimedes' approximations to $\sqrt{3}$ the reader is referred to the exhaustive paper by Dr Siegmund Günther, entitled *Die quadratischen Irrationalitäten der Alten und deren Entwickelungsmethoden* (Leipzig, 1882). The same author gives further references in his *Abriss der Geschichte der Mathematik und der Natur-wissenschaften im Altertum* forming an Appendix to Vol. v. Pt. 1 of Iwan von Müller's *Handbuch der klassischen Altertums-wissenschaft* (München, 1894).

Günther groups the different hypotheses under three general heads :

(1) those which amount to a more or less disguised use of the method of continued fractions and under which are included the solutions of De Lagny, Mollweide, Hauber, Buzengeiger, Zeuthen, P. Tannery (first solution), Heilermann;

(2) those which give the approximations in the form of a series of fractions such as $a + \dfrac{1}{q_1} + \dfrac{1}{q_1 q_2} + \dfrac{1}{q_1 q_2 q_3} + \ldots$; under this class come the solutions of Radicke, v. Pessl, Rodet (with reference to the Çulvasûtras), Tannery (second solution);

(3) those which locate the incommensurable surd between a greater and lesser limit and then proceed to draw the limits closer and closer. This class includes the solutions of Oppermann, Alexejeff, Schönborn, Hunrath, though the first two are also connected by Günther with the method of continued fractions.

Of the methods so distinguished by Günther only those need be here referred to which can, more or less, claim to rest on a historical basis in the sense of representing applications or extensions of principles laid down in the works of Greek mathematicians other than Archimedes which have come down to us. Most of these quasi-historical solutions connect themselves with the system of *side-* and *diagonal-numbers* (πλευρικοὶ and διαμετρικοὶ ἀριθμοί) explained by Theon of Smyrna (c. 130 A.D.) in a work which was intended to give so much of the principles of mathematics as was necessary for the study of the works of Plato.

The *side-* and *diagonal-numbers* are formed as follows. We start with two units, and (a) from the sum of them, (b) from the sum of twice the first unit and once the second, we form two new numbers; thus

$$1.1+1=2, \qquad 2.1+1=3.$$

Of these numbers the first is a *side-* and the second a *diagonal*-number respectively, or (as we may say)

$$a_2=2, \qquad d_2=3.$$

In the same way as these numbers were formed from $a_1=1$, $d_1=1$, successive pairs of numbers are formed from a_2, d_2, and so on, in accordance with the formula

$$a_{n+1}=a_n+d_n, \qquad d_{n+1}=2a_n+d_n,$$

whence we have

$$a_3=1.2+3=5, \qquad d_3=2.2+3=7,$$
$$a_4=1.5+7=12, \qquad d_4=2.5+7=17,$$

and so on.

Theon states, with reference to these numbers, the general proposition which we should express by the equation

$$d_n^2=2a_n^2\pm1.$$

The proof (no doubt omitted because it was well-known) is simple. For we have

$$d_n^2-2a_n^2=(2a_{n-1}+d_{n-1})^2-2(a_{n-1}+d_{n-1})^2$$
$$=2a_{n-1}^2-d_{n-1}^2$$
$$=-(d_{n-1}^2-2a_{n-1}^2)$$
$$=+(d_{n-2}^2-2a_{n-2}^2), \text{ and so on,}$$

while $d_1^2-2a_1^2=-1$; whence the proposition is established.

Cantor has pointed out that any one familiar with the truth of this proposition could not have failed to observe that, as the numbers were successively formed, the value of d_n^2/a_n^2 would approach more and more nearly to 2, and consequently the successive fractions d_n/a_n would give nearer and nearer approximations to the value of $\sqrt{2}$, or in other words that

$$\frac{1}{1}, \frac{3}{2}, \frac{7}{5}, \frac{17}{12}, \frac{41}{29}, \ldots\ldots$$

are successive approximations to $\sqrt{2}$. It is to be observed that the third of these approximations, $\frac{7}{5}$, is the Pythagorean approximation which appears to be hinted at by Plato, while the above scheme of Theon, amounting to a method of finding all the solutions in positive integers of the indeterminate equation

$$2x^2 - y^2 = \pm 1,$$

and given in a work designedly introductory to the study of Plato, distinctly suggests, as Tannery has pointed out, the probability that even in Plato's lifetime the systematic investigation of the said equation had already begun in the Academy. In this connexion Proclus' commentary on Eucl. I. 47 is interesting. It is there explained that in isosceles right-angled triangles "it is not possible to find numbers corresponding to the sides; for there is no square number which is double of a square except in the sense of *approximately* double, e.g. 7^2 is double of 5^2 less 1." When it is remembered that Theon's process has for its object the finding of any number of squares differing only by unity from double the squares of another series of numbers respectively, and that the sides of the two sets of squares are called *diagonal-* and *side*-numbers respectively, the conclusion becomes almost irresistible that Plato had such a system in mind when he spoke of ῥητὴ διάμετρος (*rational* diagonal) as compared with ἄρρητος διάμετρος (*irrational* diagonal) τῆς πεμπάδος (cf. p. lxxviii above).

One supposition then is that, following a similar line to that by which successive approximations to $\sqrt{2}$ could be obtained from the successive solutions, in rational numbers, of the indeterminate equations $2x^2 - y^2 = \pm 1$, Archimedes set himself the task of finding all the solutions, in rational numbers, of the two indeterminate equations bearing a similar relation to $\sqrt{3}$, viz.

$$x^2 - 3y^2 = 1,$$
$$x^2 - 3y^2 = -2.$$

Zeuthen appears to have been the first to connect, *eo nomine*, the ancient approximations to $\sqrt{3}$ with the solution of these equations, which are also made by Tannery the basis of his first method. But, in substance, the same method had been used as early as 1723 by De Lagny, whose hypothesis will be, for purposes of comparison, described after Tannery's which it so exactly anticipated.

Zeuthen's solution.

After recalling the fact that, even before Euclid's time, the solution of the indeterminate equation $x^2 + y^2 = z^2$ by means of the substitutions

$$x = mn, \qquad y = \frac{m^2 - n^2}{2}, \qquad z = \frac{m^2 + n^2}{2}$$

was well known, Zeuthen concludes that there could have been no difficulty in deducing from Eucl. II. 5 the identity

$$3\,(mn)^2 + \left(\frac{m^2-3n^2}{2}\right)^2 = \left(\frac{m^2+3n^2}{2}\right)^2,$$

from which, by multiplying up, it was easy to obtain the formula

$$3\,(2mn)^2 + (m^2-3n^2)^2 = (m^2+3n^2)^2.$$

If therefore one solution $m^2-3n^2=1$ was known, a second could at once be found by putting

$$x=m^2+3n^2, \qquad y=2mn.$$

Now obviously the equation

$$m^2-3n^2=1$$

is satisfied by the values $m=2$, $n=1$; hence the next solution of the equation

$$x^2-3y^2=1$$

is

$$x_1=2^2+3\,.\,1=7, \qquad y_1=2\,.\,2\,.\,1=4;$$

and, proceeding in like manner, we have any number of solutions as

$$x_2=7^2+3\,.\,4^2=97, \qquad y_2=2\,.\,7\,.\,4=56,$$

$$x_3=97^2+3\,.\,56^2=18817, \qquad y_3=2\,.\,97\,.\,56=10864,$$

and so on.

Next, addressing himself to the other equation

$$x^2-3y^2=-2,$$

Zeuthen uses the identity

$$(m+3n)^2-3\,(m+n)^2=-2\,(m^2-3n^2).$$

Thus, if we know one solution of the equation $m^2-3n^2=1$, we can proceed to substitute

$$x=m+3n, \qquad y=m+n.$$

Suppose $m=2$, $n=1$, as before; we then have

$$x_1=5, \qquad y_1=3.$$

If we put $x_2=x_1+3y_1=14$, $y_2=x_1+y_1=8$, we obtain

$$\frac{x_2}{y_2}=\frac{14}{8}=\frac{7}{4}$$

(and $m=7$, $n=4$ is seen to be a solution of $m^2-3n^2=1$).

Starting again from x_2, y_2, we have

$$x_3=38, \qquad y_3=22,$$

and

$$\frac{x_3}{y_3}=\frac{19}{11}$$

($m=19$, $n=11$ being a solution of the equation $m^2-3n^2=-2$);

$$x_4=104, \qquad y_4=60,$$

whence

$$\frac{x_4}{y_4}=\frac{26}{15}$$

(and $m=26$, $n=15$ satisfies $m^2-3n^2=1$),

$$x_5=284, \qquad y_5=164,$$

or

$$\frac{x_5}{y_5}=\frac{71}{41}.$$

Similarly $\dfrac{x_6}{y_6}=\dfrac{97}{56}$, $\dfrac{x_7}{y_7}=\dfrac{265}{153}$, and so on.

This method gives all the successive approximations to $\sqrt{3}$, taking account as it does of both the equations

$$x^2-3y^2=1,$$
$$x^2-3y^2=-2.$$

Tannery's first solution.

Tannery asks himself the question how Diophantus would have set about solving the two indeterminate equations. He takes the first equation in the generalised form

$$x^2-ay^2=1,$$

and then, assuming one solution (p, q) of the equation to be known, he supposes

$$p_1=mx-p, \quad q_1=x+q.$$

Then $\qquad p_1^2-aq_1^2\equiv m^2x^2-2mpx+p^2-ax^2-2aqx-aq^2=1,$

whence, since $p^2-aq^2=1$, by hypothesis,

$$x=2\cdot\frac{mp+aq}{m^2-a},$$

so that $\qquad p_1=\dfrac{(m^2+a)p+2amq}{m^2-a}$, $\qquad q_1=\dfrac{2mp+(m^2+a)q}{m^2-a}$,

and $p_1^2-aq_1^2=1$.

The values of p_1, q_1 so found are rational but not necessarily integral; if integral solutions are wanted, we have only to put

$$p_1=(u^2+av^2)p+2auvq, \qquad q_1=2puv+(u^2+av^2)q,$$

where (u, v) is another integral solution of $x^2-ay^2=1$.

Generally, if (p, q) be a known solution of the equation

$$x^2-ay^2=r,$$

suppose $p_1=ap+\beta q$, $q_1=\gamma p+\delta q$, and "il suffit pour déterminer a, β, γ, δ de connaître les trois groupes de solutions les plus simples et de résoudre deux couples d'équations du premier degré à deux inconnues." Thus (1) for the equation

$$x^2-3y^2=1,$$

the first three solutions are

$$(p=1, q=0), \quad (p=2, q=1), \quad (p=7, q=4),$$

whence $\qquad \left.\begin{array}{l}2=a\\1=\gamma\end{array}\right\}$ and $\left.\begin{array}{l}7=2a+\beta\\4=2\gamma+\delta\end{array}\right\}$,

so that $\qquad a=2, \beta=3, \gamma=1, \delta=2,$

and it follows that the fourth solution is given by
$$p = 2.7 + 3.4 = 26,$$
$$q = 1.7 + 2.4 = 15;$$

(2) for the equation $x^2 - 3y^2 = -2,$

the first three solutions being (1, 1), (5, 3), (19, 11), we have

$$\left.\begin{matrix}5 = a + \beta \\ 3 = \gamma + \delta\end{matrix}\right\} \text{ and } \left.\begin{matrix}19 = 5a + 3\beta \\ 11 = 5\gamma + 3\delta\end{matrix}\right\},$$

whence $a = 2$, $\beta = 3$, $\gamma = 1$, $\delta = 2$, and the next solution is given by

$$p = 2.19 + 3.11 = 71,$$
$$q = 1.19 + 2.11 = 41,$$

and so on.

Therefore, by using the two indeterminate equations and proceeding as shown, all the successive approximations to $\sqrt{3}$ can be found.

Of the two methods of dealing with the equations it will be seen that Tannery's has the advantage, as compared with Zeuthen's, that it can be applied to the solution of *any* equation of the form $x^2 - ay^2 = r$.

De Lagny's method.

The argument is this. If $\sqrt{3}$ could be exactly expressed by an improper fraction, that fraction would fall between 1 and 2, and the square of its numerator would be three times the square of its denominator. Since this is impossible, two numbers have to be sought such that the square of the greater differs as little as possible from 3 times the square of the smaller, though it may be either greater or less. De Lagny then evolved the following successive relations,

$$2^2 = 3.1^2 + 1, \quad 5^2 = 3.3^2 - 2, \quad 7^2 = 3.4^2 + 1, \quad 19^2 = 3.11^2 - 2,$$
$$26^2 = 3.15^2 + 1, \quad 71^2 = 3.41^2 - 2, \text{ etc.}$$

From these relations were derived a series of fractions greater than $\sqrt{3}$, viz. $\frac{2}{1}$, $\frac{7}{4}$, $\frac{26}{15}$, etc., and another series of fractions less than $\sqrt{3}$, viz. $\frac{5}{3}$, $\frac{19}{11}$, $\frac{71}{41}$, etc. The law of formation was found in each case to be that, if $\frac{p}{q}$ was one fraction in the series and $\frac{p'}{q'}$ the next, then

$$\frac{p'}{q'} = \frac{2p + 3q}{p + 2q}.$$

This led to the results

$$\frac{2}{1} > \frac{7}{4} > \frac{26}{15} > \frac{97}{56} > \frac{362}{209} > \frac{1351}{780} \cdots > \sqrt{3},$$

and

$$\frac{5}{3} < \frac{19}{11} < \frac{71}{41} < \frac{265}{153} < \frac{989}{571} < \frac{3691}{2131} \cdots < \sqrt{3};$$

while the law of formation of the successive approximations in each series is precisely that obtained by Tannery as the result of treating the two indeterminate equations by the Diophantine method.

Heilermann's method.

This method needs to be mentioned because it also depends upon a generalisation of the system of *side*- and *diagonal*-numbers given by Theon of Smyrna.

Theon's rule of formation was

$$S_n = S_{n-1} + D_{n-1}, \quad D_n = 2S_{n-1} + D_{n-1};$$

and Heilermann simply substitutes for 2 in the second relation any arbitrary number a, developing the following scheme,

$$S_1 = S_0 + D_0, \quad D_1 = aS_0 + D_0,$$
$$S_2 = S_1 + D_1, \quad D_2 = aS_1 + D_1,$$
$$S_3 = S_2 + D_2, \quad D_3 = aS_2 + D_2,$$
$$\vdots \qquad\qquad \vdots$$
$$S_n = S_{n-1} + D_{n-1}, \quad D_n = aS_{n-1} + D_{n-1}.$$

It follows that

$$aS_n{}^2 = aS_{n-1}{}^2 + 2aS_{n-1}D_{n-1} + aD_{n-1}{}^2,$$
$$D_n{}^2 = a^2 S_{n-1}{}^2 + 2aS_{n-1}D_{n-1} + D_{n-1}{}^2.$$

By subtraction,
$$D_n{}^2 - aS_n{}^2 = (1-a)(D_{n-1}{}^2 - aS_{n-1}{}^2)$$
$$= (1-a)^2 (D_{n-2}{}^2 - aS_{n-2}{}^2), \text{ similarly,}$$
$$= \dots\dots\dots$$
$$= (1-a)^n (D_0{}^2 - aS_0{}^2).$$

This corresponds to the most general form of the "Pellian" equation

$$x^2 - ay^2 = (\text{const.}).$$

If now we put $D_0 = S_0 = 1$, we have

$$\frac{D_n{}^2}{S_n{}^2} = a + \frac{(1-a)^{n+1}}{S_n{}^2},$$

from which it appears that, where the fraction on the right-hand side approaches zero as n increases, $\dfrac{D_n}{S_n}$ is an approximate value for $\sqrt{a}$.

Clearly in the case where $a=3$, $D_0=2$, $S_0=1$ we have

$$\frac{D_0}{S_0} = \frac{2}{1}, \quad \frac{D_1}{S_1} = \frac{5}{3}, \quad \frac{D_2}{S_2} = \frac{14}{8} = \frac{7}{4}, \quad \frac{D_3}{S_3} = \frac{19}{11}, \quad \frac{D_4}{S_4} = \frac{52}{30} = \frac{26}{15},$$

$$\frac{D_5}{S_5} = \frac{71}{41}, \quad \frac{D_6}{S_6} = \frac{194}{112} = \frac{97}{56}, \quad \frac{D_7}{S_7} = \frac{265}{153},$$

and so on.

But the method is, as shown by Heilermann, more rapid if it is used to find, not $\sqrt{a}$, but $b\sqrt{a}$, where b is so chosen as to make b^2a (which takes the place of a) somewhat near to unity. Thus suppose $a = \dfrac{27}{25}$, so that $\sqrt{a} = \dfrac{3}{5}\sqrt{3}$, and we then have (putting $D_0 = S_0 = 1$)

$$S_1 = 2, \quad D_1 = \frac{52}{25}, \text{ and } \sqrt{3} \backsim \frac{5}{3} \cdot \frac{26}{25}, \text{ or } \frac{26}{15},$$

$$S_2 = \frac{102}{25}, \quad D_2 = \frac{54+52}{25} = \frac{106}{25}, \text{ and } \sqrt{3} \backsim \frac{5}{3} \cdot \frac{106}{102}, \text{ or } \frac{265}{153},$$

$$S_3 = \frac{208}{25}, \quad D_3 = \frac{102 \cdot 27}{25 \cdot 25} + \frac{106}{25} = \frac{5404}{25 \cdot 25},$$

and
$$\sqrt{3} \backsim \frac{5404}{25 \cdot 208} \cdot \frac{5}{3}, \text{ or } \frac{1351}{780}.$$

This is one of the very few instances of success in bringing out the two Archimedean approximations in immediate sequence without any foreign values intervening. No other methods appear to connect the two values in this direct way except those of Hunrath and Hultsch depending on the formula

$$a \pm \frac{b}{2a} > \sqrt{a^2 \pm b} > a \pm \frac{b}{2a \pm 1}.$$

We now pass to the second class of solutions which develops the approximations in the form of the sum of a series of fractions, and under this head comes

Tannery's second method.

This may be exhibited by means of its application (1) to the case of the square root of a large number, e.g. $\sqrt{349450}$ or $\sqrt{571^2 + 23409}$, the first of the kind appearing in Archimedes, (2) to the case of $\sqrt{3}$.

(1) Using the formula

$$\sqrt{a^2 + b} \backsim a + \frac{b}{2a},$$

we try the effect of putting for $\sqrt{571^2 + 23409}$ the expression

$$571 + \frac{23409}{1142}.$$

It turns out that this gives correctly the integral part of the root, and we now suppose the root to be

$$571 + 20 + \frac{1}{m}.$$

Squaring and regarding $\dfrac{1}{m^2}$ as negligible, we have

$$571^2 + 400 + 22840 + \frac{1142}{m} + \frac{40}{m} = 571^2 + 23409,$$

H. A.

whence $\dfrac{1182}{m} = 169,$

and $\dfrac{1}{m} = \dfrac{169}{1182} > \dfrac{1}{7},$

so that $\sqrt{349450} > 591\dfrac{1}{7}.$

(2) Bearing in mind that

$$\sqrt{a^2+b} \backsim a + \frac{b}{2a+1},$$

we have $\sqrt{3} = \sqrt{1^2+2} \backsim 1 + \dfrac{2}{2.1+1}$

$$\backsim 1 + \frac{2}{3}, \text{ or } \frac{5}{3}.$$

Assuming then that $\sqrt{3} = \left(\dfrac{5}{3} + \dfrac{1}{m}\right)$, squaring and neglecting $\dfrac{1}{m^2}$, we obtain

$$\frac{25}{9} + \frac{10}{3m} = 3,$$

whence $m = 15$, and we get as the second approximation

$$\frac{5}{3} + \frac{1}{15}, \text{ or } \frac{26}{15}.$$

We have now $26^2 - 3.15^2 = 1,$

and can proceed to find other approximations by means of Tannery's first method.

Or we can also put $\left(1 + \dfrac{2}{3} + \dfrac{1}{15} + \dfrac{1}{n}\right)^2 = 3,$

and, neglecting $\dfrac{1}{n^2}$, we get

$$\frac{26^2}{15^2} + \frac{52}{15n} = 3,$$

whence $n = -15.52 = -780,$ and

$$\sqrt{3} \backsim \left(1 + \frac{2}{3} + \frac{1}{15} - \frac{1}{780} = \frac{1351}{780}\right).$$

It is however to be observed that this method only connects $\dfrac{1351}{780}$ with $\dfrac{26}{15}$ and not with the intermediate approximation $\dfrac{265}{153}$, to obtain which Tannery implicitly uses a particular case of the formula of Hunrath and Hultsch.

Rodet's method was apparently invented to explain the approximation in the Çulvasûtras*

$$\sqrt{2} \backsim 1 + \frac{1}{3} + \frac{1}{3.4} - \frac{1}{3.4.34};$$

* See Cantor, *Vorlesungen über Gesch. d. Math.* p. 600 sq.

but, given the approximation $\frac{4}{3}$, the other two successive approximations indicated by the formula can be obtained by the method of squaring just described* without such elaborate work as that of Rodet, which, when applied to $\sqrt{3}$, only gives the same results as the simpler method.

Lastly, with reference to the third class of solutions, it may be mentioned

(1) that Oppermann used the formula

$$\frac{a+b}{2} > \sqrt{ab} > \frac{2ab}{a+b},$$

which gave successively

$$\frac{2}{1} > \sqrt{3} > \frac{3}{2},$$

$$\frac{7}{4} > \sqrt{3} > \frac{12}{7},$$

$$\frac{97}{56} > \sqrt{3} > \frac{168}{97},$$

but only led to one of the Archimedean approximations, and that by combining the last two ratios, thus

$$\frac{97+168}{56+97} = \frac{265}{153},$$

(2) that Schönborn came somewhat near to the formula successfully used by Hunrath and Hultsch when he proved† that

$$a \pm \frac{b}{2a} > \sqrt{a^2 \pm b} > a \pm \frac{b}{2a \pm \sqrt{b}}.$$

---

* Cantor had already pointed this out in his first edition of 1880.
† *Zeitschrift für Math. u. Physik (Hist. litt. Abtheilung)* xxviii. (1883), p. 169 sq.

CHAPTER V.

ON THE PROBLEMS KNOWN AS ΝΕΥΣΕΙΣ.

THE word νεῦσις, commonly *inclinatio* in Latin, is difficult to translate satisfactorily, but its meaning will be gathered from some general remarks by Pappus having reference to the two Books of Apollonius entitled νεύσεις (now lost). Pappus says*, "A line is said to *verge* (νεύειν) towards a point if, being produced, it reach the point," and he gives, among particular cases of the general form of the problem, the following.

"Two lines being given in position, to place between them a straight line given in length and verging towards a given point."

"If there be given in position (1) a semicircle and a straight line at right angles to the base, or (2) two semicircles with their bases in a straight line, to place between the two lines a straight line given in length and verging towards a corner (γωνίαν) of a semicircle."

Thus a straight line has to be laid across two lines or curves so that it passes through a given point and the intercept on it between the lines or curves is equal to a given length†.

§ 1. The following allusions to particular νεύσεις are found in Archimedes. The proofs of Props. 5, 6, 7 of the book *On Spirals* use respectively three particular cases of the general theorem that,

* Pappus (ed. Hultsch) VII. p. 670.

† In the German translation of Zeuthen's work, *Die Lehre von den Kegelschnitten im Altertum*, νεῦσις is translated by "Einschiebung," or as we might say "insertion," but this fails to express the condition that the required line must pass through a given point, just as *inclinatio* (and for that matter the Greek term itself) fails to express the other requirement that the intercept on the line must be of given length.

if A be any point on a circle and BC any diameter, it is possible to draw through A a straight line, meeting the circle again in P and BC produced in R, such that the intercept PR is equal to any given

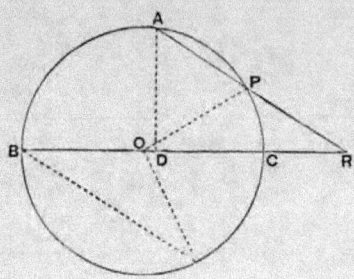

length. In each particular case the fact is merely stated as true without any explanation or proof, and

(1) Prop. 5 assumes the case where the tangent at *A* is parallel to *BC*,

(2) Prop. 6 the case where the points *A*, *P* in the figure are interchanged,

(3) Prop. 7 the case where *A*, *P* are in the relative positions shown in the figure.

Again, (4) Props. 8 and 9 each assume (as before, without proof, and without giving any solution of the implied problem) that, *if AE, BC be two chords of a circle intersecting at right angles in a point D such that BD > DC, then it is possible to draw through A another line ARP, meeting BC in R and the circle again in P, such that PR = DE.*

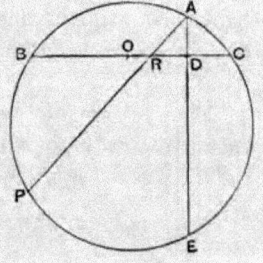

Lastly, with the assumptions in Props. 5, 6, 7 should be compared Prop. 8 of the *Liber Assumptorum*, which may well be due to Archimedes, whatever may be said of the composition of the whole book. This proposition proves that, *if in the first figure APR is so drawn that PR is equal to the radius OP, then the arc AB is three times the arc PC.* In other words, if an arc *AB* of a circle be taken subtending any angle at the centre *O*, an arc equal to one-third of the given arc can be found, *i.e. the given angle can be trisected, if only APR can be drawn through A in such a manner*

*that the intercept PR between the circle and BO produced is equal to
the radius of the circle.* Thus the *trisection of an angle* is reduced to
a νεῦσις exactly similar to those assumed as possible in Props. 6, 7
of the book *On Spirals.*

The νεύσεις so referred to by Archimedes are not, in general,
capable of solution by means of the straight line and circle alone,
as may be easily shown. Suppose in the first figure that x
represents the unknown length OR, where O is the middle point
of BC, and that k is the given length to which PR is to be equal;
also let $OD = a$, $AD = b$, $BC = 2c$. Then, whether BC be a diameter
or (more generally) any chord of the circle, we have

$$AR . RP = BR . RC,$$

and therefore $$k \sqrt{b^2 + (x - a)^2} = x^2 - c^2.$$

The resulting equation, after rationalisation, is an equation of the
fourth degree in x; or, if we denote the length of AR by y, we have,
for the determination of x and y, the two equations

$$\left.\begin{aligned} y^2 &= (x - a)^2 + b^2 \\ ky &= x^2 - c^2 \end{aligned}\right\} \quad \dots\dots\dots\dots\dots\dots(a).$$

In other words, if we have a rectangular system of coordinate
axes, the values of x and y satisfying the conditions of the problem
can be determined as the coordinates of the points of intersection of
a certain rectangular hyperbola and a certain parabola.

In one particular case, that namely in which D coincides with O
the middle point of BC, or in which A is one extremity of the
diameter bisecting BC at right angles, $a = 0$, and the equations
reduce to the single equation

$$y^2 - ky = b^2 + c^2,$$

which is a quadratic and can be geometrically solved by the
traditional method of application of areas; for, if u be substituted
for $y - k$, so that $u = AP$, the equation becomes

$$u (k + u) = b^2 + c^2,$$

and we have simply "to apply to a straight line of length k a
rectangle exceeding by a square figure and equal to a given
area $(b^2 + c^2)$."

The other νεῦσις referred to in Props. 8 and 9 can be solved in
the more general form where k, the given length to which PR
is to be equal, has any value within a certain maximum and is not

necessarily equal to DE, in exactly the same manner; and the two equations corresponding to (a) will be for the second figure

$$\left. \begin{array}{l} y^2 = (a - x)^2 + b^2 \\ ky = c^2 - x^2 \end{array} \right\} \dots\dots\dots\dots\dots\dots(\beta).$$

Here, again, the problem can be solved by the ordinary method of application of areas in the particular case where AE is the diameter bisecting BC at right angles; and it is interesting to note that this particular case appears to be assumed in a fragment of Hippocrates' *Quadrature of lunes* preserved in a quotation by Simplicius* from Eudemus' History of Geometry, while Hippocrates flourished probably as early as 450 B.C.

Accordingly we find that Pappus distinguishes different classes of νεύσεις corresponding to his classification of geometrical problems in general. According to him, the Greeks distinguished three kinds of problems, some being *plane*, others *solid*, and others *linear*. He proceeds thus †: "Those which can be solved by means of a straight line and a circumference of a circle may properly be called *plane* (ἐπίπεδα); for the lines by means of which such problems are solved have their origin in a plane. Those however which are solved by using for their discovery (εὕρεσιν) one or more of the sections of the cone have been called *solid* (στερεά); for the construction requires the use of surfaces of solid figures, namely, those of cones. There remains a third kind of problem, that which is called *linear* (γραμμικόν); for other lines [curves] besides those mentioned are assumed for the construction whose origin is more complicated and less natural, as they are generated from more irregular surfaces and intricate movements." Among other instances of the *linear* class of curves Pappus mentions spirals, the curves known as *quadratrices*, conchoids and cissoids. He adds that "it seems to be a grave error which geometers fall into whenever any one discovers the solution of a plane problem by means of conics or linear curves, or generally solves it by means of a foreign kind, as is the case, for example, (1) with the problem in the fifth Book of the Conics of Apollonius relating to the parabola ‡,

---

* Simplicius, *Comment. in Aristot. Phys.* pp. 61—68 (ed. Diels). The whole quotation is reproduced by Bretschneider, *Die Geometrie und die Geometer vor Euklides*, pp. 109—121. As regards the assumed construction see particularly p. 64 and p. xxiv of Diels' edition; cf. Bretschneider, pp. 114, 115, and Zeuthen, *Die Lehre von den Kegelschnitten im Altertum*, pp. 269, 270.

† Pappus IV. pp. 270—272.

‡ Cf. *Apollonius of Perga*, pp. cxxviii. cxxix.

and (2) when Archimedes assumes in his work on the spiral a νεῦσις of a solid character with reference to a circle; for it is possible without calling in the aid of anything solid to find the [proof of the] theorem given by the latter [Archimedes], that is, to prove that the circumference of the circle arrived at in the first revolution is equal to the straight line drawn at right angles to the initial line to meet the tangent to the spiral."

The "solid νεῦσις" referred to in this passage is that assumed to be possible in Props. 8 and 9 of the book *On Spirals*, and is mentioned again by Pappus in another place where he shows how to solve the problem by means of conics*. This solution will be given later, but, when Pappus objects to the procedure of Archimedes as unorthodox, the objection appears strained if we consider what precisely it is that Archimedes assumes. It is not the actual solution which is assumed, but only its *possibility*; and its possibility can be perceived without any use of conics. For in the particular case it is only necessary, as a condition of possibility, that DE in the second figure above should not be the *maximum* length which the intercept PR could have as APR revolves about A from the position ADE in the direction of the centre of the circle; and that DE is not the maximum length which PR can have is almost self-evident. In fact, if P, instead of moving along the circle, moved along the straight line through E parallel to BC, and if ARP moved from the position ADE in the direction of the centre, the length of PR would continually increase, and *a fortiori*, so long as P is on the arc of the circle cut off by the parallel through E to BC, PR must be greater in length than DE; and on the other hand, as ARP moves further in the direction of B, it must sometime intercept a length PR equal to DE before P reaches B, when PR vanishes. Since, then, Archimedes' method merely depends upon the theoretical possibility of a solution of the νεῦσις, and this possibility could be inferred from quite elementary considerations, he had no occasion to use conic sections for the purpose immediately in view, and he cannot fairly be said to have solved a plane problem by the use of conics.

At the same time we may safely assume that Archimedes was in possession of a solution of the νεῦσις referred to. But there is no evidence to show how he solved it, whether by means of conics, or otherwise. That he would have been *able* to effect the solution,

---

* Pappus IV. p. 298 sq.

as Pappus does, by the use of conics cannot be doubted. A precedent
for the introduction of conics where a "solid problem" had to be
solved was at hand in the determination of two mean proportionals
between two unequal straight lines by Menaechmus, the inventor of
the conic sections, who used for the purpose the intersections of a
parabola and a rectangular hyperbola. The solution of the cubic
equation on which the proposition *On the Sphere and Cylinder* II. 4
depends is also effected by means of the intersections of a parabola
with a rectangular hyperbola in the fragment given by Eutocius
and by him assumed to be the work of Archimedes himself*.

Whenever a problem did not admit of solution by means of the
straight line and circle, its solution, where possible, by means of
conics was of the greatest theoretical importance. First, the
possibility of such a solution enabled the problem to be classified
as a "solid problem"; hence the importance attached by Pappus
to solution by means of conics. But, secondly, the method had
other great advantages, particularly in view of the requirement that
the solution of a problem should be accompanied by a διορισμός
giving the criterion for the possibility of a real solution. Often too
the διορισμός involved (as frequently in Apollonius) the determination
of the number of solutions as well as the limits for their possibility.
Thus, in any case where the solution of a problem depended on the
intersections of two conics, the theory of conics afforded an effective
means of investigating διορισμοί.

§ 2. But though the solution of "solid problems" by means of
conics had such advantages, it was not the only method open to
Archimedes. An alternative would be the use of some mechanical
construction such as was often used by the Greek geometers and is
recognised by Pappus himself as a legitimate substitute for conics,
which are not easy to draw in a plane†. Thus in Apollonius'
solution of the problem of the two mean proportionals as given by
Eutocius a ruler is supposed to be moved about a point until the
points at which the ruler crosses two given straight lines at right
angles are equidistant from a certain other fixed point; and the
same construction is also given under Heron's name. Another
version of Apollonius' solution is that given by Ioannes Philoponus,
which assumes that, given a circle with diameter *OC* and two

---

* See note to *On the Sphere and Cylinder*, II. 4.
† Pappus III. p. 54.

straight lines OD, OE through O and at right angles to one another, a line can be drawn through C, meeting the circle again in F and the two lines in D, E respectively, such that the intercepts CD, FE are equal. This solution was no doubt discovered by means of the intersection of the circle with a rectangular hyperbola drawn with OD, OE as asymptotes and passing through C; and this supposition accords with Pappus' statement that Apollonius solved the problem by means of the sections of the cone*. The equivalent mechanical construction is given by Eutocius as that of Philo Byzantinus, who turns a ruler about C until CD, FE are equal†.

Now clearly a similar method could be used for the purpose of effecting a νεῦσις. We have only to suppose a ruler (or any object with a straight edge) with two marks made on it at a distance equal to the given length which the problem requires to be intercepted between two curves by a line passing through the fixed point; then, if the ruler be so moved that it always passes through the fixed point, while one of the marked points on it follows the course of one of the curves, it is only necessary to move the ruler until the second marked point falls on the other curve. Some such operation as this may have led Nicomedes to the discovery of his curve, the conchoid, which he introduced (according to Pappus) into his doubling of the cube, and by which he also trisected an angle (according to the same authority). From the fact that Nicomedes is said to have spoken disrespectfully of Eratosthenes' mechanical solution of the duplication problem, and therefore must have lived later than Eratosthenes, it is concluded that his date must have been subsequent to 200 B.C., while on the other hand he must have written earlier than 70 B.C., since Geminus knew the name of the curve about that date; Tannery places him between Archimedes and Apollonius‡. While therefore there appears to be no evidence of the use, before the time of Nicomedes, of such a mechanical method of solving a νεῦσις, the interval between Archimedes and the discovery of the conchoid can hardly have been very long. As a matter of fact, the conchoid of Nicomedes can be used to solve not only all the νεύσεις mentioned in Archimedes but any case of such a problem where one of the curves is a straight

---

* Pappus III. p. 56.
† For fuller details see *Apollonius of Perga*, pp. cxxv—cxxvii.
‡ *Bulletin des Sciences Mathématiques*, 2ᵉ série VII. p. 284.

line.   Both Pappus and Eutocius attribute to Nicomedes the invention of a machine for drawing his conchoid. AB is supposed to be

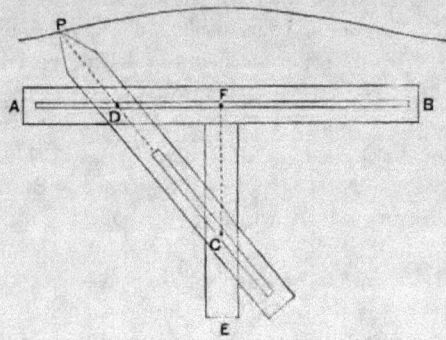

a ruler with a slot in it parallel to its length, FE a second ruler at
right angles to the first with a fixed peg in it, C.   This peg moves
in a slot made in a third ruler parallel to its length, while this
ruler has a fixed peg on it, D, in a straight line with the slot in
which C moves ; and the peg D can move along the slot in AB.   If
then the ruler PD moves so that the peg D describes the length of
the slot in AB on each side of F, the extremity of the ruler, P,
describes the curve which is called a conchoid.   Nicomedes called
the straight line AB the *ruler* (κανών), the fixed point C the *pole*
(πόλος), and the length PD the *distance* (διάστημα) ; and the
fundamental property of the curve, which in polar coordinates
would now be denoted by the equation $r = a + b \sec \theta$, is that, if
any radius vector be drawn from C to the curve, as CP, the length
intercepted on the radius vector between the curve and the straight
line AB is constant.   Thus any νεῦσις in which one of the two
given lines is a straight line can be solved by means of the
intersection of the other line with a certain conchoid whose pole
is the fixed point to which the required straight line must verge
(νεύειν).   In practice Pappus tells us that the conchoid was not
always actually drawn, but that "some," for greater convenience,
moved the ruler about the fixed point until by trial the intercept
was made equal to the given length*.

     § 3.   The following is the way in which Pappus applies
conic sections to the solution of the νεῦσις referred to in Props. 8, 9
of the book *On Spirals*.   He begins with two lemmas.

* Pappus IV. p. 246.

(1) If from a given point A any straight line be drawn meeting a straight line BC given in position in R, and if RQ be drawn perpendicular to BC and bearing a given ratio to AR, the locus of Q is a *hyperbola*.

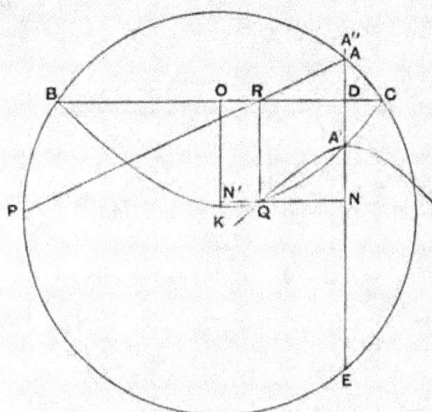

For draw AD perpendicular to BC, and on AD produced take A' such that

$$QR : RA = A'D : DA = \text{(the given ratio)}.$$

Measure DA'' along DA equal to DA'.

Then, if QN be perpendicular to AN,

$$(AR^2 - AD^2) : (QR^2 - A'D^2) = \text{(const.)},$$

or $QN^2 : A'N . A''N = \text{(const.)}$

(2) If BC be given in length, and if RQ, a straight line drawn at right angles to BC from any point R on it, be such that

$$BR . RC = k . RQ,$$

where k is a straight line of given length, then the locus of Q is a *parabola*.

Let O be the middle point of BC, and let OK be drawn at right angles to it and of such length that

$$OC^2 = k . KO.$$

Draw QN' perpendicular to OK.

Then $QN'^2 = OR^2 = OC^2 - BR . RC$

$$= k . (KO - RQ), \text{ by hypothesis,}$$

$$= k . KN'.$$

In the particular case referred to by Archimedes (with the slight generalisation that the given length k to which PR is to be equal is not necessarily equal to DE) we have

(1) the given ratio $RQ : AR$ is unity, or $RQ = AR$, whence A'' coincides with A, and, by the first lemma,

$$QN^2 = AN \cdot A'N,$$

so that Q lies on a *rectangular hyperbola*.

(2) $BR \cdot RC = AR \cdot RP = k \cdot AR = k \cdot RQ$, and, by the second lemma, Q lies on a certain *parabola*.

If now we take O as origin, OC as axis of x and OK as axis of y, and if we put $OD = a$, $AD = b$, $BC = 2c$, the hyperbola and parabola determining the position of Q are respectively denoted by the equations

$$(a - x)^2 = y^2 - b^2,$$

$$c^2 - x^2 = ky,$$

which correspond exactly to the equations (β) above obtained by purely algebraical methods.

Pappus says nothing of the διορισμός which is necessary to the complete solution of the generalised problem, the διορισμός namely which determines the *maximum* value of k for which the solution is possible. This maximum value would of course correspond to the case in which the rectangular hyperbola and the parabola touch one another. Zeuthen has shown[*] that the corresponding value of k can be determined by means of the intersection of two other hyperbolas or of a hyperbola and a parabola, and there is no doubt that Apollonius, with his knowledge of conics, and in accordance with his avowed object in giving the properties useful and necessary for διορισμοί, would have been able to work out this particular διορισμός by means of conics; but there is no evidence to show that Archimedes investigated it by the aid of conics, or indeed at all, it being clear, as shown above, that it was not necessary for his immediate purpose.

This chapter may fitly conclude with a description of (1) some important applications of νεύσεις given by Pappus, and (2) certain particular cases of the same class of problems which are *plane*, that is, can be solved by the aid of the straight line and circle only, and which were (according to Pappus) shown by the Greek geometers to be of that character.

[*] Zeuthen, *Die Lehre von den Kegelschnitten im Altertum*, pp. 273—5.

§ 4. One of the two important applications of 'solid' νεῦσεις was discovered by Nicomedes, the inventor of the conchoid, who introduced that curve for solving a νεῦσις to which he reduced the problem of *doubling the cube** or (what amounts to the same thing) the *finding of two mean proportionals between two given unequal straight lines.*

Let the given unequal straight lines be placed at right angles as *CL, LA.* Complete the parallelogram *ABCL*, and bisect *AB* at *D*, and *BC* at *E.* Join *LD* and produce it to meet *CB* produced in *H.* From *E* draw *EF* at right angles to *BC*, and take a point *F* on *EF* such that *CF* is equal to *AD.* Join *HF*, and through *C* draw *CG* parallel to *HF.* If we produce *BC* to *K*, the straight lines *CG, CK*

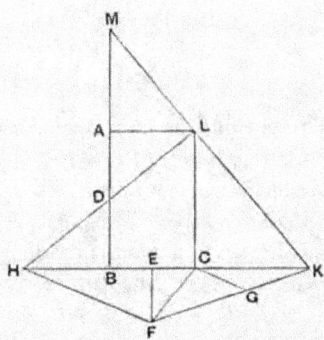

form an angle, and we now draw from the given point *F* a straight line *FGK*, meeting *CG, CK* in *G, K* respectively, such that the intercept *GK* is equal to *AD* or *FC.* (This is the νεῦσις to which the problem is reduced, and it can be solved by means of a conchoid with *F* as pole.)

Join *KL* and produce it to meet *BA* produced in *M.*

Then shall *CK, AM* be the required mean proportionals between *CL, LA,* or

$$CL : CK = CK : AM = AM : AL.$$

We have, by Eucl. II. 6,

$$BK \cdot KC + CE^2 = EK^2.$$

If we add EF^2 to each side,

$$BK \cdot KC + CF^2 = FK^2.$$

Now, by parallels,

$$MA : AB = ML : LK$$
$$= BC : CK;$$

---

* Pappus IV. p. 242 sq. and III. p. 58 sq.; Eutocius on Archimedes, *On the Sphere and Cylinder,* II. 1 (Vol. III. p. 114 sq.)

and, since $AB = 2AD$, and $BC = \frac{1}{2}HC$,

$$MA : AD = HC : CK$$
$$= FG : GK, \text{ by parallels,}$$

whence, *componendo*,

$$MD : AD = FK : GK.$$

But $GK = AD$; therefore $MD = FK$, and $MD^2 = FK^2$.

Again,　　　　　$MD^2 = BM \cdot MA + AD^2$,

and　　　　　　$FK^2 = BK \cdot KC + CF^2$, from above,

while　　　　　$MD^2 = FK^2$, and $AD^2 = CF^2$;

therefore　　　$BM \cdot MA = BK \cdot KC$.

Hence　　　　$CK : MA = BM : BK$
$$\left.\begin{array}{l} = MA : AL \\ = LC : CK \end{array}\right\}, \text{ by parallels,}$$

that is,　　　$LC : CK = CK : MA = MA : AL.$

§ 5. The second important problem which can be reduced to a 'solid' νεῦσις is the *trisection of any angle*. One method of reducing it to a νεῦσις has been mentioned above as following from Prop. 8 of the *Liber Assumptorum*. This method is not mentioned by Pappus, who describes (IV. p. 272 sq.) another way of effecting the reduction, introducing it with the words, "The earlier geometers, when they sought to solve the aforesaid problem about the [trisection of the] angle, a problem by nature 'solid,' by 'plane' methods, were unable to discover the solution; for they were not yet accustomed to the use of the sections of the cone, and were for that reason at a loss. Later, however, they trisected an angle by means of conics, having used for the discovery of it the following νεῦσις."

The νεῦσις is thus enunciated: Given a rectangle $ABCD$, let it be required to draw through A a straight line AQR, meeting CD in Q and BC produced in R, such that the intercept QR is equal to a given length, k suppose.

Suppose the problem solved, QR being equal to k. Draw DP parallel to QR and RP parallel to CD, meeting in P. Then, in the parallelogram DR, $DP = QR = k$.

Hence P lies on a *circle* with centre D and radius k.

Again, by Eucl. I. 43 relating to the complements of the parallelograms about the diagonal of the complete parallelogram,

$$BC \cdot CD = BR \cdot QD$$
$$= PR \cdot RB;$$

and, since $BC . CD$ is given, it follows that P lies on a *rectangular hyperbola* with BR, BA as asymptotes and passing through D.

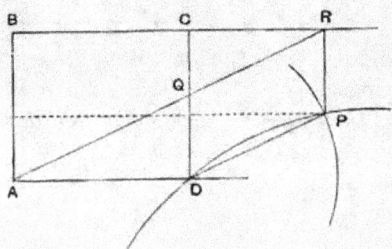

Therefore, to effect the construction, we have only to draw this rectangular hyperbola and the circle with centre D and radius equal to k. The intersection of the two curves gives the point P, and R is determined by drawing PR parallel to DC. Thus AQR is found.

[Though Pappus makes $ABCD$ a *rectangle*, the construction applies equally if $ABCD$ is any parallelogram.]

Now suppose ABC to be any acute angle which it is required to trisect. Let AC be perpendicular to BC. Complete the parallelogram $ADBC$, and produce DA.

Suppose the problem solved, and let the angle CBE be one-third of the angle ABC. Let BE meet AC in E and DA produced in F. Bisect EF in H, and join AH.

Then, since the angle ABE is equal to twice the angle EBC and, by parallels, the angles EBC, EFA are equal,

$$\angle ABE = 2 \angle AFH = \angle AHB.$$

Therefore $$AB = AH = HF,$$

and $$EF = 2HF$$

$$= 2AB.$$

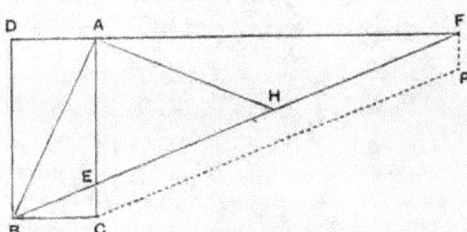

Hence, in order to trisect the angle ABC, we have only to solve the following νεῦσις: *Given the rectangle $ADBC$ whose diagonal*

is AB, to draw through B a straight line BEF, meeting AC in E and DA produced in F, such that EF may be equal to twice AB; and this νεῦσις is solved in the manner just shown.

These methods of doubling the cube and trisecting any acute angle are seen to depend upon the application of one and the same νεῦσις, which may be stated in its most general form thus. *Given any two straight lines forming an angle and any fixed point which is not on either line, it is required to draw through the fixed point a straight line such that the portion of it intercepted between the fixed lines is equal to a given length.* If AE, AC be

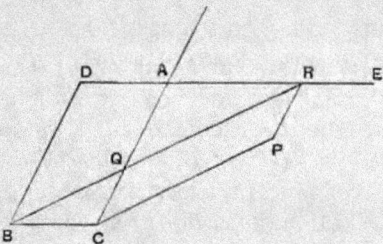

the fixed lines and B the fixed point, let the parallelogram $ACBD$ be completed, and suppose that BQR, meeting CA in Q and AE in R, satisfies the conditions of the problem, so that QR is equal to the given length. If then the parallelogram $CQRP$ is completed, we may regard P as an auxiliary point to be determined in order that the problem may be solved; and we have seen that P can be found as one of the points of intersection of (1) a circle with centre C and radius equal to k, the given length, and (2) the hyperbola which passes through C and has DE, DB for its asymptotes.

It remains only to consider some particular cases of the problem which do not require conics for their solution, but are 'plane' problems requiring only the use of the straight line and circle.

§ 6. We know from Pappus that Apollonius occupied himself, in his two Books of νεύσεις, with problems of that type which were capable of solution by '*plane*' methods. As a matter of fact, the above νεῦσις reduces to a 'plane' problem in the particular case where B lies on one of the bisectors of the angle between the two given straight lines, or (in other words) where the parallelogram $ACBD$ is a rhombus or a square. Accordingly we find Pappus enunciating, as one of the 'plane' cases which had been singled out for proof on account of their greater utility for many purposes, the following*: Given a rhombus with one side produced, to fit into the exterior angle a straight line given in length and verging to the opposite angle; and he gives later on, in his lemmas to Apollonius' work, a theorem bearing on the problem with regard to the rhombus, and (after a preliminary lemma) a solution of the νεῦσις with reference to a square.

The question therefore arises, how did the Greek geometers discover these and other particular cases, where a problem which is in general 'solid,' and therefore requires the use of conics (or a mechanical equivalent), becomes 'plane'? Zeuthen is of opinion that they were probably discovered as the result of a study of the general solution by means of conics†. I do not feel convinced of this, for the following reasons.

(1) The authenticated instances appear to be very rare in which we should be justified in assuming that the Greeks used the properties of conics, in the same way as we should combine and transform two Cartesian equations of the second degree, for the purpose of proving that the intersections of two conics also lie on certain circles or straight lines. It is true that we may reasonably infer that Apollonius discovered by a method of this sort his solution of the problem of doubling the cube where, in place of the parabola and rectangular hyperbola used by Menaechmus, he employs the same hyperbola along with the *circle* which passes through the points common to the hyperbola and parabola‡; but in the only propositions contained in his conics which offer an opportunity for making a similar reduction§, Apollonius does not make it, and is blamed by Pappus for not doing so. In the propositions referred to the feet of the normals to a parabola drawn from a given point are determined as the intersections of the parabola with a certain rectangular hyperbola, and Pappus objects

---

* Pappus vii. p. 670.

† "Mit dieser selben Aufgabe ist nämlich ein wichtiges Beispiel dafür verknüpft, dass man bemüht war solche Fälle zu entdecken, in denen Aufgaben, zu deren Lösung im allgemeinen Kegelschnitte erforderlich sind, sich mittels Zirkel und Lineal lösen lassen. Da nun das Studium der allgemeinen Lösung durch Kegelschnitte das beste Mittel gewährt solche Fälle zu entdecken, so ist es ziemlich wahrscheinlich, dass man wirklich diesen Weg eingeschlagen hat." Zeuthen, *op. cit.* p. 280.

‡ *Apollonius of Perga*, p. cxxv, cxxvi.

§ *Ibid.* p. cxxviii and pp. 182, 186 (*Conics*, v. 58, 62

to this method as an instance of discovering the solution of a 'plane' problem by means of conics*, the objection having reference to the use of a *hyperbola* where the same points could be obtained as the intersections of the parabola with a certain *circle.* Now the proof of this latter fact would present no difficulty to Apollonius, and Pappus must have been aware that it would not; if therefore he objects in the circumstances to the use of the hyperbola, it is at least arguable that he would equally have objected had Apollonius brought in the hyperbola and used its properties for the purpose of proving the problem to be 'plane' in the particular case.

(2) The solution of the general problem by means of conics brings in the auxiliary point P and the straight line CP. We should therefore naturally expect to find some trace of these in the particular solutions of the νεῦσις for a rhombus and square; but they do not appear in the corresponding demonstrations and figures given by Pappus.

Zeuthen considers that the νεῦσις with reference to a square was probably shown to be 'plane' by means of the same investigation which showed that the more general case of the rhombus was also capable of solution with the help of the straight line and circle only, i.e. by a systematic study of the general solution by means of conics. This supposition seems to him more probable than the view that the discovery of the plane construction for the square may have been accidental; for (he says) if the same problem is treated solely by the aid of elementary geometrical expedients, the discovery that it is 'plane' is by no means a simple matter†. Here, again, I am not convinced by Zeuthen's argument, as it seems to me that a simpler explanation is possible of the way in which the Greeks were led to the discovery that the particular νεῦσεις were plane. They knew in the first place that the trisection of a *right angle* was a 'plane' problem, and therefore that *half a right angle* could be trisected by means of the straight line and circle. It followed

---

* Pappus iv. p. 270. Cf. p. ciii above.

† " Die Ausführbarkeit kann dann auf die zuerst angedeutete Weise gefunden sein, die den allgemeinen Fall, wo der Winkel zwischen den gegebenen Geraden beliebig ist, in sich begreift. Dies scheint mir viel wahrscheinlicher als die Annahme, dass die Entdeckung dieser ebenen Konstruction zufällig sein sollte; denn wenn man dieselbe Aufgabe nur mittels rein elementar-geometrischer Hülfsmittel behandelt, so liegt die Entdeckung, dass sie eben ist, ziemlich fern." Zeuthen, *op. cit.* p. 282.

therefore that the corresponding νεῦσις, i.e. that for a square, was a 'plane' problem in the particular case where the given length to which the required intercept was to be equal was double of the diagonal of the square. This fact would naturally suggest the question whether the problem was still plane if k had any other value; and, when once this question was thoroughly investigated, the proof that the problem was 'plane,' and the solution of it, could hardly have evaded for long the pursuit of geometers so ingenious as the Greeks. This will, I think, be clear when the solution given by Pappus and reproduced below is examined. Again, after it had been proved that the νεῦσις with reference to a square was 'plane,' what more natural than the further inquiry as to whether the intermediate case between that of the square and parallelogram, that of the rhombus, might perhaps be a 'plane' problem?

As regards the actual solution of the plane νεύσεις with respect to the rhombus and square, i.e. the cases in general where the fixed point B lies on one of the bisectors of the angles between the two given straight lines, Zeuthen says that only in one of the cases have we a positive statement that the Greeks solved the νεῦσις by means of the circle and ruler, the case, namely, where $ACBD$ is a square[*]. This appears to be a misapprehension, for not only does Pappus mention the case of the rhombus as one of the plane νεύσεις which the Greeks had solved, but it is clear, from a proposition given by him later, how it was actually solved. The proposition is stated by Pappus to be "involved" (παραθεωρούμενον, meaning presumably "the subject of concurrent investigation") in the 8th problem of Apollonius' first Book of νεύσεις, and is enunciated in the following form[†]. *Given a rhombus AD with diameter BC produced to E, if EF be a mean proportional between BE, EC, and if a circle be described with centre E and radius EF cutting CD in K and AC produced in H, BKH shall be a straight line.* The proof is as follows.

Let the circle cut AC in L, and join HE, KE, LE. Let LK meet BC in M.

---

[*] "Indessen besitzen wir doch nur in einem einzelnen hierher gehörigen Falle eine positive Angabe darüber, dass die Griechen die Einschiebung mittels Zirkel und Lineal ausgeführt haben, wenn nämlich die gegebenen Geraden zugleich rechte Winkel bilden, $AIBC$ also ein Quadrat wird." Zeuthen, *op. cit.* p. 281.

[†] Pappus VII. p. 778.

Since, from the property of the rhombus, the angles LCM, KCM are equal, and therefore CL, CK make equal angles with the diameter FG of the circle, it follows that $CL = CK$.

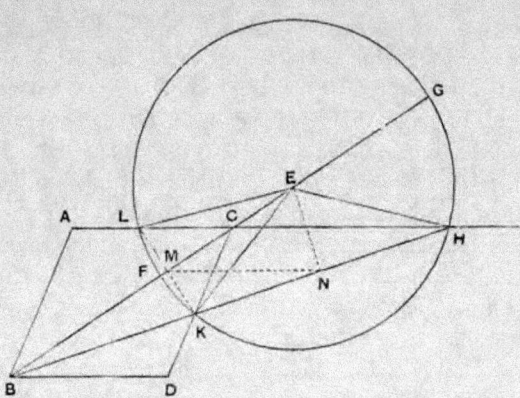

Also $EK = EL$, and CE is common to the triangles ECK, ECL. Therefore the said triangles are equal in all respects, and

$$\angle CKE = \angle CLE = \angle CHE.$$

Now, by hypothesis,

$$EB : EF = EF : EC,$$

or $$EB : EK = EK : EC \qquad \text{(since } EF = EK\text{)},$$

and the angle CEK is common to the triangles BEK, KEC; therefore the triangles BEK, KEC are similar, and

$$\angle CBK = \angle CKE$$

$$= \angle CHE, \text{ from above.}$$

Again, $$\angle HCE = \angle ACB = \angle BCK.$$

Thus in the triangles CBK, CHE two angles are equal respectively;

therefore $$\angle CEH = \angle CKB.$$

But, since $\angle CKE = \angle CHE$, from above, the points K, C, E, H are concyclic.

Hence $\angle CEH + \angle CKH = $ (two right angles).

Accordingly, since $$\angle CEH = \angle CKB,$$

$$\angle CKH + \angle CKB = \text{(two right angles)},$$

and BKH is a straight line.

Now the form of the proposition at once suggests that, in the 8th problem referred to, Apollonius had simply given a construction involving the drawing of a circle cutting CD and AC produced in the points K, H respectively, and Pappus' proof that BKH is a straight line is intended to prove that HK *verges towards* B, or (in other words) to verify that the construction given by Apollonius *solves a certain* νεῦσις *requiring BKH to be drawn so that KH is equal to a given length.*

The analysis leading to the construction must have been worked out somewhat as follows.

Suppose BKH drawn so that KH is equal to the given length k. Bisect KH at N, and draw NE at right angles to KH meeting BC produced in E.

Draw KM perpendicular to BC and produce it to meet CA in L. Then, from the property of the rhombus, the triangles KCM, LCM are equal in all respects.

Therefore $KM = ML$; and accordingly, if MN be joined, MN, LH are parallel.

Now, since the angles at M, N are right, a circle can be described about $EMKN$.

Therefore $\qquad \angle CEK = \angle MNK$, in the same segment,

$$\angle = \angle CHK, \text{ by parallels.}$$

Hence a circle can be described about $CEHK$. It follows that

$$\angle BCD = \angle CEK + \angle CKE$$
$$= \angle CHK + \angle CHE$$
$$= \angle EHK = \angle EKH.$$

Therefore the triangles EKH, DBC are similar.

Lastly, $\qquad \angle CKN = \angle CBK + \angle BCK;$

and, subtracting from these equals the equal angles EKN, BCK respectively, we have

$$\angle EKC = \angle EBK.$$

Hence the triangles EBK, EKC are similar, and

$$BE : EK = EK : EC,$$

or $\qquad\qquad BE . EC = EK^2.$

But, by similar triangles, $EK : KH = DC : CB,$

and the ratio $DC : CB$ is given, while KH is also given $(= k)$.

Therefore EK is given, and, in order to find E, we have only, in the Greek phrase, to "apply to BC a rectangle exceeding by a square figure and equal to the given area EK^2."

Thus the construction given by Apollonius was clearly the following*.

If k be the given length, take a straight line p such that

$$p : k = AB : BC.$$

Apply to BC a rectangle exceeding by a square figure and equal to the area p^2. Let $BE . EC$ be this rectangle, and with E as centre and radius equal to p describe a circle cutting AC produced in H and CD in K.

HK is then equal to k, and verges towards B, as proved by Pappus; the problem is therefore solved.

The construction used by Apollonius for the 'plane' νεῦσις with reference to the rhombus having been thus restored by means of the theorem given by Pappus, we are enabled to understand the purpose

---

* This construction was suggested to me by a careful examination of Pappus' proposition without other aid; but it is no new discovery. Samuel Horsley gives the same construction in his restoration of *Apollonii Pergaei Inclinationum libri duo* (Oxford, 1770); he explains, however, that he went astray in consequence of a mistake in the figure given in the MSS., and was unable to deduce the construction from Pappus's proposition until he was recalled to the right track by a solution of the same problem by Hugo d'Omerique. This solution appears in a work entitled, *Analysis geometrica, sive nova et vera methodus resolvendi tam problemata geometrica quam arithmeticas quaestiones*, published at Cadiz in 1698. D'Omerique's construction, which is practically identical with that of Apollonius, appears to have been evolved by means of an independent analysis of his own, since he makes no reference to Pappus, as he does in other cases where Pappus is drawn upon (e.g. when giving the construction for the case of the square attributed by Pappus to one Heraclitus). The construction differs from that given above only in the fact that the circle is merely used to determine the point K, after which BK is joined and produced to meet AC in H. Of other solutions of the same problem two may here be mentioned. (1) The solution contained in Marino Ghetaldi's posthumous work *De Resolutione et Compositione Mathematica Libri quinque* (Rome, 1630), and included among the solutions of other problems all purporting to be solved "methodo qua antiqui utebantur," is, though geometrical, entirely different from that above given, being effected by means of a reduction of the problem to a simpler plane νεῦσις of the same character as that assumed by Hippocrates in his *Quadrature of lunes*. (2) Christian Huygens (*De circuli magnitudine inventa; accedunt problematum quorundam illustrium constructiones*, Lugduni Batavorum, 1654) gave a rather complicated solution, which may be described as a generalisation of Heraclitus' solution in the case of a square.

for which Pappus, while still on the subject of the "8th problem" of Apollonius, adds a solution for the particular case of the *square* (which he calls a "problem after Heraclitus") with an introductory lemma. It seems clear that Apollonius did not treat the case of the square separately from the rhombus because the solution for the rhombus was equally applicable to the square, and this supposition is confirmed by the fact that, in setting out the main problems discussed in the νεύσεις, Pappus only mentions the rhombus and not the square. Being however acquainted with a solution by one Heraclitus of the νεῦσις relating to a square which was not on the same lines as that of Apollonius, while it was not applicable to the case of the rhombus, Pappus adds it as an alternative method for the square which is worth noting*. This is no doubt the explanation of the heading to the lemma prefixed to Heraclitus' problem which Hultsch found so much difficulty in explaining and put in brackets as an interpolation by a writer who misunderstood the figure and the object of the theorem. The words mean "Lemma useful for the [problem] with reference to squares taking the place of the rhombus" (literally "having the same property as the rhombus"), i.e. a lemma useful for Heraclitus' solution of the

---

* This view of the matter receives strong support from the following facts. In Pappus' summary (p. 670) of the contents of the νεύσεις of Apollonius "two cases" of the νεῦσις with reference to the rhombus are mentioned last among the particular problems given in the first of the two Books. As we have seen, one case (that given above) was the subject of the "8th problem" of Apollonius, and it is equally clear that the other case was dealt with in the "9th problem." The other case is clearly that in which the line to be drawn through B, instead of crossing the exterior angle of the rhombus at C, lies across the angle C itself, i.e. meets CA, CD both produced. In the former case the solution of the problem is always possible whatever be the length of k; but in the second case clearly the problem is not capable of solution if k, the given length, is less than a certain minimum. Hence the problem requires a διορισμός to determine the minimum length of k. Accordingly we find Pappus giving, after 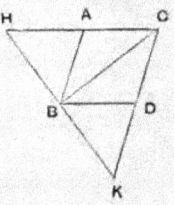 the interposition of the case of the square, a "lemma useful for the διορισμός of the 9th problem," which proves that, if $CH = CK$ and B be the middle point of HK, then HK is the least straight line which can be drawn through B to meet CH, CK. Pappus adds that the διορισμός for the rhombus is then evident; if HK be the line drawn through B perpendicular to CB and meeting CA, CD produced in H, K, then, in order that the problem may admit of solution, the given length k must be not less than HK.

νεῦσις in the particular case of a square*. The lemma is as follows.

ABCD being a square, suppose BHE drawn so as to meet CD in H and AD produced in E, and let EF be drawn perpendicular to BE meeting BC produced in F. To prove that

$$CF^2 = BC^2 + HE^2.$$

Suppose *EG* drawn parallel to *DC* meeting *CF* in *G*. Then since *BEF* is a right angle, the angles *HBC*, *FEG* are equal.

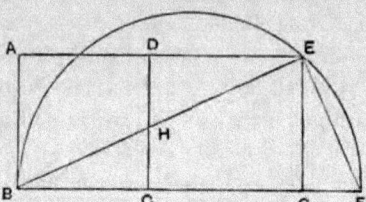

Therefore the triaugles *BCH*, *EGF* are equal in all respects, and

$$EF = BH.$$

Now　　　　　　　$$BF^2 = BE^2 + EF^2,$$

or　　　$$BC \cdot BF + BF \cdot FC = BH \cdot BE + BE \cdot EH + EF^2.$$

But, the angles *HCF*, *HEF* being right, the points *C*, *H*, *E*, *F* are concyclic, and therefore

$$BC \cdot BF = BH \cdot BE.$$

Subtracting these equals, we have

$$BF \cdot FC = BE \cdot EH + EF^2$$
$$= BE \cdot EH + BH^2$$
$$= BH \cdot HE + EH^2 + BH^2$$
$$= EB \cdot BH + EH^2$$
$$= FB \cdot BC + EH^2.$$

* Hultsch translates the words λῆμμα χρήσιμον εἰς τὸ ἐπὶ τετραγώνων ποιούντων τὰ αὐτὰ τῷ ῥόμβῳ (p. 780) thus, "Lemma utile ad *problema* de quadratis quorum summa rhombo aequalis est," and has a note in his Appendix (p. 1260) explaining what he supposes to be meant. The 'squares' he takes to be the given square and the square on the given length of the intercept, and the rhombus to be one for which he indicates a construction but which is not shown in Pappus' figure. Thus he is obliged to translate τῷ ῥόμβῳ as "*a* rhombus," which is one objection to his interpretation, while "whose squares are equal" scarcely seems a possible rendering of ποιούντων τὰ αὐτά.

Take away the common part $BC \cdot CF$, and
$$CF^2 = BC^2 + EH^2.$$

Heraclitus' analysis and construction are now as follows.

Suppose that we have drawn BHE so that HE has a given length k.

Since $CF^2 = BC^2 + EH^2$, or $BC^2 + k^2$,

and BC and k are both given,

CF is given, and therefore BF is given.

Thus the semicircle on BF as diameter is given, and therefore also E, its intersection with the given line ADE; hence BE is given.

To effect the construction, we first find a square equal to the sum of the given square and the square on k. We then produce BC to F so that CF is equal to the side of the square so found. If a semicircle be now described on BF as diameter, it will pass above D (since $CF > CD$, and therefore $BC \cdot CF > CD^2$), and will therefore meet AD produced in some point E.

Join BE meeting CD in H.

Then $HE = k$, and the problem is solved.

CHAPTER VI.

CUBIC EQUATIONS.

It has often been explained how the Greek geometers were able to solve geometrically all forms of the quadratic equation which give positive roots; while they could take no account of others because the conception of a negative quantity was unknown to them. The quadratic equation was regarded as a simple equation connecting areas, and its geometrical expression was facilitated by the methods which they possessed of transforming any rectilineal areas whatever into parallelograms, rectangles, and ultimately squares, of equal area; its solution then depended on the principle of *application of areas*, the discovery of which is attributed to the Pythagoreans. Thus any plane problem which could be reduced to the geometrical equivalent of a quadratic equation with a positive root was at once solved. A particular form of the equation was the *pure* quadratic, which meant for the Greeks the problem of finding a square equal to a given rectilineal area. This area could be transformed into a rectangle, and the general form of the equation thus became $x^2 = ab$, so that it was only necessary to find a mean proportional between a and b. In the particular case where the area was given as the sum of two or more squares, or as the difference of two squares, an alternative method depended on the Pythagorean theorem of Eucl. I. 47 (applied, if necessary, any number of times successively). The connexion between the two methods is seen by comparing Eucl. VI. 13, where the mean proportional between a and b is found, and Eucl. II. 14, where the same problem is solved without the use of proportions by means of I. 47, and where in fact the formula used is

$$x^2 = ab = \left(\frac{a+b}{2}\right)^2 - \left(\frac{a-b}{2}\right)^2.$$

The choice between the two methods was equally patent when the equation to be solved was $x^2 = pa^2$, where p is any integer; hence the *multiplication* of squares was seen to be dependent on the finding of a mean proportional. The equation $x^2 = 2a^2$ was the simplest equation of the kind, and the discovery of a geometrical construction for the side of a square equal to twice a given square was specially important, as it was the beginning of the theory of incommensurables or 'irrationals' (ἀλόγων πραγματεία) which was invented by Pythagoras. There is every reason to believe that this successful doubling of the square was what suggested the question whether a construction could not be found for the *doubling of the cube*, and the stories of the tomb erected by Minos for his son and of the oracle bidding the Delians to double a cubical altar were no doubt intended to invest the purely mathematical problem with an element of romance. It may then have been the connexion between the doubling of the square and the finding of one mean proportional which suggested the reduction of the doubling of the cube to the problem of finding *two mean proportionals* between two unequal straight lines. This reduction, attributed to Hippocrates of Chios, showed at the same time the possibility of *multiplying* the cube by any ratio. Thus, if x, y are two mean proportionals between a, b, we have

$$a : x = x : y = y : b,$$

and we derive at once

$$a : b = a^3 : x^3,$$

whence a cube (x^3) is obtained which bears to a^3 the ratio $b : a$, while any fraction $\dfrac{p}{q}$ can be transformed into a ratio between lines of which one (the consequent) is equal to the side a of the given cube. Thus the finding of two mean proportionals gives the solution of any pure cubic equation, or the equivalent of extracting the cube root, just as the single mean proportional is equivalent to extracting the square root. For suppose the given equation to be $x^3 = bcd$. We have then only to find a mean proportional a between c and d, and the equation becomes $x^3 = a^2 . b = a^3 . \dfrac{b}{a}$ which is exactly the multiplication of a cube by a ratio between lines which the two mean proportionals enable us to effect.

As a matter of fact, we do not find that the great geometers were in the habit of reducing problems to the multiplication of the cube *eo nomine*, but to the equivalent problem of the two mean proportionals; and the cubic equation $x^3 = a^2b$ is not usually stated in that form but as a proportion. Thus in the two propositions *On the Sphere and Cylinder* II. 1, 5, where Archimedes uses the two mean proportionals, it is required to find x where

$$a^2 : x^2 = x : b ;$$

he does not speak of finding the side of a cube equal to a certain parallelepiped, as the analogy of finding a square equal to a given rectangle might have suggested. So far therefore we do not find any evidence of a general system of adding and subtracting solids by transforming parallelepipeds into cubes and cubes into parallelepipeds which we should have expected to see in operation if the Greeks had systematically investigated the solution of the general form of the cubic equation by a method analogous to that of the *application of areas* employed in dealing with quadratic equations.

The question then arises, did the Greek geometers deal thus generally with the cubic equation

$$x^3 \pm ax^2 \pm Bx \pm \Gamma = 0,$$

which, on the supposition that it was regarded as an independent problem in solid geometry, would be for them a simple equation between solid figures, x and a both representing linear magnitudes, B an area (a rectangle), and Γ a volume (a parallelepiped)? And was the reduction of a problem of an order higher than that which could be solved by means of a quadratic equation to the solution of a cubic equation in the form shown above a regular and recognised method of dealing with such a problem? The only direct evidence pointing to such a supposition is found in Archimedes, who reduces the problem of dividing a sphere by a plane into two segments whose volumes are in a given ratio (*On the Sphere and Cylinder* II. 4) to the solution of a cubic equation which he states in a form equivalent to

$$4a^2 : x^2 = (3a - x) : \frac{m}{m + n} a \quad \ldots\ldots\ldots\ldots\ldots\ldots(1),$$

where a is the radius of the sphere, $m : n$ the given ratio (being a ratio between straight lines of which $m > n$), and x the height of the greater of the required segments. Archimedes explains that this is a particular case of a more general problem, to divide a straight line (a) into two parts (x, $a - x$) such that one part ($a - x$) is to another given straight line (c) as a given area (which for convenience' sake we suppose transformed into a square, b^2) is to the square on the other part (x^2), i.e. so that

$$(a - x) : c = b^2 : x^2. \quad\quad\text{............................(2)}.$$

He further explains that the equation (2) stated thus generally requires a διορισμός, i.e. that the limits for the possibility of a real solution, etc., require to be investigated, but that the particular case (with the conditions obtaining in the particular proposition) requires no διορισμός, i.e. the equation (1) will always give a real solution. He adds that "the analysis and synthesis of both these problems will be given at the end." That is, he promises to give separately a complete investigation of the equation (2), which is equivalent to the cubic equation

$$x^2 (a - x) = b^2 c \quad\quad\text{............................(3)}$$

and to apply it to the particular case (1).

Wherever the solution was given, it was temporarily lost, having apparently disappeared even before the time of Dionysodorus and Diocles (the latter of whom lived, according to Cantor, not later than about 100 B.C.); but Eutocius describes how he found an old fragment which appeared to contain the original solution of Archimedes, and gives it in full. It will be seen on reference to Eutocius' note (which I have reproduced immediately after the proposition to which it relates, *On the Sphere and Cylinder* II. 4) that the solution (the genuineness of which there seems to be no reason to doubt) was effected by means of the intersection of a parabola and a rectangular hyperbola whose equations may respectively be written thus,

$$x^2 = \frac{b^2}{a} y,$$

$$(a - x) y = ac.$$

The διορισμός takes the form of investigating the *maximum* possible value of $x^2 (a - x)$, and it is proved that this maximum value for a real solution is that corresponding to the value $x = \frac{2}{3} a$. This is established by showing that, if $b^2 c = \frac{4}{27} a^3$, the curves touch at the point for which $x = \frac{2}{3} a$. If on the other hand $b^2 c < \frac{4}{27} a^3$, it is proved that there are two real solutions. In the particular case (1) it is clear that the condition for a real solution is satisfied, for the expression in (1) corresponding to b^2c in (2) is $\dfrac{m}{m+n} 4a^3$, and it is only necessary that

$$\frac{m}{m+n} 4a^3 \not> \frac{4}{27}(3a)^3, \text{ or } 4a^3,$$

which is obviously true.

Hence it is clear that not only did Archimedes solve the cubic equation (3) by means of the intersections of two conics, but he also discussed completely the conditions under which there are 0, 1 or 2 roots lying between 0 and a. It is to be noted further that the διορισμός is similar in character to that by which Apollonius investigates the number of possible normals that can be drawn to a conic from a given point*. Lastly, Archimedes' method is seen to be an extension of that used by Menaechmus for the solution of the pure cubic equation. This can be put in the form

$$a^3 : x^3 = a : b,$$

which can again be put in Archimedes' form thus,

$$a^2 : x^2 = x : b,$$

and the conics used by Menaechmus are respectively

$$x^2 = ay, \quad xy = ab,$$

which were of course suggested by the two mean proportionals satisfying the equations

$$a : x = x : y = y : b.$$

The case above described is not the only one where we may assume Archimedes to have solved a problem by first reducing it to a cubic equation and then solving that. At the end of the preface to the book *On Conoids and Spheroids* he says that the results therein obtained may be used for discovering many theorems and problems, and, as instances of the latter, he mentions the following, "from a given spheroidal figure or conoid to cut off, by a plane drawn parallel to a given plane, a segment which shall be equal to a given cone or cylinder, or to a given sphere." Though Archimedes does not give the solutions, the following considerations may satisfy us as to his method.

(1) The case of the 'right-angled conoid' (the paraboloid of revolution) is a 'plane' problem and therefore does not concern us here.

---

* Cf. *Apollonius of Perga*, p. 168 sqq.

(2) In the case of the spheroid, the volume of the whole spheroid could be easily ascertained, and, by means of that, the ratio between the required segment and the remaining segment; after which the problem could be solved in exactly the same way as the similar one in the case of the sphere above described, since the results in *On Conoids and Spheroids*, Props. 29—32, correspond to those of *On the Sphere and Cylinder* II. 2. Or Archimedes may have proceeded in this case by a more direct method, which we may represent thus. Let a plane be drawn through the axis of the spheroid perpendicular to the given plane (and therefore to the base of the required segment). This plane will cut the elliptical base of the segment in one of its axes, which we will call $2y$. Let x be the length of the axis of the segment (or the length intercepted within the segment of the diameter of the spheroid passing through the centre of the base of the segment). Then the area of the base of the segment will vary as y^2 (since all sections of the spheroid parallel to the given plane must be similar), and therefore the volume of the *cone* which has the same vertex and base as the required segment will vary as y^2x. And the ratio of the volume of the segment to that of the cone is (*On Conoids and Spheroids*, Props. 29—32) the ratio $(3a - x) : (2a - x)$, where $2a$ is the length of the diameter of the spheroid which passes through the vertex of the segment. Therefore

$$y^2x \cdot \frac{3a - x}{2a - x} = C,$$

where C is a known volume. Further, since x, y are the coordinates of a point on the elliptical section of the spheroid made by the plane through the axis perpendicular to the cutting plane, referred to a diameter of that ellipse and the tangent at the extremity of the diameter, the ratio $y^2 : x(2a - x)$ is given. Hence the equation can be put in the form

$$x^2(3a - x) = b^2c,$$

and this again is the same equation as that solved in the fragment given by Eutocius. A διορισμός is formally necessary in this case, though it only requires the constants to be such that the volume to which the segment is to be equal must be less than that of the whole spheroid.

(3) For the 'obtuse-angled conoid' (hyperboloid of revolution) it would be necessary to use the direct method just described for

the spheroid, and, if the notation be the same, the corresponding equations will be found, with the help of *On Conoids and Spheroids*, Props. 25, 26, to be

$$y^2 x \cdot \frac{3a + x}{2a + x} = C,$$

and, since the ratio $y^2 : x\,(2a + x)$ is constant,

$$x^2\,(3a + x) = b^2 c.$$

If this equation is written in the form of a proportion like the similar one above, it becomes

$$b^2 : x^2 = (3a + x) : c.$$

There can be no doubt that Archimedes solved this equation as well as the similar one with a negative sign, i.e. he solved the two equations

$$x^3 \pm ax^2 \mp b^2 c = 0,$$

obtaining all their positive real roots. In other words, he solved completely, so far as the real roots are concerned, a cubic equation in which the term in x is absent, although the determination of the positive and negative roots of one and the same equation meant for him two separate problems. And it is clear that all cubic equations can be easily reduced to the type which Archimedes solved.

We possess one other solution of the cubic equation to which the division of a sphere into segments bearing a given ratio to one another is reduced by Archimedes. This solution is by Dionysodorus, and is given in the same note of Eutocius*. Dionysodorus does not generalise the equation, however, as is done in the fragment quoted above; he merely addresses himself to the particular case,

$$4a^2 : x^2 = (3a - x) : \frac{m}{m + n}\,a,$$

thereby avoiding the necessity for a διορισμός. The curves which he uses are the parabola

$$\frac{m}{m + n}\,a\,(3a - x) = y^2$$

and the rectangular hyperbola

$$\frac{m}{m + n}\,2a^2 = xy.$$

When we turn to Apollonius, we find him emphasising in his

---

* *On the Sphere and Cylinder* II. 4 (note at end).

preface to Book IV. of the *Conics** the usefulness of investigations
of the possible number of points in which conics may intersect one
another or circles, because "they at all events afford a more ready
means of observing some things, e.g. that several solutions are
possible, or that they are so many in number, and again that no
solution is possible"; and he shows his mastery of this method
of investigation in Book V., where he determines the number of
normals that can be drawn to a conic through any given point, the
condition that two normals through it coincide, or (in other words)
that the point lies on the evolute of the conic, and so on. For these
purposes he uses the points of intersection of a certain rectangular
hyperbola with the conic in question, and among the cases we find
(v. 51, 58, 62) some which can be reduced to cubic equations, those
namely in which the conic is a parabola and the axis of the parabola
is parallel to one of the asymptotes of the hyperbola. Apollonius
however does not bring in the cubic equation; he addresses himself
to the direct geometrical solution of the problem in hand without
reducing it to another. This is after all only natural, because the
solution necessitated the drawing of the rectangular hyperbola in
the actual figure containing the conic in question; thus, e.g. in the
case of the problem leading to a cubic equation, Apollonius can, so
to speak, compress two steps into one, and the introduction of the
cubic as such would be mere surplusage. The case was different
with Archimedes, when he had no conic in his original figure; and
the fact that he set himself to solve a cubic somewhat more general
than that actually involved in the problem made separate treatment
with a number of new figures necessary. Moreover Apollonius was
at the same time dealing, in other propositions, with cases which did
not reduce to cubics, but would, if put in an algebraical form, lead
to biquadratic equations, and these, expressed as such, would have
had no meaning for the Greeks; there was therefore the less reason
in the simpler case to introduce a subsidiary problem.

As already indicated, the cubic equation, as a subject of systematic and independent study, appears to have been lost sight of
within a century or so after the death of Archimedes. Thus Diocles,
the discoverer of the cissoid, speaks of the problem of the division of
the sphere into segments in a given ratio as having been reduced
by Archimedes "to another problem, which he does not solve in
his work on the sphere and cylinder"; and he then proceeds to

* *Apollonius of Perga*, p. lxxiii.

solve the original problem directly, without in any way bringing in the cubic. This circumstance does not argue any want of geometrical ability in Diocles; on the contrary, his solution of the original problem is a remarkable instance of dexterity in the use of conics for the solution of a somewhat complicated problem, and it proceeds on independent lines in that it depends on the intersection of an *ellipse* and a rectangular hyperbola, whereas the solutions of the cubic equation have accustomed us to the use of the *parabola* and the rectangular hyperbola. I have reproduced Diocles' solution in its proper place as part of the note of Eutocius on Archimedes' proposition; but it will, I think, be convenient to give here its equivalent in the ordinary notation of analytical geometry, in accordance with the plan of this chapter. Archimedes had proved [*On the Sphere and Cylinder* II. 2] that, if k be the height of a segment cut off by a plane from a sphere of radius a, and if h be the height of the cone standing on the same base as that of the segment and equal in volume to the segment, then

$$(3a - k) : (2a - k) = h : k.$$

Also, if h' be the height of the cone similarly related to the remaining segment of the sphere,

$$(a + k) : k = h' : (2a - k).$$

From these equations we derive

$$(h - k) : k = a : (2a - k),$$

and $$(h' - 2a + k) : (2a - k) = a : k.$$

Slightly generalising these equations by substituting for a in the third term of each proportion another length b, and adding the condition that the segments (and therefore the cones) are to bear to each other the ratio $m : n$, Diocles sets himself to solve the three equations

$$\left.\begin{array}{r}(h - k) : k = b : (2a - k) \\ (h' - 2a + k) : (2a - k) = b : k \\ h : h' = m : n\end{array}\right\} \quad \ldots\ldots\ldots\ldots(A).$$

and

Suppose $m > n$, so that $k > a$. The problem then is to divide a straight line of length $2a$ into two parts k and $(2a - k)$ of which k is the greater, and which are such that the three given equations are all simultaneously satisfied.

Imagine two coordinate axes such that the origin is the middle point of the given straight line, the axis of y is at right angles to it, and x is positive when measured along that half of the given straight line which is to contain the required point of division. Then the conics drawn by Diocles are

(1) the ellipse represented by the equation

$$(y + a - x)^2 = \frac{n}{m}\{(a+b)^2 - x^2\},$$

and (2) the rectangular hyperbola

$$(x + a)(y + b) = 2ab.$$

One intersection between these conics gives a value of x between 0 and a, and leads to the solution required. Treating the equations algebraically, and eliminating y by means of the second equation which gives

$$y = \frac{a - x}{a + x} \cdot b,$$

we obtain from the first equation

$$(a - x)^2 \left(1 + \frac{b}{a+x}\right)^2 = \frac{n}{m}\{(a+b)^2 - x^2\},$$

that is, $\qquad (a + x)^2(a + b - x) = \frac{m}{n}(a - x)^2(a + b + x) \ldots\ldots\ldots$ (B).

In other words Diocles' method is the equivalent of solving a complete cubic equation containing all the three powers of x and a constant, though no mention is made of such an equation.

To verify the correctness of the result we have only to remember that, x being the distance of the point of division from the middle point of the given straight line,

$$k = a + x, \quad 2a - k = a - x.$$

Thus, from the first two of the given equations (A) we obtain respectively

$$h = a + x + \frac{a + x}{a - x} \cdot b,$$

$$h' = a - x + \frac{a - x}{a + x} \cdot b,$$

whence, by means of the third equation, we derive

$$(a + x)^2(a + b - x) = \frac{m}{n}(a - x)^2(a + b + x),$$

which is the same equation as that found by elimination above (B).

I have purposely postponed, until the evidence respecting the Greek treatment of the cubic equation was complete, any allusion to an interesting hypothesis of Zeuthen's* which, if it could be accepted as proved, would explain some difficulties involved in Pappus' account of the orthodox classification of problems and loci. I have already quoted the passage in which Pappus distinguishes the problems which are *plane* (ἐπίπεδα), those which are *solid* (στερεά) and those which are *linear* (γραμμικά)†. Parallel to this division of problems into three orders or classes is the distinction between three classes of *loci*‡. The first class consists of *plane loci* (τόποι ἐπίπεδοι) which are exclusively straight lines and circles, the second of *solid loci* (τόποι στερεοί) which are conic sections§, and the third of *linear loci* (τόποι γραμμικοί). It is at the same time clearly implied by Pappus that problems were originally called plane, solid or linear respectively for the specific reason that they required for their solution the geometrical loci which bore the corresponding names. But there are some logical defects in the classification both as regards the problems and the loci.

(1) Pappus speaks of its being a serious error on the part of geometers to solve a plane problem by means of conics (i.e. 'solid loci') or 'linear' curves, and generally to solve a problem "by means of a foreign kind" (ἐξ ἀνοικείου γένους). If this principle were applied strictly, the objection would surely apply equally to the solution of a 'solid' problem by means of a 'linear' curve. Yet, though e.g. Pappus mentions the conchoid and the cissoid as being 'linear' curves, he does not object to their employment in the solution of the problem of the two mean proportionals, which is a 'solid' problem.

(2) The application of the term 'solid loci' to the three conic sections must have reference simply to the *definition* of the curves as sections of a solid figure, viz. the cone, and it was no doubt in contrast to the 'solid locus' that the 'plane locus' was so called. This agrees with the statement of Pappus that 'plane' problems may

* *Die Lehre von den Kegelschnitten*, p. 226 sqq.

† p. ciii.

‡ Pappus VII. pp. 652, 662.

§ It is true that Proclus (p. 394, ed. Friedlein) gives a wider definition of "solid lines" as those which arise "from some section of a solid figure, as the cylindrical helix and the conic curves"; but the reference to the cylindrical helix would seem to be due to some confusion.

properly be so called because the lines by means of which they are solved "have their origin in a plane." But, though this may be regarded as a satisfactory distinction when 'plane' and 'solid' loci are merely considered in relation to one another, it becomes at once logically defective when the third or 'linear' class is also brought in. For, on the one hand, Pappus shows how the 'quadratrix' (a 'linear' curve) can be produced by a construction in three dimensions ("by means of surface-loci," διὰ τῶν πρὸς ἐπιφανείαις τόπων); and, on the other hand, other 'linear' loci, the conchoid and cissoid, have their origin in a plane. If then Pappus' account of the origin of the terms 'plane' and 'solid' as applied to problems and loci is literally correct, it would seem necessary to assume that the third name of 'linear' problems and loci was not invented until a period when the terms 'plane' and 'solid loci' had been so long recognised and used that their origin was forgotten.

To get rid of these difficulties, Zeuthen suggests that the terms 'plane' and 'solid' were first applied to *problems*, and that they came *afterwards* to be applied to the geometrical loci which were used for the purpose of solving them. On this interpretation, when problems which could be solved by means of the straight line and circle were called 'plane,' the term is supposed to have had reference, not to any particular property of the straight line or circle, but to the fact that the problems were such as depend on an equation of a degree not higher than the second. The solution of a quadratic equation took the geometrical form of application of areas, and the term 'plane' became a natural one to apply to the class of problems so soon as the Greeks found themselves confronted with a new class of problems to which, in contrast, the term 'solid' could be applied. This would happen when the operations by which problems were reduced to applications of areas were tried upon problems which depend on the solution of a *cubic* equation. Zeuthen, then, supposes that the Greeks sought to give this equation a similar shape to that which the reduced 'plane' problem took, that is, to form a simple equation between *solids* corresponding to the cubic equation

$$x^3 + ax^2 + Bx + \Gamma = 0;$$

the term 'solid' or 'plane' being then applied according as it had been reduced, in the manner indicated, to the geometrical equivalent of a cubic or a quadratic equation.

Zeuthen further explains the term 'linear problem' as having been invented afterwards to describe the cases which, being equivalent to algebraical equations of an order higher than the third, would not admit of reduction to a simple relation between lengths, areas and volumes, and either could not be reduced to an equation at all or could only be represented as such by the use of compound ratios. The term 'linear' may perhaps have been applied because, in such cases, recourse was had to new classes of curves, directly and without any intermediate step in the shape of an equation. Or, possibly, the term may not have been used at all until a time when the original source of the names 'plane' and 'solid' problems had been forgotten.

On these assumptions, it would still be necessary to explain how Pappus came to give a more extended meaning to the term 'solid problem,' which according to him equally includes those problems which, though solved by the same method of conics as was used to solve the equivalent of cubics, do not reduce to cubic equations but to biquadratics. This is explained by the supposition that, the cubic equation having by the time of Apollonius been obscured from view owing to the attention given to the method of solution by means of conics and the discovery that the latter method was one admitting of wider application, the possibility of solution by means of conics came itself to be regarded as the criterion determining the class of problem, and the name 'solid problem' came to be used in the sense given to it by Pappus through a natural misapprehension. A similar supposition would account, in Zeuthen's view, for a circumstance which would otherwise seem strange, viz. that Apollonius does not use the expression 'solid problem,' though it might have been looked for in the preface to the fourth Book of the *Conics*. The term may have been avoided by Apollonius because it then had the more restricted meaning attributed to it by Zeuthen and therefore would not have been applicable to all the problems which Apollonius had in view.

It must be admitted that Zeuthen's hypothesis is in several respects attractive. I cannot however feel satisfied that the positive evidence in favour of it is sufficiently strong to outweigh the authority of Pappus where his statements tell the other way. To make the position clear, we have to remember that Menaechmus, the discoverer of the conic sections, was a pupil of Eudoxus who flourished about 365 B.C.; probably therefore we may place the discovery of conics at about 350 B.C. Now Aristaeus 'the elder'

wrote a book on *solid loci* (στερεοὶ τόποι) the date of which Cantor
concludes to have been about 320 B.C. Thus, on Zeuthen's hypothesis, the 'solid problems' the solution of which by means of conics
caused the latter to be called 'solid loci' must have been such as
had been already investigated and recognised as solid problems
before 320 B.C., while the definite appropriation, so to speak, of the
newly discovered curves to the service of the class of problems must
have come about in the short period between their discovery and
the date of Aristaeus' work. It is therefore important to consider
what particular problems leading to cubic equations appear to have
been the subject of speculation before 320 B.C. We have certainly
no ground for assuming that the cubic equation used by Archimedes
(*On the Sphere and Cylinder* II. 4) was one of these problems; for
the problem of cutting a sphere into segments bearing a given ratio
to one another could not have been investigated by geometers who
had not succeeded in finding the volume of a sphere and a segment
of a sphere, and we know that Archimedes was the first to discover
this. On the other hand there was the duplication of the cube, or
the solution of a pure cubic equation, which was a problem dating
from very early times. Also it is certain that the trisection of an
angle had long exercised the minds of the Greek geometers. Pappus
says that "the ancient geometers" considered this problem and first
tried to solve it, though it was by nature a solid problem (πρόβλημα
τῇ φύσει στερεὸν ὑπάρχον), by means of plane considerations (διὰ τῶν
ἐπιπέδων) but failed; and we know that Hippias of Elis invented,
about 420 B.C., a transcendental curve which was capable of being
used for two purposes, the trisection of an angle, and the quadrature
of a circle*. This curve came to be called the Quadratrix†, but, as
Deinostratus, a brother of Menaechmus, was apparently the first to
apply the curve to the quadrature of the circle‡, we may no doubt
conclude that it was originally intended for the purpose of trisecting

* Proclus (ed. Friedlein), p. 272.

† The character of the curve may be described as follows. Suppose there
are two rectangular axes Oy, Ox and that a straight line OP of a certain length
(a) revolves uniformly from a position along Oy to a position along Ox, while a
straight line remaining always parallel to Ox and passing through P in its
original position also moves uniformly and reaches Ox in the same time as the
moving radius OP. The point of intersection of this line and OP describes the
Quadratrix, which may therefore be represented by the equation

$$y/a = 2\theta/\pi.$$

‡ Pappus IV. pp. 250—2.

an angle. Seeing therefore that the Greek geometers had used their best efforts to solve this problem before the invention of conics, it may easily be that they had succeeded in reducing it to the geometrical equivalent of a cubic equation. They would not have been unequal to effecting this reduction by means of the figure of the νεῦσις given above on p. cxii. with a few lines added. The proof would of course be the equivalent of eliminating x between the two equations

$$\left.\begin{aligned} xy &= ab \\ (x-a)^2 + (y-b)^2 &= 4\,(a^2+b^2) \end{aligned}\right\} \dots\dots\dots\dots(a)$$

where $x = DF$, $y = FP = EC$, $a = DA$, $b = DB$.

The second equation gives

$$(x+a)\,(x-3a) = (y+b)\,(3b-y).$$

From the first equation it is easily seen that

$$(x+a) : (y+b) = a : y,$$

and that $\qquad (x-3a)\,y = a\,(b-3y);$

we have therefore $\qquad a^2\,(b-3y) = y^2\,(3b-y)\dots\dots\dots\dots\dots(\beta)$

[or $\qquad y^3 - 3by^2 - 3a^2y + a^2b = 0$].

If then the trisection of an angle had been reduced to the geometrical equivalent of this cubic equation, it would be natural for the Greeks to speak of it as a *solid problem*. In this respect it would be seen to be similar in character to the simpler problem of the duplication of the cube or the equivalent of a pure cubic equation; and it would be natural to see whether the transformation of volumes would enable the mixed cubic to be reduced to the form of the pure cubic, in the same way as the transformation of areas enabled the mixed quadratic to be reduced to the pure quadratic. The reduction to the pure cubic would soon be seen to be impossible, and the stereometric line of investigation would prove unfruitful and be abandoned accordingly.

The two problems of the duplication of the cube and the trisection of an angle, leading in one case to a pure cubic equation and in the other to a mixed cubic, are then the only problems leading to cubic equations which we can be certain that the Greeks had occupied themselves with up to the time of the discovery of the conic sections. Menaechmus, who discovered these, showed that they could be successfully used for finding the two mean proportionals and therefore for solving the pure cubic equation, and the

next question is whether it had been proved before the date of
Aristaeus' *Solid Loci* that the trisection of an angle could be
effected by means of the same conics, either in the form of the
νεῦσις above described directly and without the reduction to a cubic
equation, or in the form of the subsidiary cubic (β). Now (1) the
solution of the cubic would be somewhat difficult in the days when
conics were still a new thing. The solution of the equation (β) as
such would involve the drawing of the conics which we should
represent by the equations

$$xy = a^2,$$

$$bx = 3a^2 + 3by - y^2,$$

and the construction would be decidedly more difficult than that
used by Archimedes in connexion with his cubic, which only requires
the construction of the conics

$$x^2 = \frac{b^2}{a} y,$$

$$(a - x) y = ac ;$$

hence we can hardly assume that the trisection of an angle in the
form of the subsidiary *cubic equation* was solved by means of conics
before 320 B.C. (2) The angle may have been trisected by means
of conics in the sense that the νεῦσις referred to was effected by
drawing the curves (a), i.e. a rectangular hyperbola and a circle.
This could easily have been done before the date of Aristaeus; but
if the assignment of the name 'solid loci' to conics had in view their
applicability to the direct solution of the problem in this manner
without any reference to the cubic equation, or simply because
the problem had been before proved to be 'solid' by means of the
reduction to that cubic, then there does not appear to be any
reason why the Quadratrix, which had been used for the same
purpose, should not *at the time* have been also regarded as a 'solid
locus,' in which case *Aristaeus* could hardly have appropriated the
latter term, in his work, to conics alone. (3) The only remaining
alternative consistent with Zeuthen's view of the origin of the
name 'solid locus' appears to be to suppose that conics were so
called simply because they gave a means of solving *one* 'solid
problem,' viz. the doubling of the cube, and not a problem of the
more general character corresponding to a mixed cubic equation, in
which case the justification for the general name 'solid locus' could
only be admitted on the assumption that it was adopted at a time

when the Greeks were still hoping to be able to reduce the general cubic equation to the pure form. I think however that the traditional explanation of the term is more natural than this would be. Conics were the first curves of general interest for the description of which recourse to solid figures was necessary as distinct from the ordinary construction of plane figures in a plane*; hence the use of the term 'solid locus' for conics on the mere ground of their solid origin would be a natural way of describing the new class of curves in the first instance, and the term would be likely to remain in use, even when the solid origin was no longer thought of, just as the individual conics continued to be called "sections of a right-angled, obtuse-angled, and acute-angled" cone respectively.

While therefore, as I have said, the two problems mentioned *might* naturally have been called 'solid problems' before the discovery of 'solid loci,' I do not think there is sufficient evidence to show that 'solid problem' was then or later a *technical term* for a problem capable of reduction to a cubic equation in the sense of implying that the geometrical equivalent of the general cubic equation was investigated for its own sake, independently of its applications, and that it ever occupied such a recognised position in Greek geometry that a problem would be considered solved so soon as it was reduced to a cubic equation. If this had been so, and if the technical term for such a cubic was 'solid problem,' I find it hard to see how Archimedes could have failed to imply something of the kind when arriving at his cubic equation. Instead of this, his words rather suggest that he had attacked it as *res integra*. Again, if the general cubic had been regarded over any length of time as a problem of independent interest which was solved by means of the intersections of conics, the fact could hardly have been unknown to Nicoteles who is mentioned in the preface to Book IV. of the *Conics* of Apollonius as having had a controversy with Conon respecting the investigations in which the latter discussed the maximum number of points of intersection between two conics. Now Nicoteles is stated by Apollonius to have maintained that no use

---

* It is true that Archytas' solution of the problem of the two mean proportionals used a curve of double curvature drawn on a cylinder; but this was not such a curve as was likely to be investigated for itself or even to be regarded as a *locus*, strictly speaking; hence the solid origin of this isolated curve would not be likely to suggest objections to the appropriation of the term 'solid locus' to conics.

could be made of the discoveries of Conon for διορισμοί; but it seems incredible that Nicoteles could have made such a statement, even for controversial purposes, if cubic equations then formed a recognised class of problems for the discussion of which the intersections of conics were necessarily all-important.

I think therefore that the positive evidence available will not justify us in accepting the conclusions of Zeuthen except to the following extent.

1. Pappus' explanation of the meaning of the term 'plane problem' (ἐπίπεδον πρόβλημα) as used by the ancients can hardly be right. Pappus says, namely, that "problems which can be solved by means of the straight line and circle *may properly be called* plane (λέγοιτ' ἂν εἰκότως ἐπίπεδα); for the lines by means of which such problems are solved have their origin in a plane." The words "may properly be called" suggest that, so far as plane problems were concerned, Pappus was not giving the ancient definition of them, but his own inference as to why they were called 'plane.' The true significance of the term is no doubt, as Zeuthen says, not that straight lines and circles have their origin in a plane (which would be equally true of some other curves), but that the problems in question admitted of solution by the ordinary plane methods of transformation of areas, manipulation of simple equations between areas, and in particular the application of areas. In other words, plane problems were those which, if expressed algebraically, depend on equations of a degree not higher than the second.

2. When further problems were attacked which proved to be beyond the scope of the plane methods referred to, it would be found that some of such problems, in particular the duplication of the cube and the trisection of an angle, were reducible to simple equations between *volumes* instead of equations between areas; and it is quite possible that, following the analogy of the distinction existing in nature between plane figures and solid figures (an analogy which was also followed in the distinction between *numbers* as 'plane' and 'solid' expressly drawn by Euclid), the Greeks applied the term 'solid problem' to such a problem as they could reduce to an equation between volumes, as distinct from a 'plane problem' reducible to a simple equation between areas.

3. The first 'solid problem' in this sense which they succeeded

in solving was the multiplication of the cube, corresponding to the solution of a pure cubic equation in algebra, and it was found that this could be effected by means of curves obtained by making plane sections of a solid figure, namely the cone. Thus curves having a solid origin were found to solve one particular solid problem, which could not but seem an appropriate result; and hence the conic, as being the simplest curve so connected with a solid problem, was considered to be properly termed a 'solid locus,' whether because of its application or (more probably) because of its origin.

4. Further investigation showed that the general cubic equation could not be reduced, by means of stereometric methods, to the simpler form, the pure cubic; and it was found necessary to try the method of conics directly either (1) upon the derivative cubic equation or (2) upon the original problem which led to it. In practice, as e.g. in the case of the trisection of an angle, it was found that the cubic was often more difficult to solve in that manner than the original problem was. Hence the reduction of it to a cubic was dropped as an unnecessary complication, and the geometrical equivalent of a cubic equation stated as an independent problem never obtained a permanent footing as the 'solid problem' *par excellence*.

5. It followed that solution by conics came to be regarded as the criterion for distinguishing a certain class of problem, and, as conics had retained their old name of 'solid loci,' the corresponding term 'solid problem' came to be used in the wider sense in which Pappus interprets it, according to which it includes a problem depending on a biquadratic as well as a problem reducible to a cubic equation.

6. The terms 'linear problem' and 'linear locus' were then invented on the analogy of the other terms to describe respectively a problem which could not be solved by means of straight lines, circles, or conics, and a curve which could be used for solving such a problem, as explained by Pappus.

CHAPTER VII.

ANTICIPATIONS BY ARCHIMEDES OF THE INTEGRAL CALCULUS.

IT has been often remarked that, though the *method of exhaustion* exemplified in Euclid XII. 2 really brought the Greek geometers face to face with the infinitely great and the infinitely small, they never allowed themselves to use such conceptions. It is true that Antiphon, a sophist who is said to have often had disputes with Socrates, had stated[*] that, if one inscribed any regular polygon, say a square, in a circle, then inscribed an octagon by constructing isosceles triangles in the four segments, then inscribed isosceles triangles in the remaining eight segments, and so on, "until the whole area of the circle was by this means exhausted, a polygon would thus be inscribed whose sides, in consequence of their small-ness, would coincide with the circumference of the circle." But as against this Simplicius remarks, and quotes Eudemus to the same effect, that the inscribed polygon will never coincide with the circumference of the circle, even though it be possible to carry the division of the area to infinity, and to suppose that it would is to set aside a geometrical principle which lays down that magni-tudes are divisible *ad infinitum*[†]. The time had, in fact, not come for the acceptance of Antiphon's idea, and, perhaps as the result of the dialectic disputes to which the notion of the infinite gave rise, the Greek geometers shrank from the use of such expressions as infinitely great and infinitely small and substituted the idea of things *greater* or *less than any assigned magnitude*. Thus, as Hankel says[‡], they never said that a circle *is* a polygon with an infinite number of

[*] Bretschneider, p. 101.
[†] Bretschneider, p. 102.
[‡] Hankel, *Zur Geschichte der Mathematik im Alterthum und Mittelalter,* p. 123.

infinitely small sides; they always stood still before the abyss of the infinite and never ventured to overstep the bounds of clear conceptions. They never spoke of an infinitely close approximation or a limiting value of the sum of a series extending to an infinite number of terms. Yet they must have arrived practically at such a conception, e.g., in the case of the proposition that circles are to one another as the squares on their diameters, they must have been in the first instance led to infer the truth of the proposition by the idea that the circle could be regarded as the limit of an inscribed regular polygon with an indefinitely increased number of correspondingly small sides. They did not, however, rest satisfied with such an inference; they strove after an irrefragable proof, and this, from the nature of the case, could only be an indirect one. Accordingly we always find, in proofs by the method of exhaustion, a demonstration that an impossibility is involved by any other assumption than that which the proposition maintains. Moreover this stringent verification, by means of a double *reductio ad absurdum*, is repeated in every individual instance of the use of the method of exhaustion; there is no attempt to establish, in lieu of this part of the proof, any general propositions which could be simply quoted in any particular case.

The above general characteristics of the Greek method of exhaustion are equally present in the extensions of the method found in Archimedes. To illustrate this, it will be convenient, before passing to the cases where he performs genuine *integrations*, to mention his geometrical proof of the property that the area of a parabolic segment is four-thirds of the triangle with the same base and vertex. Here Archimedes *exhausts* the parabola by continually drawing, in each segment left over, a triangle with the same base and vertex as the segment. If A be the area of the triangle so inscribed in the original segment, the process gives a series of areas

$$A, \quad \tfrac{1}{4}A, \quad (\tfrac{1}{4})^2 A, \ldots$$

and the area of the segment is really the sum of the infinite series

$$A \{1 + \tfrac{1}{4} + (\tfrac{1}{4})^2 + (\tfrac{1}{4})^3 + \ldots\}.$$

But Archimedes does not express it in this way. He first proves that, if $A_1, A_2, \ldots A_n$ be any number of terms of such a series, so that $A_1 = 4A_2, A_2 = 4A_3, \ldots$, then

$$A_1 + A_2 + A_3 + \ldots + A_n + \tfrac{1}{3}A_n = \tfrac{4}{3}A_1,$$

or $\qquad A \{1 + \tfrac{1}{4} + (\tfrac{1}{4})^2 + \ldots + (\tfrac{1}{4})^{n-1} + \tfrac{1}{3}(\tfrac{1}{4})^{n-1}\} = \tfrac{4}{3}A.$

Having obtained this result, we should nowadays suppose n to increase indefinitely and should infer at once that $(\frac{1}{4})^{n-1}$ becomes indefinitely small, and that the limit of the sum on the left-hand side is the area of the parabolic segment, which must therefore be equal to $\frac{4}{3}A$. Archimedes does not avow that he inferred the result in this way; he merely *states* that the area of the segment is equal to $\frac{4}{3}A$, and then verifies it in the orthodox manner by proving that it cannot be either greater or less than $\frac{4}{3}A$.

I pass now to the extensions by Archimedes of the method of exhaustion which are the immediate subject of this chapter. It will be noticed, as an essential feature of all of them, that Archimedes takes both an inscribed figure and a circumscribed figure in relation to the curve or surface of which he is investigating the area or the solid content, and then, as it were, *compresses* the two figures into one so that they coincide with one another and with the curvilinear figure to be measured; but again it must be understood that he does not describe his method in this way or say at any time that the given curve or surface is the limiting form of the circumscribed or inscribed figure. I will take the cases in the order in which they come in the text of this book.

1.   *Surface of a sphere or spherical segment.*

The first step is to prove (*On the Sphere and Cylinder* I. 21, 22) that, if in a circle or a segment of a circle there be inscribed polygons, whose sides AB, BC, CD, ... are all equal, as shown in the respective figures, then

(*a*)   for the circle

$$(BB' + CC' + ...) : AA' = A'B : BA,$$

(*b*)   for the segment

$$(BB' + CC' + ... + KK' + LM) : AM = A'B : BA.$$

Next it is proved that, if the polygons revolve about the diameter AA', the surface described by the equal sides of the polygon in a complete revolution is [I. 24, 35]

(*a*)   equal to a circle with radius $\sqrt{AB (BB' + CC' + ... + YY')}$

or (*b*)   equal to a circle with radius $\sqrt{AB (BB' + CC' + ... + LM)}$.

Therefore, by means of the above proportions, the surfaces described by the equal sides are seen to be equal to

(a)   a circle with radius $\sqrt{AA' . A'B}$,

and (b)   a circle with radius $\sqrt{AM . A'B}$;

they are therefore respectively [I. 25, 37] *less than*

(a)   a circle with radius AA',

(b)   a circle with radius AL.

Archimedes now proceeds to take polygons circumscribed to the circle or segment of a circle (supposed in this case to be less than a semicircle) so that their sides are parallel to those of the inscribed polygons before mentioned (cf. the figures on pp. 38, 51); and he proves by like steps [I. 30, 40] that, if the polygons revolve about the diameter as before, the surfaces described by the equal sides during a complete revolution are *greater than* the same circles respectively.

Lastly, having proved these results for the inscribed and circumscribed figures respectively, Archimedes concludes and proves [I. 33, 42, 43] that the surface of the sphere or the segment of the sphere is *equal to* the first or the second of the circles respectively.

In order to see the effect of the successive steps, let us express the several results by means of trigonometry. If, in the figures on pp. 33, 47 respectively, we suppose $4n$ to be the number of sides in the polygon inscribed in the circle and $2n$ the number of the equal sides in the polygon inscribed in the segment, while in the latter case the angle AOL is denoted by a, the proportions given above are respectively equivalent to the formulae [*]

$$\sin \frac{\pi}{2n} + \sin \frac{2\pi}{2n} + \dots + \sin (2n-1) \frac{\pi}{2n} = \cot \frac{\pi}{4n},$$

and
$$\frac{2 \left\{ \sin \frac{a}{n} + \sin \frac{2a}{n} + \dots + \sin (n-1) \frac{a}{n} \right\} + \sin a}{1 - \cos a} = \cot \frac{a}{2n}.$$

Thus the two proportions give in fact a summation of the series

$$\sin \theta + \sin 2\theta + \dots + \sin (n-1) \theta$$

both generally where $n\theta$ is equal to any angle a less than π, and in the particular case where n is even and $\theta = \pi/n$.

Again, the areas of the circles which are equal to the surfaces described by the revolution of the equal sides of the *inscribed*

---

[*] These formulae are taken, with a slight modification, from Loria, *Il periodo aureo della geometria greca*, p. 108.

polygons are respectively (if a be the radius of the great circle of the sphere)

$$4\pi a^2 \sin\frac{\pi}{4n}\left\{\sin\frac{\pi}{2n}+\sin\frac{2\pi}{2n}+\dots+\sin(2n-1)\frac{\pi}{2n}\right\},\text{ or }4\pi a^2\cos\frac{\pi}{4n},$$

and

$$\pi a^2 . 2\sin\frac{a}{2n}\left[2\left\{\sin\frac{a}{n}+\sin\frac{2a}{n}+\dots+\sin(n-1)\frac{a}{n}\right\}+\sin a\right],$$

or

$$\pi a^2 . 2\cos\frac{a}{2n}(1-\cos a).$$

The areas of the circles which are equal to the surfaces described by the equal sides of the *circumscribed* polygons are obtained from the areas of the circles just given by dividing them by $\cos^2\pi/4n$ and $\cos^2 a/2n$ respectively.

Thus the results obtained by Archimedes are the same as would be obtained by taking the limiting value of the above trigonometrical expressions when n is indefinitely increased, and when therefore $\cos\pi/4n$ and $\cos a/2n$ are both unity.

But the first expressions for the areas of the circles are (when n is indefinitely increased) exactly what we represent by the integrals

$$4\pi a^2 . \tfrac{1}{2}\int_0^\pi \sin\theta\, d\theta,\text{ or }4\pi a^2,$$

and

$$\pi a^2 . \int_0^a 2\sin\theta\, d\theta,\text{ or }2\pi a^2(1-\cos a).$$

Thus Archimedes' procedure is the equivalent of a genuine integration in each case.

2. *Volume of a sphere or a sector of a sphere.*

The method does not need to be separately set out in detail here, because it depends directly on the preceding case. The investigation proceeds concurrently with that of the surface of a sphere or a segment of a sphere. The same inscribed and circumscribed figures are used, the sector of a sphere being of course compared with the solid figure made up of the figure inscribed or circumscribed to the segment and of the cone which has the same base as that figure and has its vertex at the centre of the sphere. It is then proved, (1) for the figure inscribed or circumscribed to the sphere, that its volume is equal to that of a cone with base equal to the surface of the figure and height equal to the perpendicular from the centre of the sphere on any one of the equal sides of the revolving polygon, (2) for the figure inscribed or circumscribed to the sector, that the volume is equal to that of a cone with base equal to the surface of the portion of the figure which is inscribed or circumscribed to the *segment* of the sphere included in the sector and whose height is the perpendicular from the centre on one of the equal sides of the polygon.

Thus, when the inscribed and circumscribed figures are, so to speak, compressed into one, the taking of the limit is practically the same thing in this case as in the case of the surfaces, the resulting volumes being simply the before-mentioned surfaces multiplied in each case by $\frac{1}{3}a$.

3. *Area of an ellipse.*

This case again is not strictly in point here, because it does not exhibit any of the peculiarities of Archimedes' extensions of the method of exhaustion. That method is, in fact, applied in the same manner, *mutatis mutandis*, as in Eucl. XII. 2. There is no simultaneous use of inscribed and circumscribed figures, but only the simple exhaustion of the ellipse and auxiliary circle by increasing to any desired extent the number of sides in polygons inscribed to each (*On Conoids and Spheroids*, Prop. 4).

4. *Volume of a segment of a paraboloid of revolution.*

Archimedes first states, as a Lemma, a result proved incidentally in a proposition of another treatise (*On Spirals*, Prop. 11), viz. that, if there be n terms of an arithmetical progression h, $2h$, $3h$, ..., then

$$h + 2h + 3h + \dots + nh > \tfrac{1}{2}n^2h$$
$$\text{and} \qquad h + 2h + 3h + \dots + (n-1)h < \tfrac{1}{2}n^2h \qquad \dots\dots\dots\dots(a).$$

Next he inscribes and circumscribes to the segment of the paraboloid figures made up of small cylinders (as shown in the figure of *On Conoids and Spheroids*, Props. 21, 22) whose axes lie along the axis of the segment and divide it into any number of equal parts. If c is the length of the axis AD of the segment, and if there are n cylinders in the circumscribed figure and their axes are each of length h, so that $c = nh$, Archimedes proves that

$$(1) \qquad \frac{\text{cylinder } CE}{\text{inscribed fig.}} = \frac{n^2h}{h + 2h + 3h + \dots + (n-1)h}$$
$$> 2, \text{ by the Lemma,}$$

$$\text{and } (2) \qquad \frac{\text{cylinder } CE}{\text{circumscribed fig.}} = \frac{n^2h}{h + 2h + 3h + \dots + nh}$$
$$< 2.$$

Meantime it has been proved [Props. 19, 20] that, by increasing n sufficiently, the inscribed and circumscribed figure can be made to differ by less than any assignable volume. It is accordingly concluded and proved by the usual rigorous method that

$$\text{(cylinder } CE) = 2 \text{ (segment)},$$

so that   (segment ABC) $= \frac{3}{2}$ (cone ABC).

The proof is therefore equivalent to the assertion, that if h is indefinitely diminished and n indefinitely increased, while nh remains equal to c,

$$\text{limit of } h\{h + 2h + 3h + \ldots + (n-1)h\} = \tfrac{1}{2}c^2;$$

that is, in our notation,

$$\int_0^c x\,dx = \tfrac{1}{2}c^2.$$

Thus the method is essentially the same as ours when we express the volume of the segment of the paraboloid in the form

$$\kappa \int_0^c y^2 dx,$$

where κ is a constant, which does not appear in Archimedes' result for the reason that he does not give the actual content of the segment of the paraboloid but only the ratio which it bears to the circumscribed cylinder.

5. *Volume of a segment of a hyperboloid of revolution.*

The first step in this case is to prove [*On Conoids and Spheroids,* Prop. 2] that, if there be a series of n terms,

$$ah + h^2, \quad a \cdot 2h + (2h)^2, \quad a \cdot 3h + (3h)^2, \quad \ldots \quad a \cdot nh + (nh)^2,$$

and if $(ah + h^2) + \{a \cdot 2h + (2h)^2\} + \ldots + \{a \cdot nh + (nh)^2\} = S_n,$

then $n\{a \cdot nh + (nh)^2\}/S_n < (a + nh) \Big/ \left(\dfrac{a}{2} + \dfrac{nh}{3}\right)$

and $n\{a \cdot nh + (nh)^2\}/S_{n-1} > (a + nh) \Big/ \left(\dfrac{a}{2} + \dfrac{nh}{3}\right)$ ....(β).

Next [Props. 25, 26] Archimedes draws inscribed and circumscribed figures made up of cylinders as before (figure on p. 137), and proves that, if AD is divided into n equal parts of length h, so that $nh = AD$, and if $AA' = a$, then

$$\frac{\text{cylinder } EB'}{\text{inscribed figure}} = \frac{n\{a \cdot nh + (nh)^2\}}{S_{n-1}}$$

$$> (a + nh) \bigg/ \left(\frac{a}{2} + \frac{nh}{3}\right),$$

and
$$\frac{\text{cylinder } EB'}{\text{circumscribed fig.}} = \frac{n\{a \cdot nh + (nh)^2\}}{S_n}$$

$$< (a + nh) \bigg/ \left(\frac{a}{2} + \frac{nh}{3}\right).$$

The conclusion, arrived at in the same manner as before, is that

$$\frac{\text{cylinder } EB'}{\text{segment } ABB'} = (a + nh) \bigg/ \left(\frac{a}{2} + \frac{nh}{3}\right).$$

This is the same as saying that, if $nh = b$, and if h be indefinitely diminished while n is indefinitely increased,

$$\text{limit of } n\,(ab + b^2)/S_n = (a + b) \bigg/ \left(\frac{a}{2} + \frac{b}{3}\right),$$

or
$$\text{limit of } \frac{b}{n} S_n = b^2 \left(\frac{a}{2} + \frac{b}{3}\right).$$

Now $S_n = a\,(h + 2h + \ldots + nh) + \{h^2 + (2h)^2 + \ldots + (nh)^2\}$,

so that $hS_n = ah\,(h + 2h + \ldots + nh) + h\,\{h^2 + (2h)^2 + \ldots + (nh)^2\}$.

The limit of the last expression is what we should write as

$$\int_0^b (ax + x^2)\,dx,$$

which is equal to
$$b^2 \left(\frac{a}{2} + \frac{b}{3}\right);$$

and Archimedes has given the equivalent of this integration.

6. *Volume of a segment of a spheroid.*

Archimedes does not here give the equivalent of the integration

$$\int_0^b (ax - x^2),$$

presumably because, with his method, it would have required yet another lemma corresponding to that in which the results (β) above are established.

Suppose that, in the case of a segment less than half the spheroid (figure on p. 142), $AA' = a$, $CD = \frac{1}{2}c$, $AD = b$; and let AD be divided into n equal parts of length h.

The gnomons mentioned in Props. 29, 30 are then the differences between the rectangle $cb + b^2$ and the successive rectangles

$$ch + h^2, \quad c \cdot 2h + (2h)^2, \quad \ldots \quad c \cdot (n-1)h + \{(n-1)h\}^2,$$

and in this case we have the conclusions that (if S_n be the sum of n terms of the series representing the latter rectangles)

$$\frac{\text{cylinder } EB'}{\text{inscribed figure}} = \frac{n(cb + b^2)}{n(cb + b^2) - S_n}$$

$$> (c + b) \Big/ \left(\frac{c}{2} + \frac{2b}{3}\right),$$

and

$$\frac{\text{cylinder } EB'}{\text{circumscribed fig.}} = \frac{n(cb + b^2)}{n(cb + b^2) - S_{n-1}}$$

$$< (c + b) \Big/ \left(\frac{c}{2} + \frac{2b}{3}\right),$$

and in the limit

$$\frac{\text{cylinder } EB'}{\text{segment } ABB'} = (c + b) \Big/ \left(\frac{c}{2} + \frac{2b}{3}\right).$$

Accordingly we have the limit taken of the expression

$$\frac{n(cb + b^2) - S_n}{n(cb + b^2)}, \quad \text{or} \quad 1 - \frac{S_n}{n(cb + b^2)},$$

and the integration performed is the same as that in the case of the hyperboloid above, with c substituted for a.

Archimedes discusses, as a separate case, the volume of half a spheroid [Props. 27, 28]. It differs from that just given in that c vanishes and $b = \frac{1}{2}a$, so that it is necessary to find the limit of

$$\frac{h^2 + (2h)^2 + (3h)^2 + \ldots + (nh)^2}{n(nh)^2};$$

and this is done by means of a corollary to the lemma given on pp. 107—9 [On Spirals, Prop. 10] which proves that

$$h^2 + (2h)^2 + \ldots + (nh)^2 > \frac{1}{3}n(nh)^2,$$

and

$$h^2 + (2h)^2 + \ldots + \{(n-1)h\}^2 < \frac{1}{3}n(nh)^2.$$

The limit of course corresponds to the integral

$$\int_0^b x^2 dx = \frac{1}{3}b^3.$$

7. *Area of a spiral.*

(1) Archimedes finds the area bounded by the first complete turn of a spiral and the initial line by means of the proposition just quoted, viz.

$$h^2 + (2h)^2 + \ldots + (nh)^2 > \tfrac{1}{3}n\,(nh)^2,$$

$$h^2 + (2h)^2 + \ldots + \{(n-1)\,h\}^2 < \tfrac{1}{3}n\,(nh)^2.$$

He proves [Props. 21, 22, 23] that a figure consisting of similar sectors of circles can be circumscribed about any arc of a spiral such that the area of the circumscribed figure exceeds that of the spiral by less than any assigned area, and also that a figure of the same kind can be inscribed such that the area of the spiral exceeds that of the inscribed figure by less than any assigned area. Then, lastly, he circumscribes and inscribes figures of this kind [Prop. 24]; thus e.g. in the circumscribed figure, if there are n similar sectors, the radii will be n lines forming an arithmetical progression, as h, $2h$, $3h$, ... nh, and nh will be equal to a, where a is the length intercepted on the initial line by the spiral at the end of the first turn. Since, then, similar sectors are to one another as the square of their radii, and n times the sector of radius nh or a is equal to the circle with the same radius, the first of the above formulae proves that

$$(\text{circumscribed fig.}) > \tfrac{1}{3}\pi a^2.$$

A similar procedure for the inscribed figure leads, by the use of the second formula, to the result that

$$(\text{inscribed fig.}) < \tfrac{1}{3}\pi a^2.$$

The conclusion, arrived at in the usual manner, is that

$$(\text{area of spiral}) = \tfrac{1}{3}\pi a^2\,;$$

and the proof is equivalent to taking the limit of

$$\frac{\pi}{n}\left[h^2 + (2h)^2 + \ldots + \{(n-1)\,h\}^2\right]$$

or of

$$\frac{\pi h}{a}\left[h^2 + (2h)^2 + \ldots + \{(n-1)\,h\}^2\right],$$

which last limit we should express as

$$\frac{\pi}{a}\int_0^a x^2\,dx = \tfrac{1}{3}\pi a^2$$

[It is clear that this method of proof equally gives the area bounded by the spiral and any radius vector of length b not being greater than a; for we have only to substitute $\pi b/a$ for π, and to remember that in this case $nh = b$. We thus obtain for the area

$$\frac{\pi}{a} \int_0^b x^2\, dx, \text{ or } \tfrac{1}{3}\pi b^3/a.]$$

(2) To find the area bounded by an arc on any turn of the spiral (not being greater than a complete turn) and the radii vectores to its extremities, of lengths b and c say, where $c > b$, Archimedes uses the proposition that, if there be an arithmetic progression consisting of the terms

$$b,\ b + h,\ b + 2h,\ \dots\ b + (n - 1)\,h,$$

and if $\quad S_n = b^2 + (b + h)^2 + (b + 2h)^2 + \dots + \{b + (n - 1)\,h\}^2,$

then $\quad \dfrac{(n - 1)\{b + (n-1)h\}^2}{S_n - b^2} < \dfrac{\{b + (n-1)\,h\}^2}{\{b + (n-1)h\}b + \frac{1}{3}\{(n-1)h\}^2},$

and $\quad \dfrac{(n - 1)\{b + (n-1)h\}^2}{S_{n-1}} > \dfrac{\{b + (n-1)\,h\}^2}{\{b + (n-1)h\}b + \frac{1}{3}\{(n-1)h\}^2}.$

[On Spirals, Prop. 11 and note.]

Then in Prop. 26 he circumscribes and inscribes figures consisting of similar sectors of circles, as before. There are $n - 1$ sectors in each figure and therefore n radii altogether, including both b and c, so that we can take them to be the terms of the arithmetic progression given above, where $\{b + (n - 1)\,h\} = c$. It is thus proved, by means of the above inequalities, that

$$\frac{\text{sector } OB'C}{\text{circumscribed fig.}} < \frac{\{b + (n - 1)\,h\}^2}{\{b + (n-1)h\}b + \frac{1}{3}\{(n-1)h\}^2} < \frac{\text{sector } OB'C}{\text{inscr. fig.}};$$

and it is concluded after the usual manner that

$$\frac{\text{sector } OB'C}{\text{spiral } OBC} = \frac{\{b + (n - 1)\,h\}^2}{\{b + (n-1)h\}b + \frac{1}{3}\{(n-1)h\}^2}$$

$$= \frac{c^2}{cb + \frac{1}{3}(c - b)^2}.$$

Remembering that $n - 1 = (c - b)/h$, we see that the result is the same thing as proving that, in the limit, when n becomes indefinitely great and h indefinitely small, while $b + (n-1)h = c$,

$$\text{limit of } h\left[b^2 + (b+h)^2 + \ldots + \{b + (n-2)h\}^2\right]$$
$$= (c-b)\{cb + \tfrac{1}{3}(c-b)^2\}$$
$$= \tfrac{1}{3}(c^3 - b^3);$$

that is, with our notation,

$$\int_b^c x^2\,dx = \tfrac{1}{3}(c^3 - b^3).$$

(3) Archimedes works out separately [Prop. 25], by exactly the same method, the particular case where the area is that described in any one complete turn of the spiral beginning from the initial line. This is equivalent to substituting $(n-1)a$ for b and na for c, where a is the radius vector to the end of the first complete turn of the spiral.

It will be observed that Archimedes does not use the result corresponding to

$$\int_0^c x^2\,dx - \int_h^c x^2\,dx = \int_0^b x^2\,dx.$$

8. *Area of a parabolic segment.*

Of the two solutions which Archimedes gives of the problem of squaring a parabolic segment, it is the *mechanical* solution which gives the equivalent of a genuine integration. In Props. 14, 15 of the *Quadrature of the Parabola* it is proved that, of two figures inscribed and circumscribed to the segment and consisting in each case of trapezia whose parallel sides are diameters of the parabola, the inscribed figure is less, and the circumscribed figure greater, than one-third of a certain triangle (EqQ in the figure on p. 242). Then in Prop. 16 we have the usual process which is equivalent to taking the limit when the trapezia become infinite in number and their breadth infinitely small, and it is proved that

$$\text{(area of segment)} = \tfrac{1}{3} \triangle EqQ.$$

The result is the equivalent of using the equation of the parabola referred to Qq as axis of x and the diameter through Q as axis of y, viz.

$$py = x(2a - x),$$

which can, as shown on p. 236, be obtained from Prop. 4, and finding

$$\int_0^{2a} y\,dx,$$

where y has the value in terms of x given by the equation; and of course

$$\frac{1}{p}\int_0^{2a}(2ax-x^2)\,dx=\frac{4a^3}{3p}.$$

The equivalence of the method to an integration can also be seen thus. It is proved in Prop. 16 (see figure on p. 244) that, if qE be divided into n equal parts and the construction of the proposition be made, Qq is divided at $O_1, O_2, \ldots$ into the same number of equal parts. The area of the circumscribed figure is then easily seen to be the sum of the areas of the triangles

$$QqF,\quad QR_1F_1,\quad QR_2F_2,\quad \ldots$$

that is, of the areas of the triangles

$$QqF,\quad QO_1R_1,\quad QO_2D_1,\quad \ldots$$

Suppose now that the area of the triangle QqF is denoted by Δ, and it follows that

$$\text{(circumscribed fig.)}=\Delta\left\{1+\frac{(n-1)^2}{n^2}+\frac{(n-2)^2}{n^2}+\ldots+\frac{1}{n^2}\right\}$$

$$=\frac{1}{n^2\Delta^2}\cdot\Delta\left\{\Delta^2+2^2\Delta^2+\ldots+n^2\Delta^2\right\}.$$

Similarly we obtain

$$\text{(inscribed fig.)}=\frac{1}{n^2\Delta^3}\cdot\Delta\left\{\Delta^2+2^2\Delta^2+\ldots+(n-1)^2\Delta^2\right\}.$$

Taking the limit we have, if A denote the area of the triangle EqQ, so that $A=n\Delta$,

$$\text{(area of segment)}=\frac{1}{A^2}\int_0^A\Delta^2 d\Delta$$

$$=\tfrac{1}{3}A.$$

If the conclusion be regarded in this manner, the integration is the same as that which corresponds to Archimedes' squaring of the spiral.

CHAPTER VIII.

THE TERMINOLOGY OF ARCHIMEDES.

So far as the language of Archimedes is that of Greek geometry in general, it must necessarily have much in common with that of Euclid and Apollonius, and it is therefore inevitable that the present chapter should repeat many of the explanations of terms of general application which I have already given in the corresponding chapter of my edition of Apollonius' *Conics**. But I think it will be best to make this chapter so far as possible complete and self-contained, even at the cost of some slight repetition, which will however be relieved (1) by the fact that all the particular phrases quoted by way of illustration will be taken from the text of Archimedes instead of Apollonius, and (2) by the addition of a large amount of entirely different matter corresponding to the great variety of subjects dealt with by Archimedes as compared with the limitation of the work of Apollonius to the one subject of conics.

One element of difficulty in the present case arises out of the circumstance that, whereas Archimedes wrote in the Doric dialect, the original language has been in some books completely, and in others partially, transformed into the ordinary dialect of Greek. Uniformity of dialect cannot therefore be preserved in the quotations about to be made; but I have thought it best, when explaining single words, to use the ordinary form, and, when illustrating their use by quoting phrases or sentences, to give the latter as they appear in Heiberg's text, whether in Doric or Attic in the particular case. Lest the casual reader should imagine the paroxytone words εὐθεῖαι, διάμετροι, πεσεῖται, πεσοῦνται, ἐσσεῖται, δυνάνται, ἅπτεται, καλεῖσθαι, κεῖσθαι and the like to be misprints, I add that the quotations in Doric from Heiberg's text have the unfamiliar Doric accents.

I shall again follow the plan of grouping the various technical

---

* *Apollonius of Perga*, pp. clvii—clxx.

terms under certain general headings, which will enable the Greek term corresponding to each expression in the ordinary mathematical phraseology of the present day to be readily traced wherever such a Greek equivalent exists.

Points and lines.

A *point* is σημεῖον, *the point* Β τὸ Β σημεῖον or τὸ Β simply; *a point on* (a line or curve) σημεῖον ἐπί (with gen.) or ἐν; *a point raised above* (a plane) σημεῖον μετέωρον; *any two points whatever being taken* δύο σημείων λαμβανομένων ὁποιωνοῦν.

At a point (e.g. of an angle) πρός (with dat.), *having its vertex at the centre of the sphere* κορυφὴν ἔχων πρὸς τῷ κέντρῳ τῆς σφαίρας; of lines meeting *in* a point, touching or dividing *at* a point, etc., κατά (with acc.), thus ΑΕ *is bisected at* Ζ is ἁ ΑΕ δίχα τεμνέται κατὰ τὸ Ζ; of a point falling *on* or being placed *on* another ἐπί or κατά (with acc.), thus Ζ *will fall on* Γ, τὸ μὲν Ζ ἐπὶ τὸ Γ πεσεῖται, *so that* Ε *lies on* Δ, ὥστε τὸ μὲν Ε κατὰ τὸ Δ κεῖσθαι.

Particular points are *extremity* πέρας, *vertex* κορυφή, *centre* κέντρον, *point of division* διαίρεσις, *point of meeting* σύμπτωσις, *point of section* τομή, *point of bisection* διχοτομία, *the middle point* τὸ μέσον; *the points of division* Η, Ι, Κ, τὰ τῶν διαιρεσίων σαμεῖα τὰ Η, Ι, Κ; *let* Β *be its middle point* μέσον δὲ αὐτᾶς ἔστω τὸ Β; *the point of section in which* (a circle) *cuts* ἁ τομά, καθ᾽ ἂν τέμνει.

A *line* is γραμμή, *a curved line* καμπύλη γραμμή, *a straight line* εὐθεῖα with or without γραμμή. *The straight line* ΘΙΚΛ, ἁ ΘΙΚΛ εὐθεῖα; but sometimes the older expression is used, *the straight line on which* (ἐπί with gen. or dat. of the pronoun) *are placed certain letters*, thus *let it be the straight line* Μ, ἔστω ἐφ᾽ ἇ τὸ Μ, *other straight lines* Κ, Λ, ἄλλαι γραμμαί, ἐφ᾽ ἃν τὰ Κ, Λ. *The straight lines between the points* αἱ μεταξὺ τῶν σημείων εὐθεῖαι, *of the lines which have the same extremities the straight line is the least* τῶν τὰ αὐτὰ πέρατα ἐχουσῶν γραμμῶν ἐλαχίστην εἶναι τὴν εὐθεῖαν, *straight lines cutting one another* εὐθεῖαι τεμνοῦσαι ἀλλάλας.

For points in relation to lines we have such expressions as the following: *the points* Γ, Θ, Μ *are on a straight line* ἐπ᾽ εὐθείας ἐστὶ τὰ Γ, Θ, Μ σαμεῖα, *the point of bisection of the straight line containing the centres of the middle magnitudes* ἁ διχοτομία τᾶς εὐθείας τᾶς ἐχούσας τὰ κέντρα τῶν μέσων μεγεθέων. A very characteristic phrase for *at a point which divides the straight line in such a proportion that...* is ἐπὶ τᾶς εὐθείας διαιρεθείσας ὥστε...; similarly ἐπὶ τᾶς ΧΕ

τμαθείσας οὕτως, ὥστε. *A certain point will be on the straight line…
dividing it so that…* ἐσσεῖται ἐπὶ τᾶς εὐθείας…διαιρέον οὕτως τὰν
εἰρημέναν εὐθεῖαν, ὥστε….

The middle point of a line is often elegantly denoted by an
adjective in agreement; thus *at the middle point of the segment* ἐπὶ
μέσου τοῦ τμάματος, *(a line) drawn from* Γ *to the middle point of*
EB, ἀπὸ τοῦ Γ ἐπὶ μέσαν τὰν ΕΒ ἀχθεῖσα, *drawn to the middle point of
the base* ἐπὶ μέσαν τὰν βάσιν ἀγομένα.

A straight line produced is *the (straight line) in the same straight
line with it* ἡ ἐπ᾽ εὐθείας αὐτῇ. *In the same straight line with the
axis* ἐπὶ τᾶς αὐτᾶς εὐθείας τῷ ἄξονι. Of a straight line falling *on*
another line κατά (with gen.) is used, e.g. πίπτουσι κατ᾽ αὐτῆς; ἐπί
(with acc.) is also used of a straight line *placed on* another, thus *if*
EH *be placed on* ΒΔ, τεθείσας τᾶς ΕΗ ἐπὶ τὰν ΒΔ.

For lines passing through points we find the following expressions: *will pass through* Ν, ἥξει διὰ τοῦ Ν; *will pass through the
centre* διὰ τοῦ κέντρου πορεύσεται, *will fall through* Θ πεσεῖται διὰ τοῦ
Θ, *verging towards* Β νεύουσα ἐπὶ τὸ Β, *pass through the same point*
ἐπὶ τὸ αὐτὸ σαμεῖον ἐρχόνται; *the diagonals of the parallelogram fall
(i.e. meet) at* Θ, κατὰ δὲ τὸ Θ αἱ διαμέτροι τοῦ παραλληλογράμμου
πίπτοντι; ΕΖ (passes) *through the points bisecting* ΑΒ, ΓΔ, ἐπὶ δὲ τὰν
διχοτομίαν τᾶν ΑΒ, ΓΔ ἁ ΕΖ. The verb εἰμί is also used of *passing
through*, thus ἐσσεῖται δὴ αὐτὰ διὰ τοῦ Θ.

For lines in relation to other lines we have *perpendicular to*
κάθετος ἐπί (with acc.), *parallel to* παράλληλος with dat. or παρά
(with acc.); *let* ΚΛ *be (drawn) from* Κ *parallel to* ΓΔ, ἀπὸ τοῦ Κ
παρὰ τὰν ΓΔ ἔστω ἁ ΚΛ.

Lines meeting one another συμπίπτουσαι ἀλλήλαις; *the point in
which* ΖΗ, ΜΝ *produced meet one another and* ΑΓ, τὸ σημεῖον, καθ᾽ ὃ
συμβάλλουσιν ἐκβαλλόμεναι αἱ ΖΗ, ΜΝ ἀλλήλαις τε καὶ τῇ ΑΓ; *so as
to meet the tangent* ὥστε ἐμπεσεῖν τᾷ ἐπιψανούσᾳ, *let straight lines be
drawn parallel to* ΑΓ *to meet the section of the cone* ἄχθων εὐθεῖαι
παρὰ τὰν ΑΓ ἔστε ποτὶ τὰν τοῦ κώνου τομάν, *to draw a straight line to
meet its circumference* ποτὶ τὰν περιφέρειαν αὐτοῦ ποτιβαλεῖν εὐθεῖαν,
the line drawn to meet ἁ ποτιπεσοῦσα, *let* ΑΕ, ΑΔ *be drawn from the
point* Α *to meet the spiral and produced to meet the circumference of
the circle* ποτιπιπτόντων ἀπὸ τοῦ Α σαμείου ποτὶ τὰν ἕλικα αἱ ΑΕ, ΑΔ
καὶ ἐκπιπτόντων ποτὶ τὰν τοῦ κύκλου περιφέρειαν; *until it meets* ΘΑ *in*
Ο, ἔστε κα συμπέσῃ τᾷ ΘΑ κατὰ τὸ Ο (of a circle).

(The straight line) will fall outside (i.e. will extend beyond) P, ἐκτὸς τοῦ P πεσεῖται; *will fall within the section of the figure* ἐντὸς πεσοῦνται τᾶς τοῦ σχήματος τομᾶς.

The (perpendicular) distance between (two parallel lines) AZ, BH, τὸ διάστημα τᾶν AZ, BH. Other ways of expressing *distances* are the following: *the magnitudes equidistant from the middle one* τὰ ἴσον ἀπέχοντα ἀπὸ τοῦ μέσου μεγέθεα, *are at equal distances from one another* ἴσα ἀπ' ἀλλάλων διέστακεν; *the segments (lengths) on* ΛH *equal to* N, τὰ ἐν τᾷ ΛH τμάματα ἰσομεγέθεα τᾷ N; *greater by one segment* ἑνὶ τμάματι μείζων.

The word εὐθεῖα itself is also often used in the sense of *distance*; cf. the terms πρώτη εὐθεῖα etc. in the book *On Spirals*, also ἁ εὐθεῖα ἁ μεταξὺ τοῦ κέντρου τοῦ ἁλίου καὶ τοῦ κέντρου τᾶς γᾶς *the distance between the centre of the sun and the centre of the earth.*

The word for *join* is ἐπιζευγνύω or ἐπιζεύγνυμι; *the straight line joining the points of contact* ἁ τὰς ἁφὰς ἐπιζευγνύουσα εὐθεῖα, BΔ *when joined* ἁ BΔ ἐπιζευχθεῖσα; *let EZ join the points of bisection of* AΔ, BΓ, ἁ δὲ EZ ἐπιζευγνυέτω τὰς διχοτομίας τᾶν AΔ, BΓ. In one case the word seems to be used in the sense of *drawing* simply, εἴ κα εὐθεῖα ἐπιζευχθῇ γραμμὰ ἐν ἐπιπέδῳ.

Angles.

An *angle* is γωνία, the three kinds of angles are *right* ὀρθή, *acute* ὀξεῖα, *obtuse* ἀμβλεῖα; *right-angled* etc. ὀρθογώνιος, ὀξυγώνιος, ἀμβλυγώνιος; *equiangular* ἰσογώνιος; *with an even number of angles* ἀρτιόγωνος or ἀρτιογώνιος.

At right angles to ὀρθὸς πρός (with acc.) or πρὸς ὀρθάς (with dat. following); thus *if a line be erected at right angles to the plane* γραμμᾶς ἀνεστακούσας ὀρθᾶς ποτὶ τὸ ἐπίπεδον, *the planes are at right angles to one another* ὀρθὰ ποτ' ἀλλαλά ἐντι τὰ ἐπίπεδα, *being at right angles to* ABΓ, πρὸς ὀρθὰς ὢν τῷ ABΓ; ΚΓ, ΞΛ *are at right angles to one another* ποτ' ὀρθάς ἐντι ἀλλάλαις αἱ ΚΓ, ΞΛ, *to cut at right angles* τέμνειν πρὸς ὀρθάς. The expression *making right angles with* is also used, e.g. ὀρθὰς ποιοῦσα γωνίας ποτὶ τὰν AB.

The complete expression for *the angle contained by the lines* AH, AΓ is ἁ γωνία ἁ περιεχομένα ὑπὸ τᾶν AH, AΓ; but there are a great variety of shorter expressions, γωνία itself being often understood; thus *the angles* Δ, E, A, B, αἱ Δ, E, A, B γωνίαι; *the angle at* Θ, ἁ ποτὶ τῷ Θ; *the angle contained by* AΔ, ΔZ, ἁ γωνία ἁ ὑπὸ τᾶν AΔ, ΔZ; *the angle* ΔHΓ, ἡ ὑπὸ τῶν ΔHΓ γωνία, ἡ ὑπὸ ΔHΓ (with or without γωνία).

Making the angle K *equal to the angle* Θ, γωνίαν ποιοῦσα τὰν Κ ἴσαν τᾷ Θ; *the angle into which the sun fits and which has its vertex at the eye* γωνία, εἰς ἂν ὁ ἅλιος ἐναρμόζει τὰν κορυφὰν ἔχουσαν ποτὶ τᾷ ὄψει; *of the sides subtending the right angle* (hypotenuses) τᾶν ὑπὸ τὰν ὀρθὰν γωνίαν ὑποτεινουσᾶν, *they subtend the same angle* ἐντὶ ὑπὸ τὰν αὐτὰν γωνίαν.

If a line through an angular point of a polygon divides it exactly symmetrically, *the opposite angles of the polygon,* αἱ ἀπεναντίον γωνίαι τοῦ πολυγώνου, are those answering to each other on each side of the bisecting line.

Planes and plane figures.

A *plane* ἐπίπεδον; *the plane through* BΔ, τὸ ἐπίπεδον τὸ κατὰ τὴν BΔ, or τὸ διὰ τῆς BΔ, *plane of the base* ἐπίπεδον τῆς βάσεως, *plane* (i.e. *base*) *of the cylinder* ἐπίπεδον τοῦ κυλίνδρου; *cutting plane* ἐπίπεδον τέμνον, *tangent plane* ἐπίπεδον ἐπιψαῦον; the *intersection* of planes is their *common section* κοινὴ τομή.

In the same plane as the circle ἐν τῷ αὐτῷ ἐπιπέδῳ τῷ κύκλῳ.

Let a plane be erected on ΠZ *at right angles to the plane in which* AB, ΓΔ *are* ἀπὸ τᾶς ΠZ ἐπίπεδον ἀνεστακέτω ὀρθὸν ποτὶ τὸ ἐπίπεδον τό, ἐν ᾧ ἐντι αἱ AB, ΓΔ.

The plane surface ἡ ἐπίπεδος (ἐπιφάνεια), *a plane segment* ἐπίπεδον τμῆμα, *a plane figure* σχῆμα ἐπίπεδον.

A *rectilineal figure* εὐθύγραμμον (σχῆμα), a *side* πλευρά, *perimeter* ἡ περίμετρος, *similar* ὅμοιος, *similarly situated* ὁμοίως κείμενος.

To *coincide with* (when one figure is applied to another), ἐφαρμόζειν followed by the dative or ἐπί (with acc.); *one part coincides with the other* ἐφαρμόζει τὸ ἕτερον μέρος ἐπὶ τὸ ἕτερον; *the plane through* NZ *coincides with the plane through* AΓ, τὸ ἐπίπεδον τὸ κατὰ τὰν NZ ἐφαρμόζει τῷ ἐπιπέδῳ τῷ κατὰ τὰν AΓ. The passive is also used; *if equal and similar plane figures coincide with one another* τῶν ἴσων καὶ ὁμοίων σχημάτων ἐπιπέδων ἐφαρμοζομένων ἐπ᾽ ἄλλαλα.

Triangles.

A *triangle* is τρίγωνον, the *triangles bounded by* (their three sides) τὰ περιεχόμενα τρίγωνα ὑπὸ τῶν.... A *right-angled triangle* τρίγωνον ὀρθογώνιον, *one of the sides about the right angle* μία τῶν περὶ τὴν ὀρθήν. *The triangle through the axis* (of a cone) τὸ διὰ τοῦ ἄξονος τρίγωνον.

Quadrilaterals.

A *quadrilateral* is a *four-sided* figure (τετράπλευρον) as distinguished from a *four-angled* figure, τετράγωνον, which means a *square*. A *trapezium*, τραπέζιον, is in one place more precisely described as a *trapezium having its two sides parallel* τραπέζιον τὰς δύο πλευρὰς ἔχον παραλλάλους ἀλλάλαις.

A *parallelogram* παραλληλόγραμμον; for a parallelogram *on* a straight line as base ἐπί (with gen.) is used, thus *the parallelograms on them are of equal height* ἐστὶν ἰσοΰψη τὰ παραλληλόγραμμα τὰ ἐπ' αὐτῶν. A *diagonal* of a parallelogram is διάμετρος, *the opposite sides of the parallelogram* αἱ κατ' ἐναντίον τοῦ παραλληλογράμμου πλευραί.

Rectangles.

The word generally used for a *rectangle* is χωρίον (*space* or *area*) without any further description. As in the case of angles, the *rectangles contained by straight lines* are generally expressed more shortly than by the phrase τὰ περιεχόμενα χωρία ὑπό; either χωρίον may be omitted or both χωρίον and περιεχόμενον, thus *the rectangle* ΑΓ, ΓΕ may be any of the following, τὸ ὑπὸ τῶν ΑΓ, ΓΕ, τὸ ὑπὸ ΑΓ, ΓΕ, τὸ ὑπὸ ΑΓΕ, and *the rectangle under* ΘΚ, ΑΗ is τὸ ὑπὸ τῆς ΘΚ καὶ τῆς ΑΗ. *Rectangles* Θ, Ι, Κ, Λ, χωρία ἐν οἷς τὰ (or ἐφ' ὧν ἕκαστον τῶν) Θ, Ι, Κ, Λ.

To *apply* a rectangle to a straight line (in the technical sense) is παραβάλλειν, and παραπίπτω is generally used in place of the passive; the participle παρακείμενος is also used in the sense of *applied to*. In each case applying *to* a straight line is expressed by παρά (with acc.). Examples are, *areas which we can apply to a given straight line* (i.e. which we can transform into a rectangle of the same area) χωρία, ἃ δυνάμεθα παρὰ τὰν δοθεῖσαν εὐθεῖαν παραβαλεῖν, *let a rectangle be applied to each of them* παραπεπτωκέτω παρ' ἑκάσταν αὐτᾶν χωρίον; *if there be applied to each of them a rectangle exceeding by a square figure, and the sides of the excesses exceed each other by an equal amount* (i.e. form an arithmetical progression) εἰ κα παρ' ἑκάσταν αὐτᾶν παραπέσῃ τι χωρίον ὑπερβάλλον εἴδει τετραγώνῳ, ἔωντι δὲ αἱ πλευραὶ τῶν ὑπερβλημάτων τῷ ἴσῳ ἀλλάλαν ὑπερεχούσαι.

The *rectangle applied* is παράβλημα.

Squares.

A *square* is τετράγωνον, a square *on* a straight line is a square (erected) *from* it (ἀπό). *The square on* ΓΞ, τὸ ἀπὸ τᾶς ΓΞ τετράγωνον, is shortened into τὸ ἀπὸ τᾶς ΓΞ, or τὸ ἀπὸ ΓΞ simply. *The square next in order to it* (when there are a number of squares in a row) is τὸ παρ' αὐτῷ τετράγωνον or τὸ ἐχόμενον τετράγωνον.

With reference to squares, a most important part is played by the word δύναμις and the various parts of the verb δύναμαι. δύναμις expresses a *square* (literally a *power*); thus in Diophantus it is used throughout as the technical term for the square of the unknown quantity in an algebraical equation, i.e. for x^2. In geometrical language it is the dative singular δυνάμει which is mostly used; thus a straight line is said to be *potentially equal*, δυνάμει ἴσα, to a certain rectangle where the meaning is that *the square on the straight line is equal* to the rectangle; similarly for *the square on* BA *is less than double the square on* AK we have ἡ BA ἐλάσσων ἐστὶν ἢ διπλασίων δυνάμει τῆς AK. The verb δύνασθαι (with or without ἴσον) has the sense of *being* δυνάμει ἴσα, and, when δύνασθαι is used alone, it is followed by the accusative; thus *the square* (on a straight line) *is equal to the rectangle contained by...* is (εὐθεῖα) ἴσον δύναται τῷ περιεχομένῳ ὑπό...; *let the square on the radius be equal to the rectangle* BΔ, ΔZ, ἡ ἐκ τοῦ κέντρου δυνάσθω τὸ ὑπὸ τῶν BΔZ, (*the difference*) *by which the square on* ZΓ *is greater than the square on half the other diameter* ᾧ μεῖζον δύναται ἁ ZΓ τᾶς ἡμισείας τᾶς ἑτέρας διαμέτρου.

A *gnomon* is γνώμων, and its *breadth* (πλάτος) is the breadth of each end; *a gnomon of breadth equal to* BI, γνώμων πλάτος ἔχων ἴσον τᾷ BI, (*a gnomon*) *whose breadth is greater by one segment than the breadth of the gnomon last taken away* οὗ πλάτος ἑνὶ τμάματι μεῖζον τοῦ πλάτεος τοῦ πρὸ αὐτοῦ ἀφαιρουμένου γνώμονος.

Polygons.

A *polygon* is πολύγωνον, an *equilateral* polygon is ἰσόπλευρον, a polygon *of an even number of sides* or *angles* ἀρτιόπλευρον or ἀρτιόγωνον; a polygon *with all its sides equal except* BΔ, ΔA, ἴσας ἔχον τὰς πλευρὰς χωρὶς τῶν BΔA; a polygon *with its sides, excluding the base, equal and even in number* τὰς πλευρὰς ἔχον χωρὶς τῆς βάσεως ἴσας καὶ ἀρτίους; *an equilateral polygon the number of whose sides is measured by four* πολύγωνον ἰσόπλευρον, οὗ αἱ πλευραὶ ὑπὸ τετράδος μετροῦνται, *let the number of its sides be measured by four* τὸ πλῆθος τῶν πλευρῶν μετρείσθω ὑπὸ τετράδος. A *chiliagon* χιλιάγωνον.

The straight lines subtending two sides of the polygon (i.e. joining angles next but one to each other) αἱ ὑπὸ δύο πλευρὰς τοῦ πολυγώνου ὑποτείνουσαι, *the straight line subtending one less than half the number of the sides* ἡ ὑποτείνουσα τὰς μιᾷ ἐλάσσονας τῶν ἡμίσεων.

Circles.

A *circle* is κύκλος, *the circle* Ψ is ὁ Ψ κύκλος or ὁ κύκλος ἐν ᾧ τὸ Ψ, *let the given circle be that drawn below* ἔστω ὁ δοθεὶς κύκλος ὁ ὑποκείμενος.

The *centre* is κέντρον, the *circumference* περιφέρεια, the former word having doubtless been suggested by something *stuck in* and the latter by something, e.g. a cord stretched tight, *carried round* the centre as a fixed point and describing a circle with its other extremity. Accordingly περιφέρεια is used for a circular *arc* as well as for the whole circumference; thus *the arc* BΛ is ἡ BΛ περιφέρεια, *the (part of the) circumference of the circle cut off by the same (straight line)* ἡ τοῦ κύκλου περιφέρεια ἡ ὑπὸ τῆς αὐτῆς ἀποτεμνομένη. Though the circumference of a circle is also sometimes called its *perimeter* (ἡ περίμετρος) in the treatises *On the Sphere and Cylinder* and on the *Measurement of a Circle*, the word does not seem to have been used by Archimedes himself in this sense; he speaks, however, in the *Sand-reckoner* of the *perimeter of the earth* (περίμετρος τᾶς γᾶς).

The *radius* is ἡ ἐκ τοῦ κέντρου simply, and this expression without the article is used as a predicate as if it were one word; thus *the circle whose radius is* ΘE is ὁ κύκλος οἷ ἐκ τοῦ κέντρου ἁ ΘE; BE *is a radius of the circle* ἡ δὲ BE ἐκ τοῦ κέντρου ἐστὶ τοῦ κύκλου.

A *diameter* is διάμετρος, *the circle on* ΔE *as diameter* ὁ περὶ διάμετρον τὴν ΔE κύκλος.

For drawing a *chord* of a circle there is no special technical term, but we find such phrases as the following: ἐὰν εἰς τὸν κύκλον εὐθεῖα γραμμὴ ἐμπέσῃ *if in a circle a straight line be placed*, and the chord is then *the straight line so placed* ἡ ἐμπεσοῦσα, or quite commonly ἡ ἐν τῷ κύκλῳ (εὐθεῖα) simply. For *the chord subtending one 656th part of the circumference of a circle* we have the following interesting phrase, ἁ ὑποτείνουσα ἐν τμᾶμα διαιρεθείσας τᾶς τοῦ ABΓ κύκλου περιφερείας ἐς χνϛ´.

A *segment* of a circle is τμῆμα κύκλου; sometimes, to distinguish it from a segment of a *sphere*, it is called a *plane segment* τμῆμα ἐπίπεδον. A semicircle is ἡμικύκλιον; *a segment less than a semicircle cut off by* AB, τμῆμα ἔλασσον ἡμικυκλίου ὃ ἀποτέμνει ἡ AB. *The segments on* AE, EB (*as bases*) are τὰ ἐπὶ τῶν AE, EB τμήματα; but *the semicircle on* ZH *as diameter* is τὸ ἡμικύκλιον τὸ περὶ διάμετρον τὰν ΖΗ or τὸ ἡμικύκλιον τὸ περὶ τὰν ΖΗ simply. The expression *the angle of the semicircle*, ἀ τοῦ ἡμικυκλίου (γωνία), is used of the (right) angle contained by the diameter and the arc (or tangent) at one extremity of it.

A *sector* of a circle is τομεύς or, when it is necessary to distinguish it from what Archimedes calls a 'solid sector,' ἐπίπεδος τομεὺς κύκλου a *plane sector of a circle*. The *sector including the right angle* (at the centre) is ὁ τομεὺς ὁ τὰν ὀρθὰν γωνίαν περιέχων. Either of the radii bounding a sector is called a *side* of it, πλευρά ; *each of the sectors* (is) *equal to the sector which has a side common* (*with it*) ἕκαστος τῶν τομέων ἴσος τῷ κοινὰν ἔχοντι πλευρὰν τομεῖ ; a sector is sometimes regarded as *described on* one of the bounding radii as a side, thus *similar sectors have been described on all* (*the straight lines*) ἀναγεγράφαται ἀπὸ πασᾶν ὁμοίοι τομέες.

Of polygons *inscribed in* or *circumscribed about* a circle ἐγγράφειν εἰς or ἐν and περιγράφειν περί (with acc.) are used ; we also find περιγεγραμμένος used with the simple dative, thus τὸ περιγεγραμμένον σχῆμα τῷ τομεῖ is *the figure circumscribed to the sector*. A polygon is said to be *inscribed in a segment of a circle* when the base of the segment is one side and the other sides subtend arcs making up the circumference ; thus *let a polygon be inscribed on* ΑΓ *in the segment* ΑΒΓ, ἐπὶ τῆς ΑΓ πολύγωνον ἐγγεγράφθω εἰς τὸ ΑΒΓ τμῆμα. A regular polygon is said to be *inscribed in a sector* when the two radii are two of the sides and the other sides are all equal to one another, and a similar polygon is said to be *circumscribed about a sector* when the equal sides are formed by the tangents to the arc which are respectively parallel to the equal sides of the inscribed polygon and the remaining two sides are the bounding radii produced to meet the adjacent tangents. Of a circle *circumscribed to a polygon* περιλαμβάνειν is also used ; thus πολύγωνον κύκλος περιγεγραμμένος περιλαμβανέτω περὶ τὸ αὐτὸ κέντρον γινόμενος, as we might say *let a circumscribed circle be drawn with the same centre going round the polygon*. Similarly *the circle* ΑΒΓΔ *containing the polygon* ὁ ΑΒΓΔ κύκλος ἔχων τὸ πολύγωνον.

When a polygon is inscribed in a circle, the *segments left over* between the sides of the polygon and the subtended arcs are περιλειπόμενα τμήματα ; when a polygon is circumscribed to the circle, the spaces between the two are variously called τὰ περιλειπόμενα τῆς περιγραφῆς τμήματα, τὰ περιλειπόμενα σχήματα, τὰ περιλείμματα or τὰ ἀπολείμματα.

Spheres, etc.

In connexion with a *sphere* (σφαῖρα) a number of terms are used on the analogy of the older and similar terms connected with the circle. Thus the *centre* is κέντρον, the *radius* ἡ ἐκ τοῦ κέντρου, the *diameter* ἡ διάμετρος. Two *segments*, τμήματα σφαίρας or τμήματα σφαιρικά, are formed when a sphere is cut by a plane; a *hemisphere* is ἡμισφαίριον; *the segment of the sphere at* Γ, τὸ κατὰ τὸ Γ τμῆμα τῆς σφαίρας; *the segment on the side of* ΑΒΓ, τὸ ἀπὸ ΑΒΓ τμῆμα; *the segment including the circumference* ΒΑΔ, τὸ κατὰ τὴν ΒΑΔ περιφέρειαν τμῆμα. The curved *surface* of a sphere or segment is ἐπιφάνεια; thus *of spherical segments bounded by equal surfaces the hemisphere is greatest* is τῶν τῇ ἴσῃ ἐπιφανείᾳ περιεχομένων σφαιρικῶν τμημάτων μεῖζόν ἐστι τὸ ἡμισφαίριον. The terms *base* (βάσις), *vertex* (κορυφή) and *height* (ὕψος) are also used with reference to a segment of a sphere.

Another term borrowed from the geometry of the circle is the word *sector* (τομεύς) qualified with the adjective στερεός (solid). A *solid sector* (τομεὺς στερεός) is defined by Archimedes as the figure bounded by a cone which has its vertex at the centre of a sphere and the part of the surface of the sphere within the cone. *The segment of the sphere included in the sector* is τὸ τμῆμα τῆς σφαίρας τὸ ἐν τῷ τομεῖ or τὸ κατὰ τὸν τομέα.

A *great circle of a sphere* is ὁ μέγιστος κύκλος τῶν ἐν τῇ σφαίρᾳ and often ὁ μέγιστος κύκλος alone.

Let a sphere be cut by a plane not through the centre τετμήσθω σφαῖρα μὴ διὰ τοῦ κέντρου ἐπιπέδῳ; *a sphere cut by a plane through the centre in the circle* ΕΖΗΘ, σφαῖρα ἐπιπέδῳ τετμημένη διὰ τοῦ κέντρου κατὰ τὸν ΕΖΗΘ κύκλον.

Prisms and pyramids.

A *prism* is πρῖσμα, a *pyramid* πυραμίς. As usual, ἀναγράφειν ἀπό is used of describing a prism or pyramid on a rectilineal figure as base; thus *let a prism be described on the rectilineal figure* (*as base*) ἀναγεγράφθω ἀπὸ τοῦ εὐθυγράμμου πρῖσμα, *on the polygon circumscribed about the circle* A *let a pyramid be set up* ἀπὸ τοῦ περὶ τὸν Α κύκλον περιγεγραμμένου πολυγώνου πυραμὶς ἀνεστάτω ἀναγεγραμμένη. *A pyramid with an equilateral base* ΑΒΓ is πυραμὶς ἰσόπλευρον ἔχουσα βάσιν τὸ ΑΒΓ.

The *surface* is, as usual, ἐπιφάνεια and, when any particular face or a base is excluded, some qualifying phrase has to be used.

Thus *the surface of the prism consisting of the parallelograms* (i.e. excluding the bases) ἡ ἐπιφάνεια τοῦ πρίσματος ἡ ἐκ τῶν παραλληλογράμμων συγκειμένη; *the surface (of a pyramid) excluding the base* or *the triangle* ΑΕΓ, ἡ ἐπιφάνεια χωρὶς τῆς βάσεως or τοῦ ΑΕΓ τριγώνου.

The triangles bounding the pyramid τὰ περιέχοντα τρίγωνα τὴν πυραμίδα (as distinct from the base, which may be polygonal).

Cones and solid rhombi.

The *Elements* of Euclid only introduce *right* cones, which are simply called *cones* without the qualifying adjective. A *cone* is there defined as the surface described by the revolution of a right-angled triangle about one of the sides containing the right angle. Archimedes does not define a cone, but generally describes a right cone as an *isosceles cone* (κῶνος ἰσοσκελής), though once he calls it *right* (ὀρθός). J. H. T. Müller rightly observes that the term *isosceles* applied to a cone was suggested by the analogy of the isosceles triangle, but I doubt whether such a cone was thought of (as he supposes) as one which could be described by making an isosceles triangle revolve about the perpendicular from the vertex on the base; it seems more natural to connect it with the use of the word *side* (πλευρά) by which Archimedes designates a generator of the cone, a right cone being thus directly regarded as a cone having all its *legs equal*. The latter supposition would also accord better with the term *scalene cone* (κῶνος σκαληνός) by which Apollonius denotes an oblique circular cone; such a cone could not of course be described by the revolution of a scalene *triangle*. An oblique circular cone is simply a *cone* for Archimedes, and he does not define it; but, while he speaks of *finding* a cone with a given vertex and passing through every point on a given 'section of an acute-angled cone' [ellipse], he regards the *finding* of the cone as being equivalent to finding the *circular sections*, and we may therefore conclude that he would have defined the cone in practically the same way as Apollonius does, namely as the surface described by a straight line always passing through a fixed point and moving round the circumference of any circle not in the same plane with the point.

The *vertex* of a cone is, as usual, κορυφή, the *base* βάσις, the *axis* ἄξων and the *height* ὕψος; *the cones are of the same height* εἰσὶν οἱ κῶνοι ὑπὸ τὸ αὐτὸ ὕψος. A *generator* is called a *side* (πλευρά); *if a cone be cut by a plane meeting all the generators of the cone* εἴ κα κῶνος ἐπιπέδῳ τμαθῇ συμπίπτοντι πάσαις ταῖς τοῦ κώνου πλευραῖς.

The surface of the cone excluding the base ἡ ἐπιφάνεια τοῦ κώνου χωρὶς τῆς βάσεως ; *the conical surface between (two generators)* ΑΔ, ΔΒ, κωνικὴ ἐπιφάνεια ἡ μεταξὺ τῶν ΑΔΒ.

There is no special name for what we call a *frustum of a cone* or the portion intercepted between two planes parallel to the base; the surface of such a frustum is simply *the surface of the cone between the parallel planes* ἡ ἐπιφάνεια τοῦ κώνου μεταξὺ τῶν παραλλήλων ἐπιπέδων.

A curious term is *segment of a cone* (ἀπότμαμα κώνου), which is used of the portion of any circular cone, right or oblique, cut off towards the vertex by any plane which makes an elliptic and not a circular section. With reference to a *segment of a cone* the *axis* (ἄξων) is defined as the straight line drawn from the vertex of the cone to the centre of the elliptic base.

As usual, ἀναγράφειν ἀπό is used of *describing* a cone *on* a circle as base. Similarly, a very common phrase is ἀπὸ τοῦ κύκλου κῶνος ἔστω *let there be a cone on the circle (as base)*.

A *solid rhombus* (ῥόμβος στερεός) is the figure made up of two cones having their base common, their vertices on opposite sides of it, and their axes in one straight line. A *rhombus made up of isosceles cones* ῥόμβος ἐξ ἰσοσκελῶν κώνων συγκείμενος, and the two cones are spoken of as *the cones bounding the rhombus* οἱ κῶνοι οἱ περιέχοντες τὸν ῥόμβον.

Cylinders.

A *right cylinder* is κύλινδρος ὀρθός, and the following terms apply to the cylinder as to the cone : *base* βάσις, *one base or the other* ἡ ἑτέρα βάσις, *of which the circle* ΑΒ *is a base and* ΓΔ *opposite to it* οὗ βάσις μὲν ὁ ΑΒ κύκλος, ἀπεναντίον δὲ ὁ ΓΔ ; *axis* ἄξων, *height* ὕψος, *generator* πλευρά. *The cylindrical surface cut off by (two generators)* ΑΓ, ΒΔ, ἡ ἀποτεμνομένη κυλινδρικὴ ἐπιφάνεια ὑπὸ τῶν ΑΓ, ΒΔ ; *the surface of the cylinder adjacent to the circumference* ΑΒΓ, ἡ ἐπιφάνεια τοῦ κυλίνδρου ἡ κατὰ τὴν ΑΒΓ περιφέρειαν denotes the surface of the cylinder between the two generators drawn through the extremities of the arc.

A *frustum of a cylinder* τόμος κυλίνδρου is a portion of a cylinder intercepted between two parallel sections which are elliptic and not circular, and the *axis* (ἄξων) of it is the straight line joining the centres of the two sections, which is in the same straight line with the axis of the *cylinder*.

Conic Sections.

General terms are κωνικὰ στοιχεῖα, *elements of conics*, τὰ κωνικά (*the theory of*) *conics. Any conic section* κώνου τομὴ ὁποιαοῦν. *Chords* are simply εὐθεῖαι ἐν τᾷ τοῦ κώνου τομᾷ ἀγμέναι. Archimedes never uses the word *axis* (ἄξων) with reference to a conic ; the axes are with him *diameters* (διάμετροι), and διάμετρος, when it has reference to a complete conic, is used in this sense exclusively. A *tangent* is ἐπιψαύουσα or ἐφαπτομένη (with gen.).

The separate conic sections are still denoted by the old names ; a parabola is a *section of a right-angled cone* ὀρθογωνίου κώνου τομή, a hyperbola a *section of an obtuse-angled cone* ἀμβλυγωνίου κώνου τομή, and an ellipse a *section of an acute-angled cone* ὀξυγωνίου κώνου τομή.

The parabola.

Only the *axis* of a complete parabola is called a *diameter*, and the other diameters are simply *lines parallel to the diameter*. Thus *parallel to the diameter or itself the diameter* is παρὰ τὰν διάμετρον ἢ αὐτὰ διάμετρος ; ΔΖ *is parallel to the diameter* ἁ ΔΖ παρὰ τὰν διάμετρόν ἐστι. Once the term *principal* or *original* (*diameter*) is used, ἀρχικά (sc. διάμετρος).

A *segment* of a parabola is τμῆμα, which is more fully described as *the segment bounded by a straight line and a section of a right-angled cone* τμᾶμα τὸ περιεχόμενον ὑπό τε εὐθείας καὶ ὀρθογωνίου κώνου τομᾶς. The word διάμετρος is again used with reference to a segment of a parabola in the sense of our word *axis* ; Archimedes defines the *diameter* of any segment as *the line bisecting all the straight lines (chords) drawn parallel to its base* τὰν δίχα τέμνουσαν τὰς εὐθείας πάσας τὰς παρὰ τὰν βάσιν αὐτοῦ ἀγομένας.

The part of a parabola included between two parallel chords is called a *frustum* τόμος (ἀπὸ ὀρθογωνίου κώνου τομᾶς ἀφαιρούμενος), the two chords are its *lesser* and *greater* base (ἐλάσσων and μείζων βάσις) respectively, and the line joining the middle points of the two chords is the *diameter* (διάμετρος) of the frustum.

What we call the *latus rectum* of a parabola is in Archimedes *the line which is double of the line drawn as far as the axis* ἁ διπλασία τᾶς μέχρι τοῦ ἄξονος. In this expression the *axis* (ἄξων) is the axis of the *right-angled cone* from which the curve was originally derived by means of a section perpendicular to a generator*. Or, again, the equivalent of our word *parameter* (παρ' ἃν δύνανται αἱ ἀπὸ τᾶς τομᾶς) is used by Archimedes as by Apollonius, meaning the straight line to which the rectangle which has its breadth equal to the abscissa of a point and is equal to the square of the ordinate must be applied as base. The full phrase states that the ordinates *have their squares equal to the rectangles applied to the line equal to* N (*or the parameter*) *which have as their breadth the lines which they (the ordinates) cut off from* ΔZ (*the diameter*) *towards the extremity* Δ, δύνανται τὰ παρὰ τὰν ἴσαν τᾷ N παραπίπτοντα πλάτος ἔχοντα, ᾱς αὐταὶ ἀπολαμβάνοντι ἀπὸ τᾶς ΔZ ποτὶ τὸ Δ πέρας.

Ordinates are the *lines drawn from the section to the diameter* (*of the segment*) *parallel to the base* (*of the segment*) αἱ ἀπὸ τᾶς τομᾶς ἐπὶ τὰν ΔZ ἀγομέναι παρὰ τὰν ΑΕ, or simply αἱ ἀπὸ τᾶς τομᾶς. Once also the regular phrase *drawn ordinate-wise* τεταγμένως κατηγμένη is used to describe an ordinate, as in Apollonius.

The hyperbola.

What we call the asymptotes (αἱ ἀσύμπτωτοι in Apollonius) are in Archimedes *the lines (approaching) nearest to the section of the obtuse-angled cone* αἱ ἔγγιστα τᾶς τοῦ ἀμβλυγωνίου κώνου τομᾶς.

The *centre* is not described as such, but it is *the point at which the lines nearest (to the curve) meet* τὸ σαμεῖον, καθ' ὃ αἱ ἔγγιστα συμπίπτοντι.

This is a property of the sections of obtuse-angled cones τοῦτο γάρ ἐστιν ἐν ταῖς τοῦ ἀμβλυγωνίου κώνου τομαῖς σύμπτωμα.

The ellipse.

The major and minor axes are the *greater and lesser diameters* μείζων and ἐλάσσων διάμετρος. *Let the greater diameter be* ΑΓ, διάμετρος δὲ (αὐτᾶς) ἁ μὲν μείζων ἔστω ἐφ' ᾱς τὰ Α, Γ. *The rectangle contained by the diameters* (*axes*) τὸ περιεχόμενον ὑπὸ τᾶν διαμέτρων. One axis is called *conjugate* (συζυγής) to the other: thus *let the straight line* N *be equal to half of the other diameter which is conjugate to* ΑΒ, ἁ δὲ N εὐθεῖα ἴσα ἔστω τᾷ ἡμισείᾳ τᾶς ἑτέρας διαμέτρου, ᾱ ἐστι συζυγὴς τᾷ ΑΒ.

The *centre* is here κέντρον.

---

* Cf. *Apollonius of Perga*, pp. xxiv, xxv.

Conoids and Spheroids.

There is a remarkable similarity between the language in which Archimedes describes the genesis of his solids of revolution and that used by Euclid in defining the sphere. Thus Euclid says: *when, the diameter of a semicircle remaining fixed, the semicircle revolves and returns to the same position from which it began to move, the included figure is a sphere* σφαιρά ἐστιν, ὅταν ἡμικυκλίου μενούσης τῆς διαμέτρου περιενεχθὲν τὸ ἡμικύκλιον εἰς τὸ αὐτὸ πάλιν ἀποκατασταθῇ, ὅθεν ἤρξατο φέρεσθαι, τὸ περιληφθὲν σχῆμα; and he proceeds to state that the *axis of the sphere is the fixed straight line about which the semicircle turns* ἄξων δὲ τῆς σφαίρας ἐστὶν ἡ μένουσα εὐθεῖα, περὶ ἣν τὸ ἡμικύκλιον στρέφεται. Compare with this e.g. Archimedes' definition of the *right-angled conoid* (paraboloid of revolution): *if a section of a right-angled cone, with its diameter (axis) remaining fixed, revolves and returns to the position from which it started, the figure included by the section of the right-angled cone is called a right-angled conoid, and its axis is defined as the diameter which has remained fixed*, εἴ κα ὀρθογωνίου κώνου τομὰ μενούσας τᾶς διαμέτρου περιενεχθεῖσα ἀποκατασταθῇ πάλιν, ὅθεν ὥρμασεν, τὸ περιλαφθὲν σχῆμα ὑπὸ τᾶς τοῦ ὀρθογωνίου κώνου τομᾶς ὀρθογώνιον κωνοειδὲς καλεῖσθαι, καὶ ἄξονα μὲν αὐτοῦ τὰν μεμενακοῦσαν διάμετρον καλεῖσθαι, and it will be seen that the several phrases used are practically identical with those of Euclid, except that ὥρμασεν takes the place of ἤρξατο φέρεσθαι; and even the latter phrase occurs in Archimedes' description of the genesis of the spiral later on.

The words *conoid* κωνοειδὲς (σχῆμα) and *spheroid* σφαιροειδὲς (σχῆμα) are simply adapted from κῶνος and σφαῖρα, meaning that the respective figures have the *appearance* (εἶδος) of, or resemble, cones and spheres; and in this respect the names are perhaps more satisfactory than *paraboloid, hyperboloid* and *ellipsoid*, which can only be said to resemble the respective conics in a different sense. But when κωνοειδές is qualified by the adjective *right-angled* ὀρθογώνιον to denote the paraboloid of revolution, and by ἀμβλυγώνιον *obtuse-angled* to denote the hyperboloid of revolution, the expressions are less logical, as the solids do not resemble *right-angled* and *obtuse-angled* cones respectively; in fact, since the angle between the asymptotes of the generating hyperbola may be acute, a hyperboloid of revolution would in that case more resemble an *acute*-angled cone. The terms *right-angled* and *obtuse-angled* were merely transferred to the conoids from the names for the respective conics without any more thought of their meaning.

It is unnecessary to give separately the definition of each conoid and spheroid; the phraseology is in all cases the same as that given above for the paraboloid. But it may be remarked that Archimedes does not mention the *conjugate axis* of a hyperbola or the figure obtained by causing a hyperbola to revolve about that axis; the *conjugate axis* of a hyperbola first appears in Apollonius, who was apparently the first to conceive of the two branches of a hyperbola as one curve. Thus there is only one *obtuse-angled conoid* in Archimedes, whereas there are two kinds of spheroids according as the revolution takes place about the *greater diameter* (axis) or *lesser diameter* of the generating *section of an acute-angled cone* (ellipse); the spheroid is in the former case *oblong* (παραμᾶκες σφαιροειδές) and in the latter case *flat* (ἐπιπλατὺ σφαιροειδές).

A special feature is, however, to be observed in the description of the *obtuse-angled conoid* (hyperboloid of revolution), namely that the asymptotes of the hyperbola are supposed to revolve about the axis at the same time as the curve, and Archimedes explains that *they will include an isosceles cone* (κῶνον ἰσοσκελέα περιλαψούνται), which he thereupon defines as the cone *enveloping the conoid* (περιέχων τὸ κωνοειδές). Also in a *spheroid* the term *diameter* (διάμετρος) is appropriated to *the straight line drawn through the centre at right angles to the axis* (ἁ διὰ τοῦ κέντρου ποτ᾿ ὀρθὰς ἀγομένα τῷ ἄξονι). The *centre* of a spheroid is *the middle point of the axis* τὸ μέσον τοῦ ἄξονος.

The following terms are used of all the conoids and spheroids. The *vertex* (κορυφή) is *the point at which the axis meets the surface* τὸ σαμεῖον, καθ᾿ ὃ ἀπτέται ὁ ἄξων τᾶς ἐπιφανείας, the spheroid having of course two vertices. A *segment* (τμᾶμα) is a part cut off by a plane, and the *base* (βάσις) of the segment is defined as *the plane (figure) included by the section of the conoid (or spheroid) in the cutting plane* τὸ ἐπίπεδον τὸ περιλαφθὲν ὑπὸ τᾶς τοῦ κωνοειδέος (or σφαιροειδέος) τομᾶς ἐν τῷ ἀποτέμνοντι ἐπιπέδῳ. The *vertex of a segment* is *the point at which the tangent plane parallel to the base of the segment meets the surface*, τὸ σαμεῖον, καθ᾿ ὃ ἀπτέται τὸ ἐπίπεδον τὸ ἐπιψαῦον (τοῦ κωνοειδέος). The *axis* (ἄξων) of a segment is differently defined for the three surfaces ; (*a*) in the paraboloid it is *the straight line cut off within the segment from the line drawn through the vertex of the* segment parallel to the axis of the conoid ἀ ἐναπολαφθεῖσα εὐθεῖα ἐν τῷ τμάματι ἀπὸ τᾶς ἀχθείσας διὰ τᾶς κορυφᾶς τοῦ τμάματος παρὰ τὸν ἄξονα τοῦ κωνοειδέος, (b) in the hyperboloid it is the straight line cut off within the segment from *the line drawn through the vertex of the segment and the vertex of the cone enveloping the conoid* ἀπὸ τᾶς ἀχθείσας διὰ τᾶς κορυφᾶς τοῦ τμάματος καὶ τᾶς κορυφᾶς τοῦ κώνου τοῦ περιέχοντος τὸ κωνοειδές, (c) in the spheroid it is the part similarly cut off *from the straight line joining the vertices of the two segments* into which the base divides the spheroid, ἀπὸ τᾶς εὐθείας τᾶς τᾶς κορυφὰς αὐτῶν (τῶν τμαμάτων) ἐπιζευγνυούσας.

Archimedes does not use the word *centre* with respect to the hyperboloid of revolution, but calls the centre the *vertex of the enveloping cone*. Also the *axis* of a hyperboloid or a segment is only that part of it which is within the surface. The distance between the *vertex* of the hyperboloid or segment and the *vertex of the enveloping cone* is *the line adjacent to the axis* ἀ ποτεοῦσα τῷ ἄξονι.

The following are miscellaneous expressions. *The part intercepted within the conoid of the intersection of the planes* ἀ ἐναπολαφθεῖσα ἐν τῷ κωνοειδεῖ τᾶς γενομένας τομᾶς τῶν ἐπιπέδων, *(the plane) will have cut the spheroid through its axis* τετμακὸς ἐσσεῖται τὸ σφαιροειδὲς διὰ τοῦ ἄξονος, *so that the section it makes will be a conic section* ὥστε τὰν τομὰν ποιήσει κώνου τομάν, *let two segments be cut off in any manner* ἀποτετμάσθω δύο τμάματα ὡς ἔτυχεν *or by planes drawn in any manner* ἐπιπέδοις ὁπωσοῦν ἀγμένοις.

Half the spheroid τὸ ἁμίσεον τοῦ σφαιροειδέος, *half the line joining the vertices of the segments (of a spheroid)*, i.e. what we should call a semi-diameter, ἀ ἡμισέα αὐτᾶς τᾶς ἐπιζευγνυούσας τᾶς κορυφὰς τῶν τμαμάτων.

The spiral.

We have already had, in the conoids and spheroids, instances of the evolution of figures by the motion of curves about an axis. The same sort of motion is used for the construction of solid figures inscribed in and circumscribed about a sphere, a circle and an inscribed or circumscribed polygon being made to revolve about a diameter passing through an angular point of the polygon and dividing it and the circle symmetrically. In this case, in Archimedes' phrase, *the angular points of the polygon will move along the circumferences of circles*, αἱ γωνίαι κατὰ κύκλων περιφερειῶν ἐνεχθήσονται (or οἰσθήσονται) and *the sides will move on certain cones*, or *on the surface of a cone* κατά τινων κώνων ἐνεχθήσονται or κατ᾽ ἐπιφανείας κώνου; and sometimes the angular points or the points of contact of the sides of a circumscribed polygon are said to *describe circles* γράφουσι κύκλους. *The solid figure so formed is* τὸ γενηθὲν στερεὸν σχῆμα, and *let the sphere by its revolution make a figure* περιενεχθεῖσα ἡ σφαῖρα ποιείτω σχῆμά τι.

For the construction of the spiral, however, we have a new element introduced, that of *time*, and we have two different uniform motions combined; *if a straight line in a plane turn uniformly about one extremity which remains fixed, and return to the position from which it started and if, at the same time as the line is revolving, a point move at a uniform rate along the line starting from the fixed extremity, the point will describe a spiral in the plane*, εἴ κα εὐθεῖα...ἐν ἐπιπέδῳ...μένοντος τοῦ ἑτέρου πέρατος αὐτᾶς ἰσοταχέως περιενεχθεῖσα ἀποκατασταθῇ πάλιν, ὅθεν ὥρμασεν, ἅμα δὲ τᾷ γραμμᾷ περιαγομένᾳ φερήται τι σαμεῖον ἰσοταχέως αὐτὸ ἑαυτῷ κατὰ τᾶς εὐθείας ἀρξάμενον ἀπὸ τοῦ μένοντος πέρατος, τὸ σαμεῖον ἕλικα γράψει ἐν τῷ ἐπιπέδῳ.

The *spiral (described) in the first, second,* or *any turn* is ἁ ἕλιξ ἁ ἐν τᾷ πρώτᾳ, δευτέρᾳ, or ὁποιουῦν περιφορᾷ γεγραμμένα, and the turns other than any particular ones are *the other spirals* αἱ ἄλλαι ἕλικες.

The *distance traversed* by the point along the line in any time is ἁ εὐθεῖα ἁ διανυσθεῖσα, and *the times in which the point moved over the distances* οἱ χρόνοι, ἐν οἷς τὸ σαμεῖον τὰς γραμμὰς ἐπορεύθη; *in the time in which the revolving line reaches* ΑΓ *from* ΑΒ, ἐν ᾧ χρόνῳ ἁ περιαγομένα γραμμὰ ἀπὸ τᾶς ΑΒ ἐπὶ τὰν ΑΓ ἀφικνεῖται.

The *origin of the spiral* is ἀρχὰ τᾶς ἕλικος, *the initial line* ἀρχὰ τᾶς περιφορᾶς. The distance described by the point along the line in the first complete revolution is εὐθεῖα πρώτα (*first distance*), that described during the second revolution the *second distance* εὐθεῖα δευτέρα, and so on, the distances being *called by the number of the revolutions* ὁμωνύμως ταῖς περιφοραῖς. The *first area*, χωρίον πρῶτον, is *the area bounded by the spiral described in the first revolution and by the 'first distance'* τὸ χωρίον τὸ περιλαφθὲν ὑπό τε τᾶς ἕλικος τᾶς ἐν τᾷ πρώτᾳ περιφορᾷ γραφείσας καὶ τᾶς εὐθείας, ἅ ἐστιν πρώτα; the *second area* is that bounded by the spiral in the second turn and the 'second distance,' and so on. *The area added by the spiral in any turn* is τὸ χωρίον τὸ ποτιλαφθὲν ὑπὸ τᾶς ἕλικος ἔν τινι περιφορᾷ.

The *first circle*, κύκλος πρῶτος, is the circle described with the 'first distance' as radius and the origin as centre, the *second circle* that with the origin as centre and twice the 'first distance' as radius, and so on.

Together with as many times the whole of the circumference of the circle as (is represented by) the number less by one than (that of) the revolutions μεθ' ὅλας τὰς τοῦ κύκλου περιφερείας τοσαυτάκις λαμβανομένας, ὅσος ἐστὶν ὁ ἑνὶ ἐλάσσων ἀριθμὸς τᾶν περιφορᾶν, *the circle called by the number corresponding to that of the revolutions* ὁ κύκλος ὁ κατὰ τὸν αὐτὸν ἀριθμὸν λεγόμενος ταῖς περιφοραῖς.

With reference to any radius vector, the side which is in the direction of the revolution is *forward* τὰ προαγούμενα, the other *backward* τὰ ἑπόμενα.

Tangents, etc.

Though the word ἅπτομαι is sometimes used in Archimedes of a line *touching* a curve, its general meaning is not to *touch* but simply to *meet*; e.g. the axis of a conoid or spheroid *meets* (ἅπτεται) the surface in the vertex. (The word is also often used elsewhere than in Archimedes of points *lying on* a locus; e.g. in Pappus, p. 664, *the point will lie on a straight line given in position* ἅψεται τὸ σημεῖον θέσει δεδομένης εὐθείας.)

To *touch* a curve or surface is generally ἐφάπτεσθαι or ἐπιψαύειν (with gen.). A *tangent* is ἐφαπτομένη or ἐπιψαύουσα (sc. εὐθεῖα) and a *tangent plane* ἐπιψαῦον ἐπίπεδον. *Let tangents be drawn to the circle* ΑΒΓ, τοῦ ΑΒΓ κύκλου ἐφαπτόμεναι ἤχθωσαν; *if straight lines be drawn touching the circles* ἐὰν ἀχθῶσίν τινες ἐπιψαύουσαι τῶν κύκλων. The full phrase of *touching without cutting* is sometimes found in Archimedes; *if a plane touch (any of) the conoidal figures without cutting the conoid* εἴ κα τῶν κωνοειδέων σχημάτων ἐπίπεδον ἐφαπτῆται μὴ τέμνον τὸ κωνοειδές. The simple word ψαύειν is occasionally used (participially), *the tangent planes* τὰ ἐπίπεδα τὰ ψαύοντα.

To touch *at a point* is expressed by κατά (with acc.); *the points at which the sides...touch (or meet) the circle* σημεῖα, καθ' ἃ ἅπτονται τοῦ κύκλου αἱ πλευραί.... *Let them touch the circle at the middle points of the circumferences cut off by the sides of the inscribed polygon* ἐπιψανέτωσαν τοῦ κύκλου κατὰ μέσα τῶν περιφερειῶν τῶν ἀποτεμνομένων ὑπὸ τοῦ ἐγγεγραμμένου πολυγώνου πλευρῶν.

The distinction between ἐπιψαύειν and ἅπτομαι is well brought out in the following sentence; *but that the planes touching the spheroid meet its surface at one point only we shall prove* ὅτι δὲ τὰ ἐπιψαύοντα ἐπίπεδα τοῦ σφαιροειδέος καθ᾽ ἓν μόνον ἅπτονται σαμεῖον τᾶς ἐπιφανείας αὐτοῦ δειξοῦμες.

The point of contact ἡ ἀφή.

Tangents *drawn from* (a point) ἀγμέναι ἀπό; we find also the elliptical expression ἀπὸ τοῦ Ξ ἐφαπτέσθω ἡ ΟΞΤ, *let* ΟΞΠ *be the tangent from* Ξ, where, in the particular case, Ξ is on the circle.

Constructions.

The richness of the Greek language in expressions for constructions is forcibly illustrated by the variety of words which may be used (with different shades of meaning) for *drawing* a line. Thus we have in the first place ἄγω and the compounds διάγω (of drawing a line *through* a figure, with εἰς or ἐν following, of *producing* a plane *beyond* a figure, or of drawing a line *in* a plane), κατάγω (used of drawing an ordinate *down* from a point on a conic), προσάγω (of drawing a line *to meet* another). As an alternative to προσάγω, προσβάλλω is also used; and προσπίπτω may take the place of the passive of either verb. To *produce* is ἐκβάλλω, and the same word is also used of a plane drawn *through* a point or *through* a straight line; an alternative for the passive is supplied by ἐκπίπτω. Moreover πρόσκειμαι is an alternative word for *being produced* (literally *being added*).

In the vast majority of cases constructions are expressed by the elegant use of the perfect imperative passive (with which may be classed such forms as γεγονέτω from γίγνομαι, ἔστω from εἰμί, and κείσθω from κεῖμαι), or occasionally the aorist imperative passive. The great variety of the forms used will be understood from the following specimens. *Let* ΒΓ *be made* (or *supposed*) *equal to* Δ, κείσθω τῷ Δ ἴσον τὸ ΒΓ; *let it be drawn* ἤχθω, *let a straight line be drawn in it* (a chord of a circle) διήχθω τις εἰς αὐτὸν εὐθεῖα, *let* ΚΜ *be drawn equal to...* ἴση κατήχθω ἡ ΚΜ, *let it be joined* ἐπεζεύχθω, *let* ΚΛ *be drawn to meet* προσβεβλήσθω ἡ ΚΛ, *let them be produced* ἐκβεβλήσθωσαν, *suppose them found* εὑρήσθωσαν, *let a circle be set out* ἐκκείσθω κύκλος, *let it be taken* εἰλήφθω, *let* Κ, Η *be taken* ἔστωσαν εἰλημμέναι αἱ Κ, Η, *let a circle* Ψ *be taken* λελάφθω κύκλος ἐν ᾧ τὸ Ψ, *let it be cut* τετμήσθω, *let it be divided* διαιρήσθω (διῃρήσθω); *let one cone be cut by a plane parallel to the base and produce the section* ΕΖ, τμηθήτω ὁ ἕτερος κῶνος ἐπιπέδῳ παραλλήλῳ τῇ βάσει καὶ ποιείτω τομὴν τὴν ΕΖ, *let* ΤΖ *be cut off* ἀπολελάφθω ἁ ΤΖ; *let* (such an angle) *be left and let it be* ΝΗΓ, λελείφθω καὶ ἔστω ἡ ὑπὸ ΝΗΓ, *let a figure be made* γεγενήσθω σχῆμα, *let the sector be made* ἔστω γεγενημένος ὁ τομεύς, *let cones be described on the circles* (*as bases*) ἀναγεγράφθωσαν ἀπὸ τῶν κύκλων κῶνοι, ἀπὸ τοῦ κύκλου κῶνος ἔστω, *let it be inscribed* or *circumscribed* ἐγγεγράφθω (or ἐγγεγραμμένον ἔστω), περιγεγράφθω; *let an area* (*equal to that*) *of* AB *be applied to* ΛΗ, παραβεβλήσθω παρὰ τὰν ΛΗ τὸ χωρίον τοῦ AB; *let a segment of a circle be described on* ΘΚ, ἐπὶ τῆς ΘΚ κύκλου τμῆμα ἐφεστάσθω, *let the circle be completed* ἀναπεπληρώσθω ὁ κύκλος, *let* ΝΞ (*a parallelogram*) *be completed* συμπεπληρώσθω τὸ ΝΞ, *let it be made* πεποιήσθω, *let the rest of the construction be the same as before* τὰ ἄλλα κατεσκευάσθω τὸν αὐτὸν τρόπον τοῖς πρότερον. *Suppose it done* γεγονέτω.

Another method is to use the passive imperative of νοέω (*let it be conceived*). *Let straight lines be conceived to be drawn* νοείσθωσαν εὐθεῖαι ἠγμέναι, *let the sphere be conceived to be cut* νοείσθω ἡ σφαῖρα τετμημένη, *let a figure* (*generated*) *from the inscribed polygon be conceived as inscribed in the sphere* ἀπὸ τοῦ πολυγώνου τοῦ ἐγγραφομένου νοείσθω τι εἰς τὴν σφαῖραν ἐγγραφὲν σχῆμα. Sometimes the participle for *drawn* is left out; thus ἀπ' αὐτοῦ νοείσθω ἐπιφάνεια *let a surface be conceived* (*generated*) *from it*.

The active is much more rarely used; but we find (1) ἐάν with subjunctive, *if we cut* ἐὰν τέμωμεν, *if we draw* ἐὰν ἀγάγωμεν, *if you produce* ἐὰν ἐκβάλῃς; (2) the participle, *it is possible to inscribe...and* (*ultimately*) *to leave* δυνατόν ἐστιν ἐγγράφοντα...λείπειν, *if we continually circumscribe polygons, bisecting the remaining circumferences and drawing tangents, we shall* (*ultimately*) *leave* ἀεὶ δὴ περιγράφοντες πολύγωνα δίχα τεμνομένων τῶν περιλειπομένων περιφερειῶν καὶ ἀγομένων ἐφαπτομένων λείψομεν, *it is possible, if we take the area..., to inscribe* λαβόντα (or λαμβάνοντα) τὸ χωρίον...δυνατόν ἐστιν...ἐγγράψαι; (3) the first person singular, *I take two straight lines* λαμβάνω δύο εὐθείας, *I took a straight line* ἔλαβόν τινα εὐθεῖαν; *I draw* ΘΜ *from* Θ *parallel to* ΑΖ, ἄγω ἀπὸ τοῦ Θ τὰν ΘΜ παράλληλον τᾷ ΑΖ, *having drawn* ΓΚ *perpendicular, I cut off* ΑΚ *equal to* ΓΚ ἀγαγὼν κάθετον τὰν ΓΚ τᾷ ΓΚ ἴσαν ἀπέλαβον τὰν ΑΚ, *I inscribed a solid figure...and circumscribed another* ἐνέγραψα σχῆμα στερεόν...καὶ ἄλλο περιέγραψα.

The genitive of the passive participle is used absolutely, εὑρεθέντος δή *it being supposed found*, ἐγγραφέντος δή (*the figure*) *being inscribed*.

To make a figure similar to one (*and equal to another*) ὁμοιῶσαι, *to find experimentally* ὀργανικῶς λαβεῖν, *to cut into unequal parts* εἰς ἄνισα τέμνειν.

Operations (addition, subtraction, etc.).

1. *Addition, and sums, of magnitudes.*

To *add* is προστίθημι, for the passive of which πρόσκειμαι is often used; thus *one segment being added* ἑνὸς τμάματος ποτιτεθέντος, *the added (straight line)* ἁ ποτικειμένα, *let the common* ΗΑ, ΖΓ *be added* κοιναὶ προσκείσθωσαν αἱ ΗΑ, ΖΓ; the words are generally followed by πρός (with acc. of the thing added *to*), but sometimes by the dative, *that to which the addition was made* ᾧ ποτετέθη.

For *being added together* we have συντίθεσθαι; thus *being added to itself* συντιθέμενον αὐτὸ ἑαυτῷ, *added together* ἐς τὸ αὐτὸ συντεθέντα, *added to itself (continually)* ἐπισυντιθέμενον ἑαυτῷ.

Sums are commonly expressed for two magnitudes by συναμφό-τερος used in the following different ways; *the sum of* ΒΑ, ΑΛ συναμφότερος ἡ ΒΑΛ, *the sum of* ΔΓ, ΓΒ συναμφότερος ἡ ΔΓ, ΓΒ, *the sum of the area and the circle* τὸ συναμφότερον ὅ τε κύκλος καὶ τὸ χωρίον. Again for *sums* in general we have such expressions as *the line which is equal to both the radii* ἡ ἴση ἀμφοτέραις ταῖς ἐκ τοῦ κέντρου, *the line equal to (the sum of) all the lines joining* ἡ ἴση πάσαις ταῖς ἐπιζευγνυούσαις. Also *all the circles* οἱ πάντες κύκλοι means *the sum of* all the circles; and σύγκειται ἐκ is used for *is equal to the sum of* (two other magnitudes).

To denote *plus* μετά (with gen.) and σύν are used; *together with the bases* μετὰ τῶν βάσεων, *together with half the base of the segment* σὺν τῇ ἡμισείᾳ τῆς τοῦ τμήματος βάσεως; τε and καί also express the same thing, and the participle of προσλαμβάνω gives another way of describing *having something added to it*; thus *the squares on (all) the lines equal to the greatest together with the square on the greatest...* is τὰ τετράγωνα τὰ ἀπὸ τᾶν ἰσᾶν τᾷ μεγίστᾳ ποτιλαμβάνοντα τό τε ἀπὸ τᾶς μεγίστας τετράγωνον....

2. *Subtraction and differences.*

To *subtract from* is ἀφαιρεῖν ἀπό; *if (the rhombus) be conceived as taken away* ἐὰν νοηθῇ ἀφῃρημένος, *let the segments be subtracted* ἀφαιρεθέντων τὰ τμήματα. *Terms common to each side in an equation* are κοινά; *the squares are common to both (sides)* κοινά ἐντι ἑκατέρων τὰ τετράγωνα. Then *let the common area be subtracted* is κοινὸν ἀφῃρήσθω τὸ χωρίον, and so on; *the remainder* is denoted by the adjective λοιπός, e.g. *the conical surface remaining* λοιπὴ ἡ κωνικὴ ἐπιφάνεια.

The *difference* or *excess* is ὑπεροχή, or more fully the *excess by*

which (one magnitude) exceeds (another) ὑπεροχή, ἧ ὑπερέχει... or
ὑπεροχά, ᾇ μείζων ἐστί.... The *excess* is also expressed by means of
the verb ὑπερέχειν alone; *let the difference by which the said triangles
exceed the triangle* ΑΔΓ *be* Θ, ᾧ δὴ ὑπερέχει τὰ εἰρημένα τρίγωνα τοῦ
ΑΔΓ τριγώνου ἔστω τὸ Θ, *to exceed by less than the excess of the cone*
Ψ *over the half of the spheroid* ὑπερέχειν ἐλάσσονι ἢ ᾧ (or ἁλίκῳ)
ὑπερέχει ὁ Ψ κῶνος τοῦ ἡμίσεος τοῦ σφαιροειδέος (where ᾧ ὑπερέχει may
also be omitted). Again the *excess* may be ᾧ μείζων ἐστί. The
opposite to ὑπερέχει is λείπεται (with gen.).

Equal to twice a certain excess ἴσα δυσὶν ὑπεροχαῖς, with which
equal to one excess, ἴσα μιᾷ ὑπεροχᾷ, is contrasted.

The following sentence practically states the equivalent of an
algebraical equation; *the rectangle under* ΖΗ, ΞΔ *exceeds the rectangle under* ΖΕ, ΕΔ *by the (sum of) the rectangle contained by* ΞΔ,
ΕΗ *and the rectangle under* ΖΕ, ΞΕ, ὑπερέχει τὸ ὑπὸ τᾶν ΖΗ, ΞΔ τοῦ
ὑπὸ τᾶν ΖΕ, ΕΔ τῷ τε ὑπὸ τᾶν ΞΔ, ΕΗ περιεχομένῳ καὶ τῷ ὑπὸ τᾶν ΖΕ,
ΞΕ. Similarly *twice* ΡΗ *together with* ΠΣ *is (equal to) the sum of*
ΣΡ, ΡΠ, δύο μὲν αἱ ΡΗ μετὰ τᾶς ΠΣ συναμφότερός ἐστιν ἁ ΣΡΠ.

3. *Multiplication.*

To *multiply* is πολλαπλασιάζω; *multiply one another* (of numbers)
πολλαπλασιάζειν ἀλλάλους; to multiply *by* a number is expressed by
the dative; *let* Δ *be multiplied by* Θ πεπολλαπλασιάσθω ὁ Δ τῷ Θ.

Multiplied into is sometimes ἐπί (with acc.); thus *the rectangle*
ΠΘ, ΘΛ *into* ΘΛ (i.e. a solid figure) is τὸ ὑπὸ τῶν ΗΘ, ΘΛ ἐπὶ
τὴν ΘΛ.

4. *Division.*

To *divide* διαιρεῖν; *let it be divided into three equal parts at the
points* Κ, Θ, διῃρήσθω εἰς τρία ἴσα κατὰ τὰ Κ, Θ σαμεῖα; *to be divisible
by* μετρεῖσθαι ὑπό.

Proportions.

A *ratio* is λόγος, *proportional* is expressed by the phrase *in
proportion* ἀνάλογον, and a *proportion* is ἀναλογία. We find in
Archimedes some uses of the verb λέγω which seem to throw light
on the definition found in Euclid of the *relation* or *ratio* between
two magnitudes. One passage (*On Conoids and Spheroids*, Prop. 1)
says *if the terms similarly placed have, two and two, the same ratio
and the first magnitudes are taken in relation to some other magnitudes in any ratios whatever* εἴ κα κατὰ δύο τὸν αὐτὸν λόγον ἔχωντι

τὰ ὁμοίως τεταγμένα, λεγήται δὲ τὰ πρῶτα μεγέθεα ποτί τινα ἄλλα μεγέθεα...ἐν λόγοις ὁποιοισοῦν, *if* A, B... *be in relation to* N, Ξ... *but* Z *be not in relation to anything* (i.e. has no term corresponding to it) εἴ κα... τὰ μὲν A, B,... λεγώνται ποτὶ τὰ N, Ξ,... τὸ δὲ Z μηδὲ ποθ' ἓν λεγήται.

A *mean proportional between* is μέση ἀνάλογον τῶν..., *is a mean proportional between* μέσον λόγον ἔχει τῆς...καὶ τῆς..., *two mean proportionals* δύο μέσαι ἀνάλογον *with or without* κατὰ τὸ συνεχές *in continued proportion.*

If three straight lines be proportional ἐὰν τρεῖς εὐθεῖαι ἀνάλογον ὦσι, *a fourth proportional* τετάρτα ἀνάλογον, *if four straight lines be proportional in continued proportion* εἴ κα τέσσαρες γραμμαὶ ἀνάλογον ἔωντι ἐν τᾷ συνεχεῖ ἀναλογίᾳ, *at the point dividing (the line) in the said proportion* κατὰ τὰν ἀνάλογον τομὰν τᾷ εἰρημένᾳ.

The *ratio of one straight line to another* is e.g. ὁ τῆς PΛ πρὸς ΛΧ λόγος or ὁ (λόγος), ὃν ἔχει ἡ PΛ πρὸς τὴν ΛΧ ; *the ratio of the bases* ὁ τῶν βασίων λόγος ; *has the ratio of 5 to 2* λόγον ἔχει, ὃν πέντε πρὸς δύο.

For *having the same ratio as* we find the following constructions. *Have the same ratio to one another as the bases* τὸν αὐτὸν ἔχοντι λόγον ποτ' ἀλλάλους ταῖς βάσεσιν, *as the squares on the radii* ὃν αἱ ἐκ τῶν κέντρων δυνάμει ; TΔ *has to* PZ *the (linear) ratio which the square on* TΔ *has to the square on* H, ὃν ἔχει λόγον ἡ TΔ πρὸς τὴν Η δυνάμει, τοῦτον ἔχει τὸν λόγον ἡ TΔ πρὸς PZ μήκει. *Is divided in the same ratio* εἰς τὸν αὐτὸν λόγον τέτμηται, or simply ὁμοίως ; *will divide the diameter in the proportion of the successive odd numbers, unity corresponding to the (part) adjacent to the vertex of the segment* τὰν διάμετρον τεμοῦντι εἰς τοὺς τῶν ἑξῆς περισσῶν ἀριθμῶν λόγους, ἑνὸς λεγομένου ποτὶ τᾷ κορυφᾷ τοῦ τμάματος.

To have a less (or greater) ratio than is ἔχειν λόγον ἐλάσσονα (or μείζονα) with the genitive of the second ratio or a phrase introduced by ἤ ; *to have a less ratio than the greater magnitude has to the less*, ἔχειν λόγον ἐλάσσονα ἢ τὸ μεῖζον μέγεθος πρὸς τὸ ἔλασσον.

For *duplicate, triplicate* etc. ratios we have the following expressions : *has the triplicate ratio of the same ratio* τριπλασίονα λόγον ἔχει τοῦ αὐτοῦ λόγου, *has the duplicate ratio of* ΕΛ *to* ΑΚ διπλασίονα λόγον ἔχει ἥπερ ἡ ΕΛ πρὸς ΑΚ, *are in the triplicate ratio of the diameters in the bases* ἐν τριπλασίονι λόγῳ εἰσὶ τῶν ἐν ταῖς βάσεσι διαμέτρων, *sesquialterate ratio* ἡμιόλιος λόγος. With these expressions must be contrasted the use of *double, quadruple* etc.

ratio in the sense of a simple multiple by 2, 4 etc., e.g. *if any number of areas be placed in order, each being four times the next* εἰ κα χωρία τεθέωντι ἐξῆς ὁποσαοῦν ἐν τῷ τετραπλασίονι λόγῳ.

The ordinary expression for a proportion is *as* A *is to* B *so is* Γ *to* Δ, ὡς ἡ A πρὸς τὴν B, οὕτως ἡ Γ πρὸς τὴν Δ. *Let* ΔE *be made so that* ΔE *is to* ΓE *as the sum of* ΘA, AE *is to* AE, πεποιήσθω, ὡς συναμφότερος ἡ ΘA, AE πρὸς τὴν AE, οὕτως ἡ ΔE πρὸς ΓE. *The antecedents* are τὰ ἡγούμενα, *the consequents* τὰ ἑπόμενα.

For *reciprocally proportional* the parts of ἀντιπέπονθα are used; the *bases are reciprocally proportional to the heights* ἀντιπεπόνθασιν αἱ βάσεις ταῖς ὕψεσιν, *to be reciprocally in the same proportion* ἀντιπεπονθέμεν κατὰ τὸν αὐτὸν λόγον.

A *ratio compounded of* is λόγος συνημμένος (or συγκείμενος) ἔκ τε τοῦ...καὶ τοῦ...; *the ratio of* PΛ *to* ΛX *is equal to that compounded of* ὁ τῆς PΛ πρὸς ΛX λόγος συνῆπται ἐκ.... Two other expressions for compounded ratios are ὁ τοῦ ἀπὸ AΘ πρὸς τὸ ἀπὸ BΘ καὶ ὁ (or προσλαβὼν τὸν) τῆς AΘ πρὸς ΘB, *the ratio of the square on* AΘ *to the square on* BΘ *multiplied by the ratio of* AΘ *to* ΘB.

The technical terms for transforming such a proportion as $a : b = c : d$ are as follows:

1. ἐναλλάξ *alternately* (usually called *permutando* or *alternando*) means transforming the proportion into $a : c = b : d$.

2. ἀνάπαλιν *reversely* (usually *invertendo*), $b : a = d : c$.

3. σύνθεσις λόγου is *composition of a ratio* by which the ratio $a : b$ becomes $a + b : b$. The corresponding Greek term to *componendo* is συνθέντι, which means no doubt literally "to one who has compounded," i.e. "if we compound," the ratios. Thus συνθέντι denotes the inference that $a + b : b = c + d : d$. κατὰ σύνθεσιν is also used in the same sense by Archimedes.

4. διαίρεσις λόγου signifies the *division of a ratio* in the sense of *separation* or *subtraction* by which $a : b$ becomes $a - b : b$. Similarly διελόντι (or κατὰ διαίρεσιν) denotes the inference that $a - b : b = c - d : d$. The translation *dividendo* is therefore somewhat misleading.

5. ἀναστροφὴ λόγου *conversion of a ratio* and ἀναστρέψαντι correspond respectively to the ratio $a : a - b$ and to the inference that $a : a - b = c : c - d$.

6. δι' ἴσου *ex aequali* (sc. *distantia*) is applied e.g. to the inference from the proportions

$$a : b : c : d \text{ etc.} = A : B : C : D \text{ etc.}$$

that $\qquad\qquad a : d = A : D.$

When this dividing-out of ratios takes place between proportions with corresponding terms placed crosswise, it is described as δι' ἴσου ἐν τῇ τεταραγμένῃ ἀναλογίᾳ, *ex aequali in disturbed proportion* or ἀνομοίως τῶν λόγων τεταγμένων *the ratios being dissimilarly placed*; this is the case e.g. when we have two proportions

$$a : b = B : C,$$
$$b : c = A : B,$$

and we infer that $\qquad a : c = A : C.$

Arithmetical terms.

Whole *multiples* of any magnitude are generally described as *the double* of, *the triple* of etc., ὁ διπλάσιος, ὁ τριπλάσιος κ.τ.λ., following the gender of the particular magnitude; thus *the (surface which is) four times the greatest circle in the sphere* ἡ τετραπλασία τοῦ μεγίστου κύκλου τῶν ἐν τῇ σφαίρᾳ; *five times the sum of* AB, BE *together with ten times the sum of* ΓB, BΔ, ἁ πενταπλασία συναμφοτέρου τᾶς AB, BE μετὰ τᾶς δεκαπλασίας συναμφοτέρου τᾶς ΓB, BΔ. *The same multiple as* τοσαυταπλασίων...ὁσαπλασίων ἐστί, or ἰσάκις πολλαπλασίων...καί. The general word for a multiple of is πολλαπλάσιος or πολλαπλασίων, which may be qualified by any expression denoting the *number of times* multiplied; thus *multiplied by the same number* πολλαπλάσιος τῷ αὐτῷ ἀριθμῷ, *multiples according to the successive numbers* πολλαπλάσια κατὰ τοὺς ἑξῆς ἀριθμούς.

Another method is to use the adverbial forms *twice* δίς, *thrice* τρίς, etc., which are either followed by the nominative, e.g. *twice* EΔ δὶς ἡ EΔ, or constructed with a participle, e.g. *twice taken* δὶς λαμβανόμενος or δὶς εἰρημένος; *together with twice the whole circumference of the circle* μεθ' ὅλας τᾶς τοῦ κύκλου περιφερείας δὶς λαμβανομένας. Similarly *the same number of times (the said circumference) as is expressed by the number one less than (that of) the revolutions* τοσαυτάκις λαμβανομένας, ὅσος ἐστὶν ὁ ἑνὶ ἐλάσσων ἀριθμὸς τᾶν περιφορᾶν. An interesting phrase is the following, *as many times as the line* ΓΔ *is contained* (literally *added together*) *in* AΔ, *so many times let the time* ZH *be contained in the time* ΛH, ὁσάκις συγκεῖται ἁ ΓΔ γραμμὰ ἐν τῷ ΑΔ, τοσαυτάκις συγκείσθω ὁ χρόνος ὁ ΖΗ ἐν τῷ χρόνῳ τῷ ΛΗ.

Submultiples are denoted by the ordinal number followed by μέρος; *one-seventh* is ἕβδομον μέρος and so on, *one-half* being however ἥμισυς. When the denominator is a large number, a circumlocutory phrase is used; thus *less than* $\frac{1}{164}$ *th part of a right angle* ἐλάττων ἢ διαιρεθείσας τὰς ὀρθὰς εἰς ρξδ´ τούτων ἐν μέρος.

When the numerator of a fraction is not unity, it is expressed by the ordinal number, and the denominator by a compound substantive denoting such and such a submultiple; e.g. *two-thirds* δύο τριταμόρια, *three-fifths* τρία πεμπταμόρια.

There are two improper fractions which have special names, thus *one-and-a-half of* is ἡμιόλιος, *one-and-a-third of* ἐπίτριτος. Where a number is partly integral and partly fractional, the integer is first stated and the fraction follows introduced by καὶ ἔτι or καί *and besides*. The phrases used to express the fact that *the circumference of a circle is less than* $3\frac{1}{7}$ *but greater than* $3\frac{10}{71}$ *times its diameter* deserve special notice; (1) παντὸς κύκλου ἡ περίμετρος τῆς διαμέτρου τριπλασίων ἐστί, καὶ ἔτι ὑπερέχει ἐλάσσονι μὲν ἢ ἑβδόμῳ μέρει τῆς διαμέτρου, μείζονι δὲ ἢ δέκα ἑβδομηκοστομόνοις, and (2) τριπλασίων ἐστὶ καὶ ἐλάσσονι μὲν ἢ ἑβδόμῳ μέρει, μείζονι δὲ ἢ ι´ οα´´ μείζων. We also have the phrase for the first part ἐλάσσων ἢ τριπλασίων καὶ ἑβδόμῳ μέρει μείζων.

To *measure* μετρεῖν, *common measure* κοινὸν μέτρον, *commensurable*, *incommensurable* σύμμετρος, ἀσύμμετρος.

Mechanical terms.

Mechanics τὰ μηχανικά, *weight* βάρος; *centre of gravity* κέντρον τοῦ βάρεος with another genitive of the body or magnitude; in the plural we have either τὰ κέντρα αὐτῶν τοῦ βάρεος or τὰ κέντρα τῶν βαρέων. κέντρον is also used alone.

A *lever* ζυγός or ζύγιον, *the horizon* ὁ ὁρίζων; *in a vertical line* is represented by *perpendicularly* κατὰ κάθετον, thus *the point of suspension and the centre of gravity of the body suspended are in a vertical line* κατὰ κάθετόν ἐστι τό τε σαμεῖον τοῦ κρεμαστοῦ καὶ τὸ κέντρον τοῦ βάρεος τοῦ κρεμαμένου. Of suspension *from* or *at* ἐκ or κατά (with acc.) is used. *Let the triangle be suspended from the points* Β, Γ, κρεμάσθω τὸ τρίγωνον ἐκ τῶν Β, Γ σαμείων; *if the suspension of the triangle* ΒΔΓ *at* Β, Γ *be set free, and it be suspended at* Ε, *the triangle remains in its position* εἰ κα τοῦ ΒΔΓ τριγώνου ἁ μὲν κατὰ τὰ Β, Γ κρέμασις λυθῇ, κατὰ δὲ τὸ Ε κρεμασθῇ, μένει τὸ τρίγωνον, ὡς νῦν ἔχει.

To *incline towards* ῥέπειν ἐπί (acc.); *to be in equilibrium* ἰσορροπεῖν, *they will be in equilibrium with* Δ *held fast* κατεχομένου τοῦ Δ ἰσορροπήσει, *they will be in equilibrium at* Δ (i.e. will balance about Δ) κατὰ τὸ Δ ἰσορρυπησοῦντι; AB *is too great to balance* Γ μεῖζόν ἐστι τὸ AB ἢ ὥστε ἰσορροπεῖν τῷ Γ. The adjective for *in equilibrium* is ἰσορρεπής; *let it be in equilibrium with the triangle* ΓΔΗ, ἰσορρεπὲς ἔστω τῷ ΓΔΗ τριγώνῳ. To balance *at certain distances* (from the point of support or the centre of gravity of a system) is ἀπό τινων μακέων ἰσορροπεῖν.

Theorems, problems, etc.

A *theorem* θεώρημα (from θεωρεῖν *to investigate*); a *problem* πρόβλημα, with which the following expressions may be compared, *the* (*questions*) *propounded concerning the figures* τὰ προβεβλημένα περὶ τῶν σχημάτων, *these things are propounded for investigation* προβάλλεται τάδε θεωρῆσαι; also πρόκειμαι takes the place of the passive, *which it was proposed* (or *required*) *to find* ὅπερ προέκειτο εὑρεῖν.

Another similar word is ἐπίταγμα, *direction* or *requirement*; thus *the theorems and directions necessary for the proofs of them* τὰ θεωρήματα καὶ τὰ ἐπιτάγματα τὰ χρείαν ἔχοντα εἰς τὰς ἀποδείξίας αὐτῶν, *in order that the requirement may be fulfilled* ὅπως γένηται τὸ ἐπιταχθέν (or ἐπίταγμα). To *satisfy the requirement* is ποιεῖν τὸ ἐπίταγμα (either e.g. of lines in a figure, or of the person solving the problem).

After the setting out (ἔκθεσις) in any proposition there follows the short statement of what it is required to prove or to do. In the former case (that of a *theorem*) Archimedes uses one of three expressions δεικτέον *it is required to prove*, λέγω or φαμὶ δή *I assert* or *say*; and in the second case (that of a problem) δεῖ δή *it is required* (to do so and so).

In a problem the *analysis* ἀνάλυσις and *synthesis* σύνθεσις are distinguished, the latter being generally introduced with the words *the synthesis of the problem will be as follows* συντεθήσεται τὸ πρόβλημα οὕτως. The parts of the verb ἀναλύειν are similarly used; thus *the analysis and synthesis of each of these* (*problems*) *will be given at the end* ἑκάτερα δὲ ταῦτα ἐπὶ τέλει ἀναλυθήσεταί τε καὶ συντεθήσεται.

A notable term in connexion with problems is the διορισμός (*determination*), which means the determination of the limits within which a solution is possible*. If a solution is always possible, the problem *does not involve a* διορισμός, οὐκ ἔχει διορισμόν ; otherwise it does involve it, ἔχει διορισμόν.

Data and hypotheses.

For *given* some part of the verb δίδωμι is used, generally the participle δοθείς, but sometimes δεδομένος and once or twice διδόμενος. *Let a circle be given* δεδόσθω κύκλος, *given two unequal magnitudes* δύο μεγεθῶν ἀνίσων δοθέντων, *each of the two lines* ΓΔ, ΕΖ *is given* ἐστὶν δοθεῖσα ἑκατέρα τῶν ΓΔ, ΕΖ, *the same ratio as the given one* λόγος ὁ αὐτὸς τῷ δοθέντι. Similar expressions are *the assigned ratio* ὁ ταχθεὶς λόγος, *the given area* τὸ προτεθὲν (or προκείμενον) χωρίον.

Given in position θέσει simply (sc. δεδομένη).

Of *hypotheses* the parts of the verb ὑποτίθεμαι and (for the passive) ὑπόκειμαι are used ; *with the same suppositions* τῶν αὐτῶν ὑποκειμένων, *let the said suppositions be made* ὑποκείσθω τὰ εἰρημένα, *we make these suppositions* ὑποτιθέμεθα τάδε.

Where in a *reductio ad absurdum* the original hypothesis is referred to, and generally where an earlier step is quoted, the past tense of the verb is used ; *but it was not* (so) οὐκ ἦν δέ, *for it was less* ἦν γὰρ ἐλάσσων, *they were proved equal* ἀπεδείχθησαν ἴσοι, *for this has been proved to be possible* δεδείκται γὰρ τοῦτο δυνατὸν ἐόν. Where a hypothesis is thus quoted, the past tense of ὑπόκειμαι has various constructions after it, (1) an adjective or participle, ΑΖ, ΒΗ *were supposed equal* ἴσαι ὑπέκειντο αἱ ΑΖ, ΒΗ, *it is by hypothesis a tangent* ὑπέκειτο ἐπιψαύουσα, (2) an infinitive, *for by hypothesis it does not cut* ὑπέκειτο γὰρ μὴ τέμνειν, *the axis is by hypothesis not at right angles to the parallel planes* ὑπέκειτο ὁ ἄξων μὴ εἶμεν ὀρθὸς ποτὶ τὰ παράλλαλα ἐπίπεδα, (3) *the plane is supposed to have been drawn through the centre* τὸ ἐπίπεδον ὑπόκειται διὰ τοῦ κέντρου ἄχθαι.

Supposing it found εὑρεθέντος absolutely. *Suppose it done* γεγονέτω.

The usual idiomatic use of εἰ δὲ μή after a negative statement may be mentioned ; *it will not meet the surface in another point, otherwise...* οὐ γὰρ ἅψεται κατ' ἄλλο σαμεῖον τᾶς ἐπιφανείας· εἰ δὲ μή....

* Cf. *Apollonius of Perga*, p. lxx, note.

Inferences, and adaptation to different cases.

The usual equivalent for *therefore* is ἄρα; οὖν and τοίνυν are generally used in a somewhat weaker sense to mark the starting-point of an argument, thus ἐπεὶ οὖν may be translated as *since, then*. *Since* is ἐπεί, *because* διότι.

πολλῷ μᾶλλον *much more then* is apparently not used in Archimedes, who has πολλῷ alone; thus *much less then is the ratio of the circumscribed figure to the inscribed than that of* K *to* H πολλῷ ἄρα τὸ περιγραφὲν πρὸς τὸ ἐγγραφὲν ἐλάσσονα λόγον ἔχει τοῦ, ὃν ἔχει ἡ Κ πρὸς H.

διά with the accusative is a common way of expressing the reason why; *because the cone is isosceles* διὰ τὸ ἰσοσκελῆ εἶναι τὸν κῶνον, *for the same reason* διὰ ταὐτά.

διά with the genitive expresses the *means* by which a proposition is proved; *by means of the construction* διὰ τῆς κατασκευῆς, *by the same means* διὰ τῶν αὐτῶν, *by the same method* διὰ τοῦ αὐτοῦ τρόπου.

Whenever this is the case, the surface is greater ὅταν τοῦτο ᾖ, μείζων γίνεται ἡ ἐπιφάνεια..., *if this is the case, the angle* BAΘ *is equal...,* εἰ δὲ τοῦτο, ἴσα ἐστὶν ἁ ὑπὸ BAΘ γωνία..., *which is the same thing as showing that...* ὃ ταὐτόν ἐστι τῷ δεῖξαι, ὅτι....

Similarly for the sector ὁμοίως δὲ καὶ ἐπὶ τοῦ τομέως, *the proof is the same as (that used to show) that* ἁ αὐτὰ ἀπόδειξις ἅπερ καὶ ὅτι, *the proof that...is the same* ἁ αὐτὰ ἀπόδειξίς ἐστι καὶ διότι..., *the same argument holds for all rectilineal figures inscribed in the segments in the recognised manner (see p. 204)* ἐπὶ πάντων εὐθυγράμμων τῶν ἐγγραφομένων ἐς τὰ τμάματα γνωρίμως ὁ αὐτὸς λόγος; *it will be possible, having proved it for a circle, to transfer the same argument in the case of the sector* ἔσται ἐπὶ κύκλου δείξαντα μεταγαγεῖν τὸν ὅμοιον λόγον καὶ ἐπὶ τοῦ τομέως; *the rest will be the same, but it will be the lesser of the diameters which will be intercepted within the spheroid (instead of the greater)* τὰ μὲν ἄλλα τὰ αὐτὰ ἐσσεῖται, τᾶν δὲ διαμέτρων ἁ ἐλάσσων ἐσσεῖται ἁ ἐναπολαφθεῖσα ἐν τῷ σφαιροειδεῖ; *it will make no difference whether...or...* διοίσει δὲ οὐδέν, εἴτε...εἴτε....

Conclusions.

The proposition is therefore obvious, or is proved δῆλον οὖν ἐστι (or δέδεικται) τὸ προτεθέν; *similarly* φανερὸν οὖν ἐστιν, ὃ ἔδει δεῖξαι, and ἔδει δὲ τοῦτο δεῖξαι. *Which is absurd, or impossible* ὅπερ ἄτοπον, or ἀδύνατον.

A curious use of two negatives is contained in the following:

οὐκ ἄρα οὐκ ἔστι κέντρον τοῦ βάρεος τοῦ ΔΕΖ τριγώνου τὸ Ν σαμεῖον. ἔστιν ἄρα, *therefore it is not possible that the point* Ν *should not be the centre of gravity of the triangle* ΔΕΖ. *It must therefore be so.*

Thus a rhombus will have been formed ἔσται δὴ γεγονὼς ῥόμβος ; *two unequal straight lines have been found satisfying the requirement* εὑρημέναι εἰσὶν ἄρα δύο εὐθεῖαι ἄνισοι ποιοῦσαι τὸ ἐπίταγμα.

Direction, concavity, convexity.

In the same direction ἐπὶ τὰ αὐτά, *in the other direction* ἐπὶ τὰ ἕτερα, *concave in the same direction* ἐπὶ τὰ αὐτὰ κοίλη ; *in the same direction as* ἐπὶ τὰ αὐτά *with the dative or* ἐφ᾽ ἅ, *thus in the same direction as the vertex of the cone* ἐπὶ τὰ αὐτὰ τᾷ τοῦ κώνου κορυφᾷ, *drawn in the same direction as (that of) the convex side of it* ἐπὶ τὰ αὐτὰ ἀγομέναι, ἐφ᾽ ἅ ἐντι τὰ κυρτὰ αὐτοῦ. For *on the same side of* ἐπὶ τὰ αὐτά is followed by the genitive, *they fall on the same side of the line* ἐπὶ τὰ αὐτὰ πίπτουσι τῆς γραμμῆς.

On each side of ἐφ᾽ ἑκάτερα (with gen.); *on each side of the plane of the base* ἐφ᾽ ἑκάτερα τοῦ ἐπιπέδου τῆς βάσεως.

Miscellaneous.

Property σύμπτωμα. *Proceeding thus continually,* ἀεὶ τοῦτο ποιοῦντες, ἀεὶ τούτου γενομένου, or τούτου ἑξῆς γινομένου. *In the elements* ἐν τῇ στοιχειώσει.

One special difference between our terminology and the Greek is that whereas we speak of *any* circle, *any* straight line and the like, the Greeks say *every* circle, *every* straight line, etc. Thus *any pyramid is one third part of the prism with the same base as the pyramid and equal height* πᾶσα πυραμὶς τρίτον μέρος ἐστὶ τοῦ πρίσματος τοῦ τὰν αὐτὰν βάσιν ἔχοντος τᾷ πυραμίδι καὶ ὕψος ἴσον. *I define the diameter of any segment as* διάμετρον καλέω παντὸς τμάματος. *To exceed any assigned (magnitude) of those which are comparable with one another* ὑπερέχειν παντὸς τοῦ προτεθέντος τῶν πρὸς ἄλληλα λεγομένων.

Another noteworthy difference is illustrated in the last sentence. The Greeks did not speak as we do of *a* given area, *a* given ratio etc., but of *the* given area, *the* given ratio, and the like. Thus *It is possible...to leave certain segments less than a given area* δυνατόν ἐστιν...λείπειν τινα τμήματα, ἅπερ ἔσται ἐλάσσονα τοῦ προκειμένου χωρίου; *to divide a given sphere by a plane so that the segments have to one another an assigned ratio* τὰν δοθεῖσαν σφαῖραν ἐπιπέδῳ τεμεῖν, ὥστε τὰ τμάματα αὐτᾶς ποτ᾽ ἄλλαλα τὸν ταχθέντα λόγον ἔχειν.

Magnitudes in *arithmetical progression* are said *to exceed each other by an equal (amount)*; *if there be any number of magnitudes in arithmetical progression* εἰ κα ἔωντι μεγέθεα ὁποσαοῦν τῷ ἴσῳ ἀλλάλων ὑπερέχοντα. The *common difference* is the *excess* ὑπεροχά, and the terms collectively are spoken of as *the magnitudes exceeding by the equal (difference)* τὰ τῷ ἴσῳ ὑπερέχοντα. *The least term* is τὸ ἐλάχιστον, *the greatest term* τὸ μέγιστον. The *sum of the terms* is expressed by πάντα τὰ τῷ ἴσῳ ὑπερέχοντα.

Terms of a *geometrical progression* are simply *in (continued) proportion* ἀνάλογον, the *series* is then ἡ ἀναλογία, the *proportion*, and a term of the series is τὶς τῶν ἐν τᾷ αὐτᾷ ἀναλογίᾳ. *Numbers in geometrical progression beginning from unity* are ἀριθμοὶ ἀνάλογον ἀπὸ μονάδος. *Let the term* Λ *of the progression be taken which is distant the same number of terms from* Θ *as* Δ *is distant from unity* λελάφθω ἐκ τᾶς ἀναλογίας ὁ Λ ἀπέχων ἀπὸ τοῦ Θ τοσούτους, ὅσους ὁ Δ ἀπὸ μονάδος ἀπέχει.

THE WORKS OF
ARCHIMEDES.

ON THE SPHERE AND CYLINDER.

BOOK I.

"ARCHIMEDES to Dositheus greeting.

On a former occasion I sent you the investigations which I had up to that time completed, including the proofs, showing that any segment bounded by a straight line and a section of a right-angled cone [a parabola] is four-thirds of the triangle which has the same base with the segment and equal height. Since then certain theorems not hitherto demonstrated (ἀνελέγκτων) have occurred to me, and I have worked out the proofs of them. They are these: first, that the surface of any sphere is four times its greatest circle (τοῦ μεγίστου κύκλου); next, that the surface of any segment of a sphere is equal to a circle whose radius (ἡ ἐκ τοῦ κέντρου) is equal to the straight line drawn from the vertex (κορυφή) of the segment to the circumference of the circle which is the base of the segment; and, further, that any cylinder having its base equal to the greatest circle of those in the sphere, and height equal to the diameter of the sphere, is itself [i.e. in content] half as large again as the sphere, and its surface also [including its bases] is half as large again as the surface of the sphere. Now these properties were all along naturally inherent in the figures referred to (αὐτῇ τῇ φύσει προυπῆρχεν περὶ τὰ εἰρημένα σχήματα), but remained unknown to those who were before my time engaged in the study of geometry. Having, however, now discovered that the properties are true of these figures, I cannot feel any hesitation in setting them side by side both with my former investigations and with those of the theorems of Eudoxus on solids which are held to be most irrefragably established, namely, that any pyramid is one third part of the prism which has the same base with the pyramid and equal height, and that any cone is one third part of the cylinder which has the same base with the cone and equal height. For, though these properties also were naturally inherent in the figures all along, yet they were in fact unknown to all the many able geometers who lived before Eudoxus, and had not been observed by any one. Now, however, it will be open to those who possess the requisite ability to examine these discoveries of mine. They ought to have been published while Conon was still alive, for I should conceive that he would best have been able to grasp them and to pronounce upon them the appropriate verdict; but, as I judge it well to communicate them to those who are conversant with mathematics, I send them to you with the proofs written out, which it will be open to mathematicians to examine. Farewell.

I first set out the axioms* and the assumptions which I have used for the proofs of my propositions.

DEFINITIONS.

1. There are in a plane certain terminated bent lines (καμπύλαι γραμμαὶ πεπερασμέναι)†, which either lie wholly on the same side of the straight lines joining their extremities, or have no part of them on the other side.

2. I apply the term **concave in the same direction** to a line such that, if any two points on it are taken, either all the straight lines connecting the points fall on the same side of the line, or some fall on one and the same side while others fall on the line itself, but none on the other side.

---

* Though the word used is ἀξιώματα, the "axioms" are more of the nature of definitions; and in fact Eutocius in his notes speaks of them as such (ὅροι).

† Under the term *bent line* Archimedes includes not only curved lines of continuous curvature, but lines made up of any number of lines which may be either straight or curved.

3. Similarly also there are certain terminated surfaces, not themselves being in a plane but having their extremities in a plane, and such that they will either be wholly on the same side of the plane containing their extremities, or have no part of them on the other side.

4. I apply the term **concave in the same direction** to surfaces such that, if any two points on them are taken, the straight lines connecting the points either all fall on the same side of the surface, or some fall on one and the same side of it while some fall upon it, but none on the other side.

5. I use the term **solid sector**, when a cone cuts a sphere, and has its apex at the centre of the sphere, to denote the figure comprehended by the surface of the cone and the surface of the sphere included within the cone.

6. I apply the term **solid rhombus,** when two cones with the same base have their apices on opposite sides of the plane of the base in such a position that their axes lie in a straight line, to denote the solid figure made up of both the cones.

ASSUMPTIONS.

1. *Of all lines which have the same extremities the straight line is the least*.

* This well-known Archimedean assumption is scarcely, as it stands, a *definition* of a straight line, though Proclus says [p. 110 ed. Friedlein] "Archimedes defined (ὡρίσατο) the straight line as the least of those [lines] which have the same extremities. For because, as Euclid's definition says, ἐξ ἴσου κεῖται τοῖς ἐφ᾽ ἑαυτῆς σημείοις, it is in consequence the least of those which have the same extremities." Proclus had just before [p. 109] explained Euclid's definition, which, as will be seen, is different from the ordinary version given in our text-books; a straight line is not "that which lies evenly between its extreme points," but "that which ἐξ ἴσου τοῖς ἐφ᾽ ἑαυτῆς σημείοις κεῖται." The words of Proclus are, "He [Euclid] shows by means of this that the straight line alone [of all lines] occupies a distance (κατέχειν διάστημα) equal to that between the points on it. For, as far as one of its points is removed from another, so great is the length (μέγεθος) of the straight line of which the points are the extremities; and this is the meaning of τὸ ἐξ ἴσου κεῖσθαι τοῖς ἐφ᾽ ἑαυτῆς σημείοις. But, if you take two points on a circumference or any other line, the distance cut off between them along the line is greater than the interval separating them; and this is the case with every line except the straight line." It appears then from this that Euclid's definition should be understood in a sense very like that of

2. Of other lines in a plane and having the same extremities, [any two] such are unequal whenever both are concave in the same direction and one of them is either wholly included between the other and the straight line which has the same extremities with it, or is partly included by, and is partly common with, the other; and that [line] which is included is the lesser [of the two].

3. Similarly, of surfaces which have the same extremities, if those extremities are in a plane, the plane is the least [in area].

4. Of other surfaces with the same extremities, the extremities being in a plane, [any two] such are unequal whenever both are concave in the same direction and one surface is either wholly included between the other and the plane which has the same extremities with it, or is partly included by, and partly common with, the other; and that [surface] which is included is the lesser [of the two in area].

5. Further, of unequal lines, unequal surfaces, and unequal solids, the greater exceeds the less by such a magnitude as, when added to itself, can be made to exceed any assigned magnitude among those which are comparable with [it and with] one another*.

These things being premised, *if a polygon be inscribed in a circle, it is plain that the perimeter of the inscribed polygon is less than the circumference of the circle;* for each of the sides of the polygon is less than that part of the circumference of the circle which is cut off by it."

Archimedes' assumption, and we might perhaps translate as follows, "A straight line is that which extends equally (ἐξ ἴσου κεῖται) with the points on it," or, to follow Proclus' interpretation more closely, "A straight line is that which represents equal extension with [the distances separating] the points on it."

* With regard to this assumption compare the Introduction, chapter III. § 2.

Proposition 1.

If a polygon be circumscribed about a circle, the perimeter of the circumscribed polygon is greater than the perimeter of the circle.

Let any two adjacent sides, meeting in A, touch the circle at P, Q respectively.

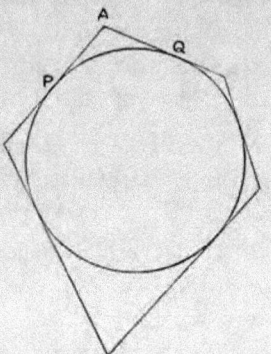

Then [*Assumptions*, 2]

$$PA + AQ > (\text{arc } PQ).$$

A similar inequality holds for each angle of the polygon; and, by addition, the required result follows.

Proposition 2.

Given two unequal magnitudes, it is possible to find two unequal straight lines such that the greater straight line has to the less a ratio less than the greater magnitude has to the less.

Let AB, D represent the two unequal magnitudes, AB being the greater.

Suppose BC measured along BA equal to D, and let GH be any straight line.

Then, if CA be added to itself a sufficient number of times, the sum will exceed D. Let AF be this sum, and take E on GH produced such that GH is the same multiple of HE that AF is of AC.

Thus $EH : HG = AC : AF.$

But, since $AF > D$ (or CB),

$$AC : AF < AC : CB.$$

Therefore, *componendo*,

$$EG : GH < AB : D.$$

Hence EG, GH are two lines satisfying the given condition.

Proposition 3.

Given two unequal magnitudes and a circle, it is possible to inscribe a polygon in the circle and to describe another about it so that the side of the circumscribed polygon may have to the side of the inscribed polygon a ratio less than that of the greater magnitude to the less.

Let A, B represent the given magnitudes, A being the greater.

Find [Prop. 2] two straight lines F, KL, of which F is the greater, such that
$$F : KL < A : B \dots\dots\dots\dots\dots\dots(1).$$

Draw LM perpendicular to LK and of such length that $KM = F$.

In the given circle let CE, DG be two diameters at right angles. Then, bisecting the angle DOC, bisecting the half again, and so on, we shall arrive ultimately at an angle (as NOC) less than twice the angle LKM.

Join NC, which (by the construction) will be the side of a regular polygon inscribed in the circle. Let OP be the radius of the circle bisecting the angle NOC (and therefore bisecting NC at right angles, in H, say), and let the tangent at P meet OC, ON produced in S, T respectively.

Now, since $\angle CON < 2 \angle LKM$,

$$\angle HOC < \angle LKM,$$

and the angles at H, L are right;

$$\text{therefore } MK : LK > OC : OH$$
$$> OP : OH.$$

Hence $\qquad ST : CN < MK : LK$
$$< F : LK;$$

therefore, *a fortiori*, by (1),

$$ST : CN < A : B.$$

Thus two polygons are found satisfying the given condition.

Proposition 4.

Again, given two unequal magnitudes and a sector, it is possible to describe a polygon about the sector and to inscribe another in it so that the side of the circumscribed polygon may have to the side of the inscribed polygon a ratio less than the greater magnitude has to the less.

[The "inscribed polygon" found in this proposition is one which has for two sides the two radii bounding the sector, while the remaining sides (the number of which is, by construction, some power of 2) subtend equal parts of the arc of the sector; the "circumscribed polygon" is formed by the tangents parallel to the sides of the inscribed polygon and by the two bounding radii produced.]

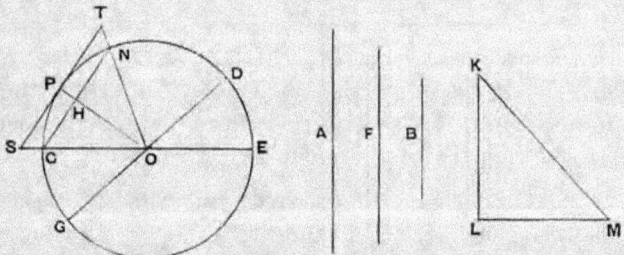

In this case we make the same construction as in the last proposition except that we bisect the angle COD of the sector, instead of the right angle between two diameters, then bisect the half again, and so on. The proof is exactly similar to the preceding one.

Proposition 5.

Given a circle and two unequal magnitudes, to describe a polygon about the circle and inscribe another in it, so that the circumscribed polygon may have to the inscribed a ratio less than the greater magnitude has to the less.

Let A be the given circle and B, C the given magnitudes, B being the greater.

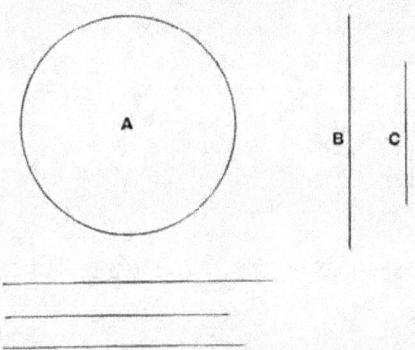

Take two unequal straight lines D, E, of which D is the greater, such that $D : E < B : C$ [Prop. 2], and let F be a mean proportional between D, E, so that D is also greater than F.

Describe (in the manner of Prop. 3) one polygon about the circle, and inscribe another in it, so that the side of the former has to the side of the latter a ratio less than the ratio $D : F$.

Thus the duplicate ratio of the side of the former polygon to the side of the latter is less than the ratio $D^2 : F^2$.

But the said duplicate ratio of the sides is equal to the ratio of the areas of the polygons, since they are similar;

therefore the area of the circumscribed polygon has to the area of the inscribed polygon a ratio less than the ratio $D^2 : F^2$, or $D : E$, and *a fortiori* less than the ratio $B : C$.

Proposition 6.

"Similarly we can show that, *given two unequal magnitudes and a sector, it is possible to circumscribe a polygon about the sector and inscribe in it another similar one so that the circumscribed may have to the inscribed a ratio less than the greater magnitude has to the less.*

And it is likewise clear that, *if a circle or a sector, as well as a certain area, be given, it is possible, by inscribing regular polygons in the circle or sector, and by continually inscribing such in the remaining segments, to leave segments of the circle or sector which are [together] less than the given area.* For this is proved in the *Elements* [Eucl. XII. 2].

But it is yet to be proved that, *given a circle or sector and an area, it is possible to describe a polygon about the circle or sector, such that the area remaining between the circumference and the circumscribed figure is less than the given area.*"

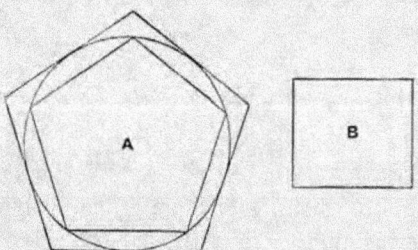

The proof for the circle (which, as Archimedes says, can be equally applied to a sector) is as follows.

Let A be the given circle and B the given area.

Now, there being two unequal magnitudes $A + B$ and A, let a polygon (C) be circumscribed about the circle and a polygon (I) inscribed in it [as in Prop. 5], so that

$$C : I < A + B : A \quad \dots\dots\dots\dots\dots(1).$$

The circumscribed polygon (C) shall be that required.

For the circle (A) is greater than the inscribed polygon (I).

Therefore, from (1), *a fortiori*,

$$C : A < A + B : A,$$

whence

$$C < A + B,$$

or

$$C - A < B.$$

Proposition 7.

If in an isosceles cone [i.e. *a right circular cone*] *a pyramid be inscribed having an equilateral base, the surface of the pyramid excluding the base is equal to a triangle having its base equal to the perimeter of the base of the pyramid and its height equal to the perpendicular drawn from the apex on one side of the base.*

Since the sides of the base of the pyramid are equal, it follows that the perpendiculars from the apex to all the sides of the base are equal; and the proof of the proposition is obvious.

Proposition 8.

If a pyramid be circumscribed about an isosceles cone, the surface of the pyramid excluding its base is equal to a triangle having its base equal to the perimeter of the base of the pyramid and its height equal to the side [i.e. *a generator*] *of the cone.*

The base of the pyramid is a polygon circumscribed about the circular base of the cone, and the line joining the apex of the cone or pyramid to the point of contact of any side of the polygon is perpendicular to that side. Also all these perpendiculars, being generators of the cone, are equal; whence the proposition follows immediately.

Proposition 9.

If in the circular base of an isosceles cone a chord be placed, and from its extremities straight lines be drawn to the apex of the cone, the triangle so formed will be less than the portion of the surface of the cone intercepted between the lines drawn to the apex.

Let ABC be the circular base of the cone, and O its apex.

Draw a chord AB in the circle, and join OA, OB. Bisect the arc ACB in C, and join AC, BC, OC.

Then $\triangle OAC + \triangle OBC > \triangle OAB.$

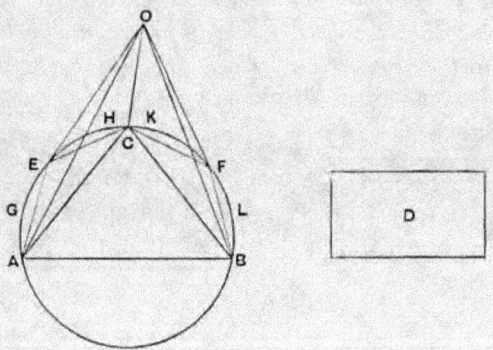

Let the excess of the sum of the first two triangles over the third be equal to the area D.

Then D is either less than the sum of the segments AEC, CFB, or not less.

I.   Let D be not less than the sum of the segments referred to.

We have now two surfaces

(1) that consisting of the portion $OAEC$ of the surface of the cone together with the segment AEC, and

(2) the triangle OAC;

and, since the two surfaces have the same extremities (the perimeter of the triangle OAC), the former surface is greater than the latter, which is *included* by it [*Assumptions*, 3 or 4].

Hence   (surface $OAEC$) + (segment AEC) > $\triangle OAC$.

Similarly (surface $OCFB$) + (segment CFB) > $\triangle OBC$.

Therefore, since D is not less than the sum of the segments, we have, by addition,

(surface $OAECFB$) + D > $\triangle OAC + \triangle OBC$

$$> \triangle OAB + D, \text{ by hypothesis.}$$

Taking away the common part D, we have the required result.

II.   Let D be less than the sum of the segments AEC, CFB.

If now we bisect the arcs AC, CB, then bisect the halves, and so on, we shall ultimately leave segments which are together less than D.   [Prop. 6]

Let AGE, EHC, CKF, FLB be those segments, and join OE, OF.

Then, as before,

(surface $OAGE$) + (segment AGE) > $\triangle OAE$

and   (surface $OEHC$) + (segment EHC) > $\triangle OEC$.

Therefore (surface $OAGHC$) + (segments AGE, EHC)

$$> \triangle OAE + \triangle OEC$$

$$> \triangle OAC, \text{ a fortiori.}$$

Similarly for the part of the surface of the cone bounded by OC, OB and the arc CFB.

Hence, by addition,

(surface $OAGEHCKFLB$) + (segments AGE, EHC, CKF, FLB)

$$> \triangle OAC + \triangle OBC$$

$$> \triangle OAB + D, \text{ by hypothesis.}$$

But the sum of the segments is less than D, and the required result follows.

Proposition 10.

If in the plane of the circular base of an isosceles cone two tangents be drawn to the circle meeting in a point, and the points of contact and the point of concourse of the tangents be respectively joined to the apex of the cone, the sum of the two triangles formed by the joining lines and the two tangents are together greater than the included portion of the surface of the cone.

Let ABC be the circular base of the cone, O its apex, AD, BD the two tangents to the circle meeting in D. Join OA, OB, OD.

Let ECF be drawn touching the circle at C, the middle point of the arc ACB, and therefore parallel to AB. Join OE, OF.

Then $$ED + DF > EF,$$
and, adding $AE + FB$ to each side,
$$AD + DB > AE + EF + FB.$$

Now OA, OC, OB, being generators of the cone, are equal, and they are respectively perpendicular to the tangents at A, C, B.

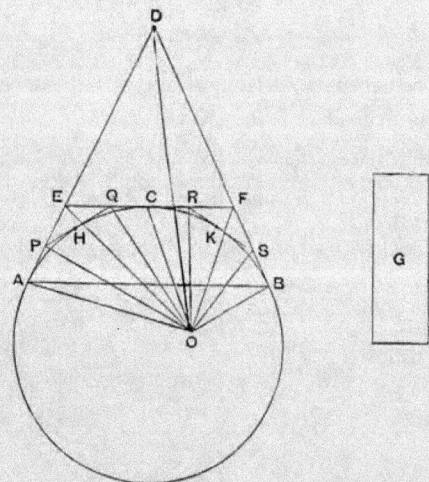

It follows that

$$\triangle OAD + \triangle ODB > \triangle OAE + \triangle OEF + \triangle OFB.$$

Let the area G be equal to the excess of the first sum over the second.

G is then either less, or not less, than the sum of the spaces $EAHC$, $FCKB$ remaining between the circle and the tangents, which sum we will call L.

I.   Let G be not less than L.

We have now two surfaces

(1) that of the pyramid with apex O and base $AEFB$, excluding the face OAB,

(2) that consisting of the part $OACB$ of the surface of the cone together with the segment ACB.

These two surfaces have the same extremities, viz. the perimeter of the triangle OAB, and, since the former *includes* the latter, the former is the greater [*Assumptions*, 4].

That is, the surface of the pyramid exclusive of the face OAB is greater than the sum of the surface $OACB$ and the segment ACB.

Taking away the segment from each sum, we have

$$\triangle OAE + \triangle OEF + \triangle OFB + L > \text{the surface } OAHCKB.$$

And G is not less than L.

It follows that

$$\triangle OAE + \triangle OEF + \triangle OFB + G,$$

which is by hypothesis equal to $\triangle OAD + \triangle ODB$, is greater than the same surface.

II.   Let G be less than L.

If we bisect the arcs AC, CB and draw tangents at their middle points, then bisect the halves and draw tangents, and so on, we shall lastly arrive at a polygon such that the sum of the parts remaining between the sides of the polygon and the circumference of the segment is less than G.

Let the remainders be those between the segment and the polygon $APQRSB$, and let their sum be M. Join OP, OQ, etc.

Then, as before,

$$\triangle OAE + \triangle OEF + \triangle OFB > \triangle OAP + \triangle OPQ + \ldots + \triangle OSB.$$

Also, as before,

(surface of pyramid $OAPQRSB$ excluding the face OAB)
> the part $OACB$ of the surface of the cone together with the segment ACB.

Taking away the segment from each sum,

$$\triangle OAP + \triangle OPQ + \ldots + M > \text{the part } OACB \text{ of the}$$
surface of the cone.

Hence, *a fortiori*,

$$\triangle OAE + \triangle OEF + \triangle OFB + G,$$

which is by hypothesis equal to

$$\triangle OAD + \triangle ODB,$$

is greater than the part $OACB$ of the surface of the cone.

Proposition 11.

If a plane parallel to the axis of a right cylinder cut the cylinder, the part of the surface of the cylinder cut off by the plane is greater than the area of the parallelogram in which the plane cuts it.

Proposition 12.

If at the extremities of two generators of any right cylinder tangents be drawn to the circular bases in the planes of those bases respectively, and if the pairs of tangents meet, the parallelograms formed by each generator and the two corresponding tangents respectively are together greater than the included portion of the surface of the cylinder between the two generators.

[The proofs of these two propositions follow exactly the methods of Props. 9, 10 respectively, and it is therefore unnecessary to reproduce them.]

" From the properties thus proved it is clear (1) that, *if a pyramid be inscribed in an isosceles cone, the surface of the pyramid excluding the base is less than the surface of the cone* [*excluding the base*], and (2) that, *if a pyramid be circumscribed about an isosceles cone, the surface of the pyramid excluding the base is greater than the surface of the cone excluding the base.*

" It is also clear from what has been proved both (1) that, *if a prism be inscribed in a right cylinder, the surface of the prism made up of its parallelograms* [*i.e. excluding its bases*] *is less than the surface of the cylinder excluding its bases*, and (2) that, *if a prism be circumscribed about a right cylinder, the surface of the prism made up of its parallelograms is greater than the surface of the cylinder excluding its bases.*"

Proposition 13.

The surface of any right cylinder excluding the bases is equal to a circle whose radius is a mean proportional between the side [*i.e. a generator*] *of the cylinder and the diameter of its base.*

Let the base of the cylinder be the circle A, and make CD equal to the diameter of this circle, and EF equal to the height of the cylinder.

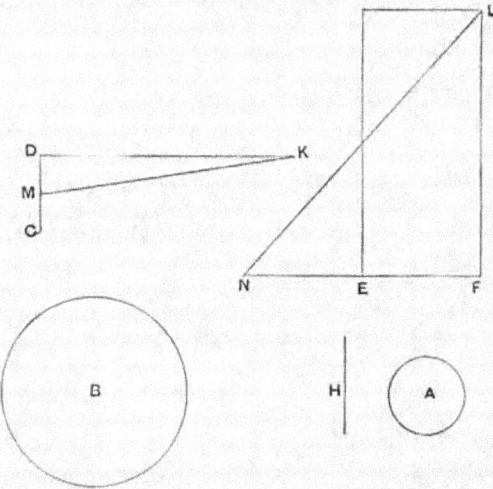

Let H be a mean proportional between CD, EF, and B a circle with radius equal to H.

Then the circle B shall be equal to the surface of the cylinder (excluding the bases), which we will call S.

For, if not, B must be either greater or less than S.

I. Suppose $B < S$.

Then it is possible to circumscribe a regular polygon about B, and to inscribe another in it, such that the ratio of the former to the latter is less than the ratio $S : B$.

Suppose this done, and circumscribe about A a polygon similar to that described about B; then erect on the polygon about A a prism of the same height as the cylinder. The prism will therefore be circumscribed to the cylinder.

Let KD, perpendicular to CD, and FL, perpendicular to EF, be each equal to the perimeter of the polygon about A. Bisect CD in M, and join MK.

Then $\triangle KDM = $ the polygon about A.

Also $\square\, EL = $ surface of prism (excluding bases).

Produce FE to N so that $FE = EN$, and join NL.

Now the polygons about A, B, being similar, are in the duplicate ratio of the radii of A, B.

Thus

$$\triangle KDM : (\text{polygon about } B) = MD^2 : H^2$$
$$= MD^2 : CD \,.\, EF$$
$$= MD : NF$$
$$= \triangle KDM : \triangle LFN$$

$$(\text{since } DK = FL).$$

Therefore $(\text{polygon about } B) = \triangle LFN$
$$= \square\, EL$$
$$= (\text{surface of prism about } A),$$

from above.

But $(\text{polygon about } B) : (\text{polygon in } B) < S : B$.

Therefore

(surface of prism about A) : (polygon in B) $< S : B$,

and, alternately,

(surface of prism about A) : $S <$ (polygon in B) : B;

which is impossible, since the surface of the prism is greater than S, while the polygon inscribed in B is less than B.

Therefore $B \not< S$.

II. Suppose $B > S$.

Let a regular polygon be circumscribed about B and another inscribed in it so that

(polygon about B) : (polygon in B) $< B : S$.

Inscribe in A a polygon similar to that inscribed in B, and erect a prism on the polygon inscribed in A of the same height as the cylinder.

Again, let DK, FL, drawn as before, be each equal to the perimeter of the polygon inscribed in A.

Then, in this case,

$$\triangle KDM > \text{(polygon inscribed in } A)$$

(since the perpendicular from the centre on a side of the polygon is less than the radius of A).

Also $\triangle LFN = \square EL = $ surface of prism (excluding bases).

Now

(polygon in A) : (polygon in B) $= MD^2 : H^2$,

$= \triangle KDM : \triangle LFN$, as before.

And $\triangle KDM > $ (polygon in A).

Therefore

$\triangle LFN$, or (surface of prism) $>$ (polygon in B).

But this is impossible, because

(polygon about B) : (polygon in B) $< B : S$,

$<$ (polygon about B) : S, a *fortiori*,

so that   (polygon in B) $> S$,

$>$ (surface of prism), a *fortiori*.

Hence B is neither greater nor less than S, and therefore

$$B = S.$$

Proposition 14.

The surface of any isosceles cone excluding the base is equal to a circle whose radius is a mean proportional between the side of the cone [a generator] and the radius of the circle which is the base of the cone.

Let the circle A be the base of the cone; draw C equal to the radius of the circle, and D equal to the side of the cone, and let E be a mean proportional between C, D.

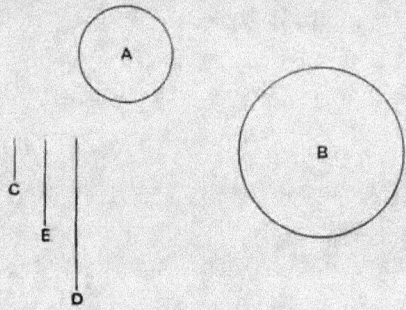

Draw a circle B with radius equal to E.

Then shall B be equal to the surface of the cone (excluding the base), which we will call S.

If not, B must be either greater or less than S.

I. Suppose $B < S$.

Let a regular polygon be described about B and a similar one inscribed in it such that the former has to the latter a ratio less than the ratio $S : B$.

Describe about A another similar polygon, and on it set up a pyramid with apex the same as that of the cone.

Then   (polygon about A) : (polygon about B)

$= C^2 : E^2$

$= C : D$

$=$ (polygon about A) : (surface of pyramid excluding base).

2—2

Therefore

(surface of pyramid) = (polygon about B).

Now   (polygon about B) : (polygon in B) < S : B.

Therefore

(surface of pyramid) : (polygon in B) < S : B,

which is impossible, (because the surface of the pyramid is greater than S, while the polygon in B is less than B).

Hence $B \not< S$.

II. Suppose $B > S$.

Take regular polygons circumscribed and inscribed to B such that the ratio of the former to the latter is less than the ratio $B : S$.

Inscribe in A a similar polygon to that inscribed in B, and erect a pyramid on the polygon inscribed in A with apex the same as that of the cone.

In this case

(polygon in A) : (polygon in B) = $C^2 : E^2$

$$= C : D$$

> (polygon in A) : (surface of pyramid excluding base).

This is clear because the ratio of C to D is greater than the ratio of the perpendicular from the centre of A on a side of the polygon to the perpendicular from the apex of the cone on the same side*.

Therefore

(surface of pyramid) > (polygon in B).

But   (polygon about B) : (polygon in B) < B : S.

Therefore, *a fortiori*,

(polygon about B) : (surface of pyramid) < B : S;

which is impossible.

Since therefore B is neither greater nor less than S,

$$B = S.$$

---

* This is of course the geometrical equivalent of saying that, if α, β be two angles each less than a right angle, and $\alpha > \beta$, then $\sin \alpha > \sin \beta$.

Proposition 15.

The surface of any isosceles cone has the same ratio to its base as the side of the cone has to the radius of the base.

By Prop. 14, the surface of the cone is equal to a circle whose radius is a mean proportional between the side of the cone and the radius of the base.

Hence, since circles are to one another as the squares of their radii, the proposition follows.

Proposition 16.

If an isosceles cone be cut by a plane parallel to the base, the portion of the surface of the cone between the parallel planes is equal to a circle whose radius is a mean proportional between (1) *the portion of the side of the cone intercepted by the parallel planes and* (2) *the line which is equal to the sum of the radii of the circles in the parallel planes.*

Let OAB be a triangle through the axis of a cone, DE its intersection with the plane cutting off the frustum, and OFC the axis of the cone.

Then the surface of the cone OAB is equal to a circle whose radius is equal to $\sqrt{OA \cdot AC}$.   [Prop. 14.]

Similarly the surface of the cone ODE is equal to a circle whose radius is equal to $\sqrt{OD \cdot DF}$.

And the surface of the frustum is equal to the difference between the two circles.

Now
$$OA \cdot AC - OD \cdot DF = DA \cdot AC + OD \cdot AC - OD \cdot DF.$$

But $OD \cdot AC = OA \cdot DF,$

since $OA : AC = OD : DF.$

Hence $OA \cdot AC - OD \cdot DF = DA \cdot AC + DA \cdot DF$
$$= DA \cdot (AC + DF).$$

And, since circles are to one another as the squares of their radii, it follows that the difference between the circles whose radii are $\sqrt{OA \cdot AC}$, $\sqrt{OD \cdot DF}$ respectively is equal to a circle whose radius is $\sqrt{DA \cdot (AC + DF)}$.

Therefore the surface of the frustum is equal to this circle.

Lemmas.

"1. *Cones having equal height have the same ratio as their bases; and those having equal bases have the same ratio as their heights* *.

2. *If a cylinder be cut by a plane parallel to the base, then, as the cylinder is to the cylinder, so is the axis to the axis †.*

3. *The cones which have the same bases as the cylinders* [and equal height] *are in the same ratio as the cylinders.*

4. *Also the bases of equal cones are reciprocally proportional to their heights; and those cones whose bases are reciprocally proportional to their heights are equal ‡.*

5. *Also the cones, the diameters of whose bases have the same ratio as their axes, are to one another in the triplicate ratio of the diameters of the bases §.*

And all these propositions have been proved by earlier geometers."

---

* Euclid xii. 11. "Cones and cylinders of equal height are to one another as their bases."

Euclid xii. 14. "Cones and cylinders on equal bases are to one another as their heights."

† Euclid xii. 13. "If a cylinder be cut by a plane parallel to the opposite planes [the bases], then, as the cylinder is to the cylinder, so will the axis be to the axis."

‡ Euclid xii. 15. "The bases of equal cones and cylinders are reciprocally proportional to their heights; and those cones and cylinders whose bases are reciprocally proportional to their heights are equal."

§ Euclid xii. 12. "Similar cones and cylinders are to one another in the triplicate ratio of the diameters of their bases."

Proposition 17.

If there be two isosceles cones, and the surface of one cone be equal to the base of the other, while the perpendicular from the centre of the base [of the first cone] on the side of that cone is equal to the height [of the second], the cones will be equal.

Let OAB, DEF be triangles through the axes of two cones respectively, C, G the centres of the respective bases, GH the

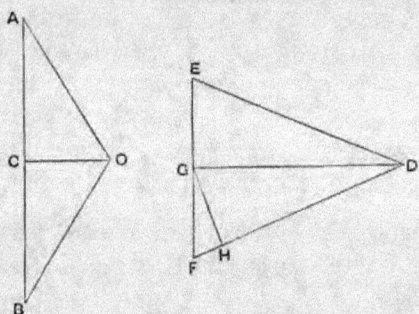

perpendicular from G on FD; and suppose that the base of the cone OAB is equal to the surface of the cone DEF, and that $OC = GH$.

Then, since the base of OAB is equal to the surface of DEF,

(base of cone OAB) : (base of cone DEF)

$$= \text{(surface of } DEF) : \text{(base of } DEF)$$
$$= DF : FG \qquad\qquad \text{[Prop. 15]}$$
$$= DG : GH, \text{ by similar triangles,}$$
$$= DG : OC.$$

Therefore the bases of the cones are reciprocally proportional to their heights; whence the cones are equal. [*Lemma* 4.]

Proposition 18.

Any solid rhombus consisting of isosceles cones is equal to the cone which has its base equal to the surface of one of the cones composing the rhombus and its height equal to the perpendicular drawn from the apex of the second cone to one side of the first cone.

Let the rhombus be $OABD$ consisting of two cones with apices O, D and with a common base (the circle about AB as diameter).

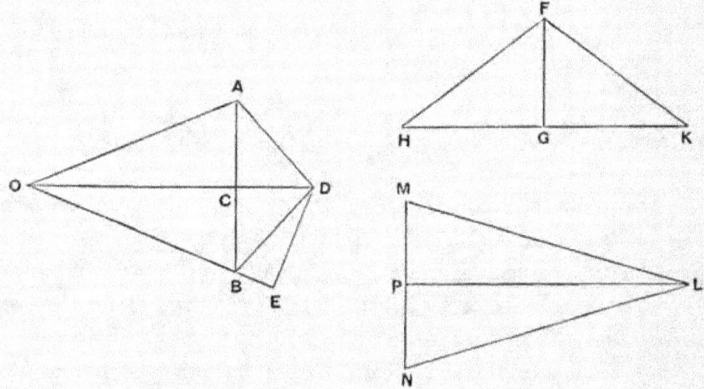

Let FHK be another cone with base equal to the surface of the cone OAB and height FG equal to DE, the perpendicular from D on OB.

Then shall the cone FHK be equal to the rhombus.

Construct a third cone LMN with base (the circle about MN) equal to the base of OAB and height LP equal to OD.

Then, since $$LP = OD,$$
$$LP : CD = OD : CD.$$

But [*Lemma* 1] $OD : CD = $ (rhombus $OADB$) : (cone DAB),

and $$LP : CD = (\text{cone } LMN) : (\text{cone } DAB).$$

It follows that

$$(\text{rhombus } OADB) = (\text{cone } LMN) \dots\dots\dots\dots\dots (1).$$

Again, since $AB = MN$, and

$$\text{(surface of } OAB) = \text{(base of } FHK),$$

(base of FHK) : (base of LMN)

$$= \text{(surface of } OAB) : \text{(base of } OAB)$$
$$= OB : BC \qquad\qquad \text{[Prop. 15]}$$
$$= OD : DE, \text{ by similar triangles,}$$
$$= LP : FG, \text{ by hypothesis.}$$

Thus, in the cones FHK, LMN, the bases are reciprocally proportional to the heights.

Therefore the cones FHK, LMN are equal,

and hence, by (1), the cone FHK is equal to the given solid rhombus.

Proposition 19.

If an isosceles cone be cut by a plane parallel to the base, and on the resulting circular section a cone be described having as its apex the centre of the base [of the first cone], and if the rhombus so formed be taken away from the whole cone, the part remaining will be equal to the cone with base equal to the surface of the portion of the first cone between the parallel planes and with height equal to the perpendicular drawn from the centre of the base of the first cone on one side of that cone.

Let the cone OAB be cut by a plane parallel to the base in the circle on DE as diameter. Let C be the centre of the base of the cone, and with C as apex and the circle about DE as base describe a cone, making with the cone ODE the rhombus $ODCE$.

Take a cone FGH with base equal to the surface of the frustum $DABE$ and height equal to the perpendicular (CK) from C on AO.

Then shall the cone FGH be equal to the difference between the cone OAB and the rhombus $ODCE$.

Take (1) a cone LMN with base equal to the surface of the cone OAB, and height equal to CK,

(2) a cone PQR with base equal to the surface of the cone ODE and height equal to CK.

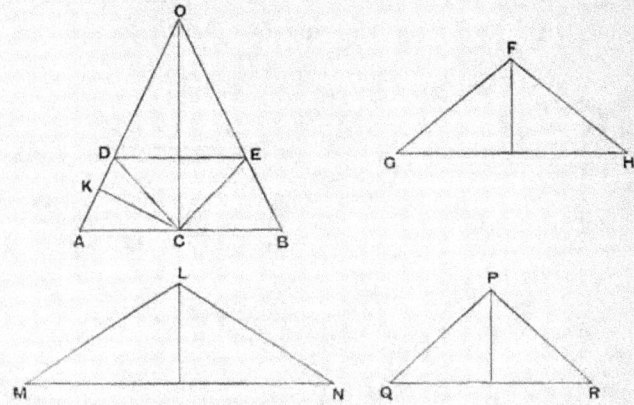

Now, since the surface of the cone OAB is equal to the surface of the cone ODE together with that of the frustum $DABE$, we have, by the construction,

(base of LMN) = (base of FGH) + (base of PQR)

and, since the heights of the three cones are equal,

(cone LMN) = (cone FGH) + (cone PQR).

But the cone LMN is equal to the cone OAB [Prop. 17], and the cone PQR is equal to the rhombus $ODCE$ [Prop. 18].

Therefore (cone OAB) = (cone FGH) + (rhombus $ODCE$), and the proposition is proved.

Proposition 20.

*If one of the two isosceles cones forming a rhombus be cut by a plane parallel to the base and on the resulting circular section a cone be described having the same apex as the second cone, and if the resulting rhombus be taken from the whole rhombus, the remainder will be equal to the cone with base equal to the surface of the portion of the cone between the parallel planes and with height equal to the perpendicular drawn from the apex of the second * cone to the side of the first cone.*

---

* There is a slight error in Heiberg's translation " prioris coni " and in the corresponding note, p. 93. The perpendicular is not drawn from the apex of the cone which is cut by the plane but from the apex of the other.

Let the rhombus be $OACB$, and let the cone OAB be cut by a plane parallel to its base in the circle about DE as diameter. With this circle as base and C as apex describe a cone, which therefore with ODE forms the rhombus $ODCE$.

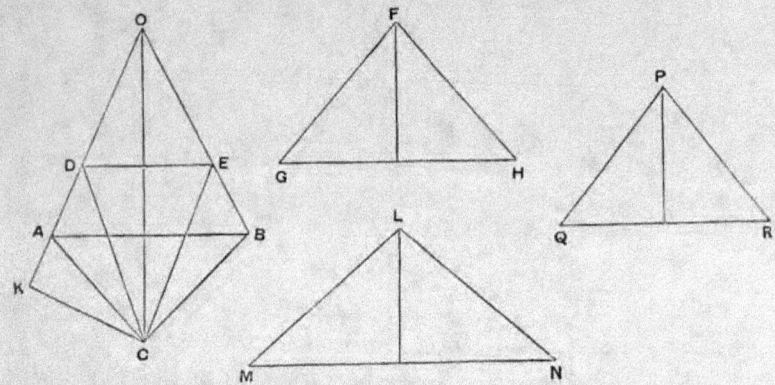

Take a cone FGH with base equal to the surface of the frustum $DABE$ and height equal to the perpendicular (CK) from C on OA.

The cone FGH shall be equal to the difference between the rhombi $OACB$, $ODCE$.

For take (1) a cone LMN with base equal to the surface of OAB and height equal to CK,

(2) a cone PQR, with base equal to the surface of ODE, and height equal to CK.

Then, since the surface of OAB is equal to the surface of ODE together with that of the frustum $DABE$, we have, by construction,

(base of LMN) = (base of PQR) + (base of FGH),

and the three cones are of equal height;

therefore   (cone LMN) = (cone PQR) + (cone FGH).

But the cone LMN is equal to the rhombus $OACB$, and the cone PQR is equal to the rhombus $ODCE$ [Prop. 18].

Hence the cone FGH is equal to the difference between the two rhombi $OACB$, $ODCE$.

Proposition 21.

A regular polygon of an even number of sides being inscribed in a circle, as $ABC...A'...C'B'A$, so that AA' is a diameter, if two angular points next but one to each other, as B, B', be joined, and the other lines parallel to BB' and joining pairs of angular points be drawn, as CC', $DD'...$, then

$$(BB' + CC' + ...) : AA' = A'B : BA.$$

Let BB', CC', DD',... meet AA' in F, G, H,...; and let CB', DC',... be joined meeting AA' in K, L,... respectively.

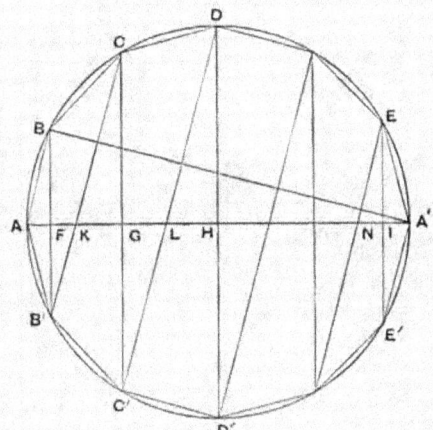

Then clearly CB', DC',... are parallel to one another and to AB.

Hence, by similar triangles,

$$BF : FA = B'F : FK$$
$$= CG : GK$$
$$= C'G : GL$$
$$\cdots\cdots\cdots$$
$$= E'I : IA';$$

and, summing the antecedents and consequents respectively, we have

$$(BB' + CC' + ...) : AA' = BF : FA$$
$$= A'B : BA.$$

Proposition 22.

If a polygon be inscribed in a segment of a circle LAL' so that all its sides excluding the base are equal and their number even, as $LK...A...K'L'$, A being the middle point of the segment, and if the lines BB', CC',... parallel to the base LL' and joining pairs of angular points be drawn, then

$$(BB' + CC' + ... + LM) : AM = A'B : BA,$$

where M is the middle point of LL' and AA' is the diameter through M.

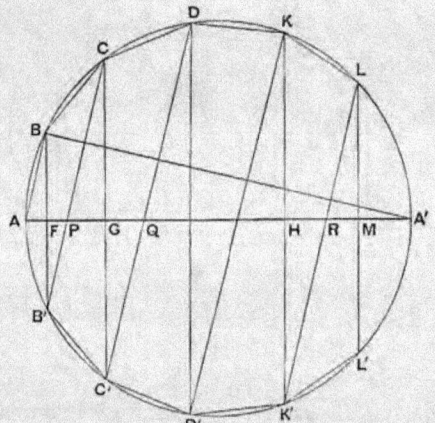

Joining CB', DC',...LK', as in the last proposition, and supposing that they meet AM in P, Q,...R, while BB', CC',..., KK' meet AM in F, G,... H, we have, by similar triangles,

$$BF : FA = B'F : FP$$
$$= CG : PG$$
$$= C'G : GQ$$
$$............$$
$$= LM : RM;$$

and, summing the antecedents and consequents, we obtain

$$(BB' + CC' + \ldots + LM) : AM = BF : FA$$
$$= A'B : BA.$$

Proposition 23.

Take a great circle $ABC\ldots$ of a sphere, and inscribe in it a regular polygon whose sides are a multiple of four in number. Let AA', MM' be diameters at right angles and joining opposite angular points of the polygon.

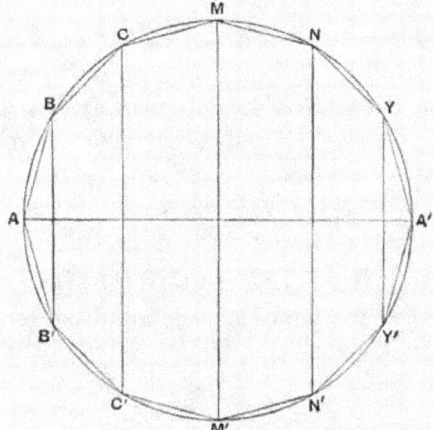

Then, if the polygon and great circle revolve together about the diameter AA', the angular points of the polygon, except A, A', will describe circles on the surface of the sphere at right angles to the diameter AA'. Also the sides of the polygon will describe portions of conical surfaces, e.g. BC will describe a surface forming part of a cone whose base is a circle about CC' as diameter and whose apex is the point in which CB, $C'B'$ produced meet each other and the diameter AA'.

Comparing the hemisphere MAM' and that half of the figure described by the revolution of the polygon which is included in the hemisphere, we see that the surface of the hemisphere and the surface of the inscribed figure have the same boundaries in one plane (viz. the circle on MM' as diameter), the former surface entirely includes the latter, and they are both concave in the same direction.

Therefore [*Assumptions*, 4] the surface of the hemisphere is greater than that of the inscribed figure; and the same is true of the other halves of the figures.

Hence *the surface of the sphere is greater than the surface described by the revolution of the polygon inscribed in the great circle about the diameter of the great circle.*

Proposition 24.

If a regular polygon $AB...A'...B'A$, the number of whose sides is a multiple of four, be inscribed in a great circle of a sphere, and if BB' subtending two sides be joined, and all the other lines parallel to BB' and joining pairs of angular points be drawn, then the surface of the figure inscribed in the sphere by the revolution of the polygon about the diameter AA' is equal to a circle the square of whose radius is equal to the rectangle

$$BA \, (BB' + CC' + ...).$$

The surface of the figure is made up of the surfaces of parts of different cones.

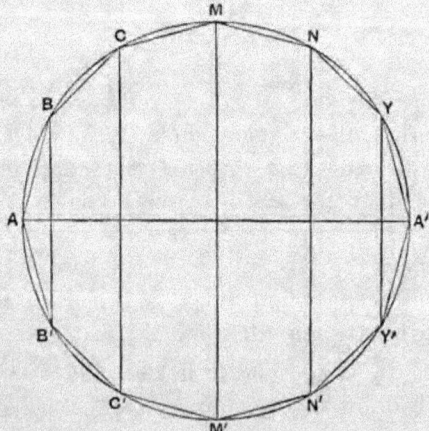

Now the surface of the cone ABB' is equal to a circle whose radius is $\sqrt{BA \cdot \frac{1}{2}BB'}$.　　　　　　　　　　　　　[Prop. 14]

The surface of the frustum $BB'C'C$ is equal to a circle of
radius $\sqrt{BC \cdot \frac{1}{2}(BB' + CC')}$,   [Prop. 16]
and so on.

It follows, since $BA = BC = \ldots$, that the whole surface is
equal to a circle whose radius is equal to

$$\sqrt{BA\,(BB' + CC' + \ldots + MM' + \ldots + Y Y')}.$$

Proposition 25.

*The surface of the figure inscribed in a sphere as in the last
propositions, consisting of portions of conical surfaces, is less than
four times the greatest circle in the sphere.*

Let $AB\ldots A'\ldots B'A$ be a regular polygon inscribed in a
great circle, the number of its sides being a multiple of four.

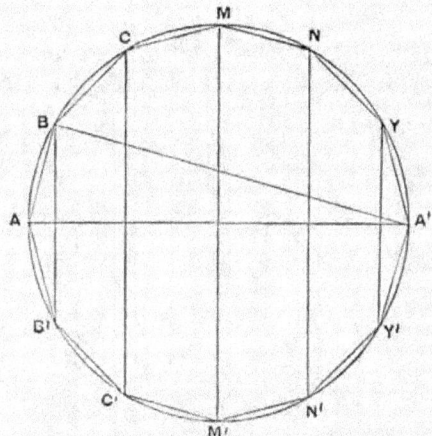

As before, let BB' be drawn subtending two sides, and
$CC',\ldots YY'$ parallel to BB'.

Let R be a circle such that the square of its radius is equal
to

$$AB\,(BB' + CC' + \ldots + YY'),$$

so that the surface of the figure inscribed in the sphere is equal
to R.   [Prop. 24]

Now

$$(BB' + CC' + \ldots + YY') : AA' = A'B : AB, \quad [\text{Prop. 21}]$$

whence $\quad AB (BB' + CC' + \ldots + YY') = AA' \cdot A'B.$

Hence $\qquad (\text{radius of } R)^2 = AA' \cdot A'B$

$$< AA'^2.$$

Therefore the surface of the inscribed figure, or the circle R, is less than four times the circle $AMA'M'$.

Proposition 26.

The figure inscribed as above in a sphere is equal [in volume] to a cone whose base is a circle equal to the surface of the figure inscribed in the sphere and whose height is equal to the perpendicular drawn from the centre of the sphere to one side of the polygon.

Suppose, as before, that $AB...A'...B'A$ is the regular polygon inscribed in a great circle, and let BB', CC', ... be joined.

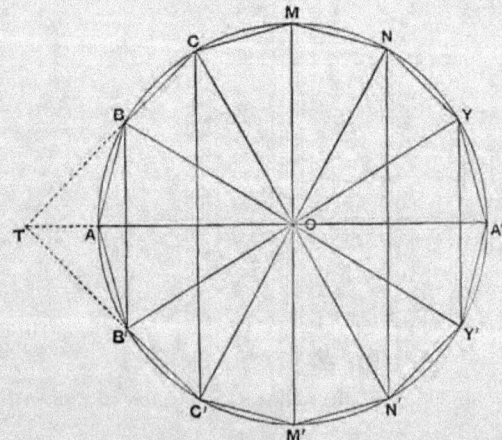

With apex O construct cones whose bases are the circles on BB', CC', ... as diameters in planes perpendicular to AA'.

Then $OBAB'$ is a solid rhombus, and its volume is equal to a cone whose base is equal to the surface of the cone ABB' and whose height is equal to the perpendicular from O on AB [Prop. 18]. Let the length of the perpendicular be p.

Again, if CB, $C'B'$ produced meet in T, the portion of the solid figure which is described by the revolution of the triangle BOC about AA' is equal to the difference between the rhombi $OCTC'$ and $OBTB'$, i.e. to a cone whose base is equal to the surface of the frustum $BB'C'C$ and whose height is p [Prop. 20].

Proceeding in this manner, and adding, we prove that, since cones of equal height are to one another as their bases, the volume of the solid of revolution is equal to a cone with height p and base equal to the sum of the surfaces of the cone BAB', the frustum $BB'C'C$, etc., i.e. a cone with height p and base equal to the surface of the solid.

Proposition 27.

The figure inscribed in the sphere as before is less than four times the cone whose base is equal to a great circle of the sphere and whose height is equal to the radius of the sphere.

By Prop. 26 the volume of the solid figure is equal to a cone whose base is equal to the surface of the solid and whose height is p, the perpendicular from O on any side of the polygon. Let R be such a cone.

Take also a cone S with base equal to the great circle, and height equal to the radius, of the sphere.

Now, since the surface of the inscribed solid is less than four times the great circle [Prop. 25], the base of the cone R is less than four times the base of the cone S.

Also the height (p) of R is less than the height of S.

Therefore the volume of R is less than four times that of S; and the proposition is proved.

Proposition 28.

Let a regular polygon, whose sides are a multiple of four in number, be circumscribed about a great circle of a given sphere, as $AB...A'...B'A$; and about the polygon describe another circle, which will therefore have the same centre as the great circle of the sphere. Let AA' bisect the polygon and cut the sphere in a, a'.

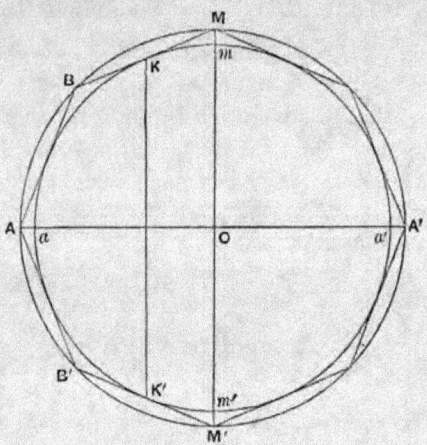

If the great circle and the circumscribed polygon revolve together about AA', the great circle will describe the surface of a sphere, the angular points of the polygon except A, A' will move round the surface of a larger sphere, the points of contact of the sides of the polygon with the great circle of the inner sphere will describe circles on that sphere in planes perpendicular to AA', and the sides of the polygon themselves will describe portions of conical surfaces. *The circumscribed figure will thus be greater than the sphere itself.*

Let any side, as BM, touch the inner circle in K, and let K' be the point of contact of the circle with $B'M'$.

Then the circle described by the revolution of KK' about AA' is the boundary in one plane of two surfaces

(1) the surface formed by the revolution of the circular segment KaK', and

3—2

(2) the surface formed by the revolution of the part
$KB...A...B'K'$ of the polygon.

Now the second surface entirely includes the first, and they
are both concave in the same direction;

therefore [*Assumptions,* 4] the second surface is greater
than the first.

The same is true of the portion of the surface on the opposite
side of the circle on KK' as diameter.

Hence, adding, we see that *the surface of the figure
circumscribed to the given sphere is greater than that of the
sphere itself.*

Proposition 29.

*In a figure circumscribed to a sphere in the manner shown
in the previous proposition the surface is equal to a circle the
square on whose radius is equal to* $AB(BB' + CC' + ...)$.

For the figure circumscribed to the sphere is inscribed in a
larger sphere, and the proof of Prop. 24 applies.

Proposition 30.

*The surface of a figure circumscribed as before about a sphere
is greater than four times the great circle of the sphere.*

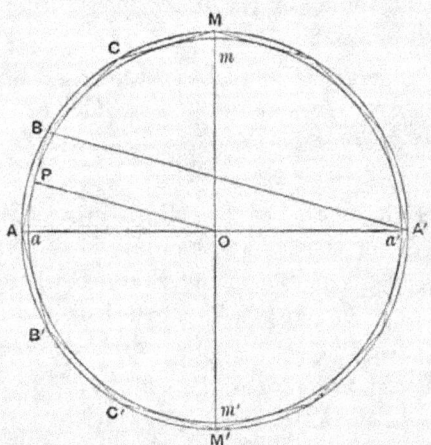

Let $AB...A'...B'A$ be the regular polygon of $4n$ sides which by its revolution about AA' describes the figure circumscribing the sphere of which $ama'm'$ is a great circle. Suppose aa', AA' to be in one straight line.

Let R be a circle equal to the surface of the circumscribed solid.

Now $(BB' + CC' + ...) : AA' = A'B : BA$, [as in Prop. 21]

so that $AB(BB' + CC' + ...) = AA' . A'B$.

Hence (radius of R) $= \sqrt{AA' . A'B}$ [Prop. 29]

$$> A'B.$$

But $A'B = 2OP$, where P is the point in which AB touches the circle $ama'm'$.

Therefore (radius of R) > (diameter of circle $ama'm'$);

whence R, and therefore the surface of the circumscribed solid, is greater than four times the great circle of the given sphere.

Proposition 31.

The solid of revolution circumscribed as before about a sphere is equal to a cone whose base is equal to the surface of the solid and whose height is equal to the radius of the sphere.

The solid is, as before, a solid inscribed in a larger sphere; and, since the perpendicular on any side of the revolving polygon is equal to the radius of the inner sphere, the proposition is identical with Prop. 26.

COR. *The solid circumscribed about the smaller sphere is greater than four times the cone whose base is a great circle of the sphere and whose height is equal to the radius of the sphere.*

For, since the surface of the solid is greater than four times the great circle of the inner sphere [Prop. 30], the cone whose base is equal to the surface of the solid and whose height is the radius of the sphere is greater than four times the cone of the same height which has the great circle for base. [*Lemma* 1.]

Hence, by the proposition, the volume of the solid is greater than four times the latter cone.

Proposition 32.

If a regular polygon with 4n sides be inscribed in a great circle of a sphere, as ab...a'...b'a, and a similar polygon AB...A'...B'A be described about the great circle, and if the polygons revolve with the great circle about the diameters aa', AA' respectively, so that they describe the surfaces of solid figures inscribed in and circumscribed to the sphere respectively, then

(1) *the surfaces of the circumscribed and inscribed figures are to one another in the duplicate ratio of their sides, and*

(2) *the figures themselves [i.e. their volumes] are in the triplicate ratio of their sides.*

(1) Let AA', aa' be in the same straight line, and let $MmOm'M'$ be a diameter at right angles to them.

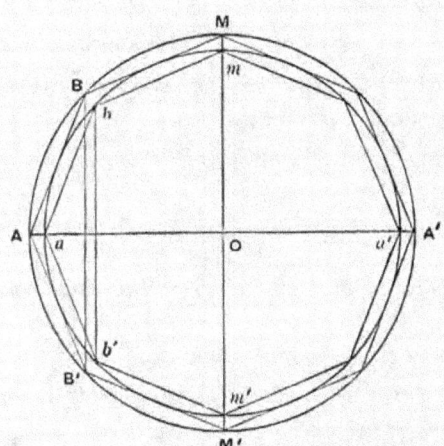

Join BB', CC', ... and bb', cc', ... which will all be parallel to one another and MM'.

Suppose R, S to be circles such that

$$R = \text{(surface of circumscribed solid)},$$
$$S = \text{(surface of inscribed solid)}.$$

Then   (radius of $R)^2 = AB\,(BB' + CC' + ...)$ [Prop. 29]

(radius of $S)^2 = ab\,(bb' + cc' + ...).$ [Prop. 24]

And, since the polygons are similar, the rectangles in these two equations are similar, and are therefore in the ratio of

$$AB^2 : ab^2.$$

Hence

(surface of circumscribed solid) : (surface of inscribed solid)

$$= AB^2 : ab^2.$$

(2)   Take a cone V whose base is the circle R and whose height is equal to Oa, and a cone W whose base is the circle S and whose height is equal to the perpendicular from O on ab, which we will call p.

Then V, W are respectively equal to the volumes of the circumscribed and inscribed figures.   [Props. 31, 26]

Now, since the polygons are similar,

$AB : ab = Oa : p$

     $= $ (height of cone V) : (height of cone W);

and, as shown above, the bases of the cones (the circles R, S) are in the ratio of AB^2 to ab^2.

Therefore $V : W = AB^3 : ab^3.$

Proposition 33.

The surface of any sphere is equal to four times the greatest circle in it.

Let C be a circle equal to four times the great circle.

Then, if C is not equal to the surface of the sphere, it must either be less or greater.

I.   Suppose C less than the surface of the sphere.

It is then possible to find two lines β, γ, of which β is the greater, such that

$$\beta : \gamma < (\text{surface of sphere}) : C. \qquad [\text{Prop. 2}]$$

Take such lines, and let δ be a mean proportional between them.

Suppose similar regular polygons with $4n$ sides circumscribed about and inscribed in a great circle such that the ratio of their sides is less than the ratio $\beta : \delta$.   [Prop. 3]

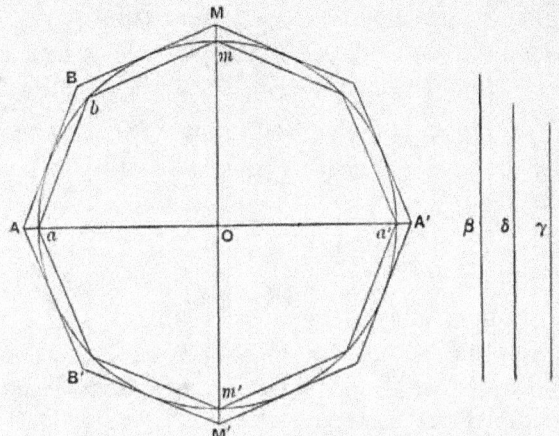

Let the polygons with the circle revolve together about a diameter common to all, describing solids of revolution as before.

Then (surface of outer solid) : (surface of inner solid)

$$= \text{(side of outer)}^2 : \text{(side of inner)}^2 \quad \text{[Prop. 32]}$$
$$< \beta^2 : \delta^2, \text{ or } \beta : \gamma$$
$$< \text{(surface of sphere)} : C, \text{ a fortiori.}$$

But this is impossible, since the surface of the circumscribed solid is greater than that of the sphere [Prop. 28], while the surface of the inscribed solid is less than C [Prop. 25].

Therefore C is not less than the surface of the sphere.

II.   Suppose C greater than the surface of the sphere.

Take lines β, γ, of which β is the greater, such that

$$\beta : \gamma < C : \text{(surface of sphere)}.$$

Circumscribe and inscribe to the great circle similar regular polygons, as before, such that their sides are in a ratio less than that of β to δ, and suppose solids of revolution generated in the usual manner.

Then, in this case,

(surface of circumscribed solid) : (surface of inscribed solid)

$< C$: (surface of sphere).

But this is impossible, because the surface of the circumscribed solid is greater than C [Prop. 30], while the surface of the inscribed solid is less than that of the sphere [Prop. 23].

Thus C is not greater than the surface of the sphere.

Therefore, since it is neither greater nor less, C is equal to the surface of the sphere.

Proposition 34.

Any sphere is equal to four times the cone which has its base equal to the greatest circle in the sphere and its height equal to the radius of the sphere.

Let the sphere be that of which $ama'm'$ is a great circle.

If now the sphere is not equal to four times the cone described, it is either greater or less.

I.   If possible, let the sphere be greater than four times the cone.

Suppose V to be a cone whose base is equal to four times the great circle and whose height is equal to the radius of the sphere.

Then, by hypothesis, the sphere is greater than V; and two lines β, γ can be found (of which β is the greater) such that

$\beta : \gamma <$ (volume of sphere) : V.

Between β and γ place two arithmetic means δ, ϵ.

As before, let similar regular polygons with sides $4n$ in number be circumscribed about and inscribed in the great circle, such that their sides are in a ratio less than $\beta : \delta$.

Imagine the diameter aa' of the circle to be in the same straight line with a diameter of both polygons, and imagine the latter to revolve with the circle about aa', describing the surfaces of two solids of revolution. The volumes of these solids are therefore in the triplicate ratio of their sides.   [Prop. 32]

Thus (vol. of outer solid) : (vol. of inscribed solid)

$$< \beta^3 : \delta^3, \text{ by hypothesis,}$$

$$< \beta : \gamma, \text{ } a \text{ } fortiori \text{ (since } \beta : \gamma > \beta^3 : \delta^3)^*,$$

$$< \text{(volume of sphere)} : V, \text{ } a \text{ } fortiori.$$

But this is impossible, since the volume of the circumscribed

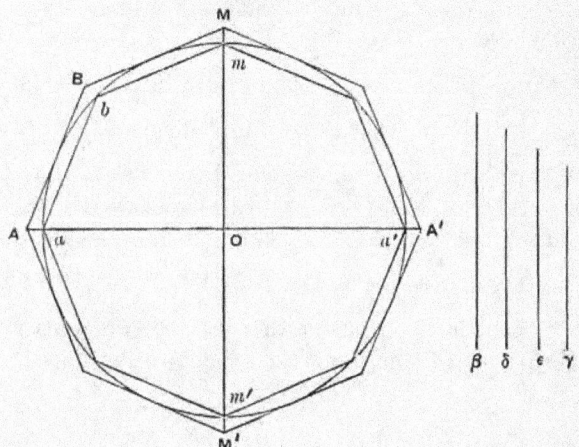

* That $\beta : \gamma > \beta^3 : \delta^3$ is assumed by Archimedes. Eutocius proves the property in his commentary as follows.

Take x such that $\beta : \delta = \delta : x.$

Thus $\beta - \delta : \beta = \delta - x : \delta$

and, since $\beta > \delta$, $\beta - \delta > \delta - x$.

But, by hypothesis, $\beta - \delta = \delta - \epsilon.$

Therefore $\delta - \epsilon > \delta - x,$

or $x > \epsilon.$

Again, suppose $\delta : x = x : y,$

and, as before, we have $\delta - x > x - y,$

so that, a fortiori, $\delta - \epsilon > x - y.$

Therefore $\epsilon - \gamma > x - y;$

and, since $x > \epsilon$, $y > \gamma$.

Now, by hypothesis, β, δ, x, y are in continued proportion; therefore $\beta^3 : \delta^3 = \beta : y$

$$< \beta : \gamma.$$

solid is greater than that of the sphere [Prop. 28], while the volume of the inscribed solid is less than V [Prop. 27].

Hence the sphere is not greater than V, or four times the cone described in the enunciation.

II.   If possible, let the sphere be less than V.

In this case we take β, γ (β being the greater) such that

$$\beta : \gamma < V : (\text{volume of sphere}).$$

The rest of the construction and proof proceeding as before, we have finally

$$(\text{volume of outer solid}) : (\text{volume of inscribed solid})$$

$$< V : (\text{volume of sphere}).$$

But this is impossible, because the volume of the outer solid is greater than V [Prop. 31, Cor.], and the volume of the inscribed solid is less than the volume of the sphere.

Hence the sphere is not less than V.

Since then the sphere is neither less nor greater than V, it is equal to V, or to four times the cone described in the enunciation.

COR.   From what has been proved it follows that *every cylinder whose base is the greatest circle in a sphere and whose height is equal to the diameter of the sphere is* $\frac{3}{2}$ *of the sphere, and its surface together with its bases is* $\frac{3}{2}$ *of the surface of the sphere.*

For the cylinder is three times the cone with the same base and height [Eucl. XII. 10], i.e. six times the cone with the same base and with height equal to the radius of the sphere.

But the sphere is four times the latter cone [Prop. 34]. Therefore the cylinder is $\frac{3}{2}$ of the sphere.

Again, the surface of a cylinder (excluding the bases) is equal to a circle whose radius is a mean proportional between the height of the cylinder and the diameter of its base [Prop. 13].

In this case the height is equal to the diameter of the base and therefore the circle is that whose radius is the diameter of the sphere, or a circle equal to four times the great circle of the sphere.

Therefore the surface of the cylinder with the bases is equal to six times the great circle.

And the surface of the sphere is four times the great circle [Prop. 33]; whence

(surface of cylinder with bases) = $\frac{3}{2}$. (surface of sphere).

Proposition 35.

If in a segment of a circle LAL' (where A is the middle point of the arc) a polygon LK...A...K'L' be inscribed of which LL' is one side, while the other sides are 2n in number and all equal, and if the polygon revolve with the segment about the diameter AM, generating a solid figure inscribed in a segment of a sphere, then the surface of the inscribed solid is equal to a circle the square on whose radius is equal to the rectangle

$$AB\left(BB' + CC' + \dots + KK' + \frac{LL'}{2}\right).$$

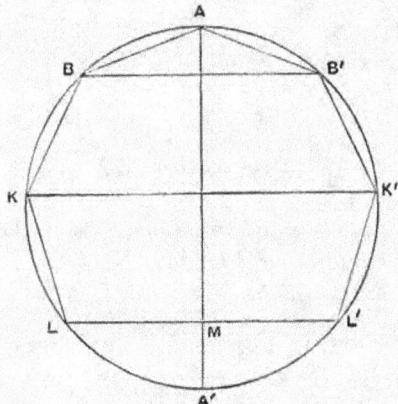

The surface of the inscribed figure is made up of portions of surfaces of cones.

If we take these successively, the surface of the cone BAB' is equal to a circle whose radius is

$$\sqrt{AB \cdot \tfrac{1}{2}BB'}. \qquad\qquad \text{[Prop. 14]}$$

The surface of the frustum of a cone $BCC'B'$ is equal to a circle whose radius is

$$\sqrt{AB \cdot \frac{BB' + CC'}{2}}; \qquad\qquad \text{[Prop. 16]}$$

and so on.

Proceeding in this way and adding, we find, since circles are to one another as the squares of their radii, that the surface of the inscribed figure is equal to a circle whose radius is

$$\sqrt{AB\left(BB' + CC' + \ldots + KK' + \frac{LL'}{2}\right)}.$$

Proposition 36.

The surface of the figure inscribed as before in the segment of a sphere is less than that of the segment of the sphere.

This is clear, because the circular base of the segment is a common boundary of each of two surfaces, of which one, the segment, includes the other, the solid, while both are concave in the same direction [*Assumptions*, 4].

Proposition 37.

The surface of the solid figure inscribed in the segment of the sphere by the revolution of $LK \ldots A \ldots K'L'$ about AM is less than a circle with radius equal to AL.

Let the diameter AM meet the circle of which LAL' is a segment again in A'. Join $A'B$.

As in Prop. 35, the surface of the inscribed solid is equal to a circle the square on whose radius is

$$AB(BB' + CC' + \ldots + KK' + LM).$$

But this rectangle $= A'B . AM$ [Prop. 22]

$< A'A . AM$

$< AL^2.$

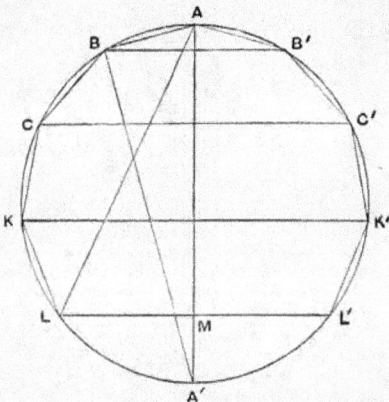

Hence the surface of the inscribed solid is less than the circle whose radius is AL.

Proposition 38.

The solid figure described as before in a segment of a sphere less than a hemisphere, together with the cone whose base is the base of the segment and whose apex is the centre of the sphere, is equal to a cone whose base is equal to the surface of the inscribed solid and whose height is equal to the perpendicular from the centre of the sphere on any side of the polygon.

Let O be the centre of the sphere, and p the length of the perpendicular from O on AB.

Suppose cones described with O as apex, and with the circles on BB', CC',... as diameters as bases.

Then the rhombus $OBAB'$ is equal to a cone whose base is equal to the surface of the cone BAB', and whose height is p.

[Prop. 18]

Again, if CB, $C'B'$ meet in T, the solid described by the triangle BOC as the polygon revolves about AO is the difference between the rhombi $OCTC'$ and $OBTB'$, and is therefore equal to a cone whose base is equal to the surface of the frustum $BCC'B'$ and whose height is p. [Prop. 20]

Similarly for the part of the solid described by the triangle COD as the polygon revolves; and so on.

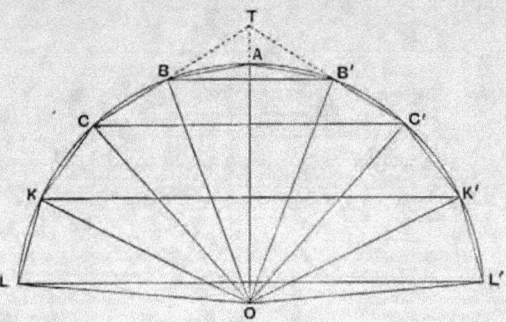

Hence, by addition, the solid figure inscribed in the segment together with the cone OLL' is equal to a cone whose base is the surface of the inscribed solid and whose height is p.

COR. *The cone whose base is a circle with radius equal to AL and whose height is equal to the radius of the sphere is greater than the sum of the inscribed solid and the cone OLL'.*

For, by the proposition, the inscribed solid together with the cone OLL' is equal to a cone with base equal to the surface of the solid and with height p.

This latter cone is less than a cone with height equal to OA and with base equal to the circle whose radius is AL, because the height p is less than OA, while the surface of the solid is less than a circle with radius AL. [Prop. 37]

Proposition 39.

Let lal' be a segment of a great circle of a sphere, being less than a semicircle. Let O be the centre of the sphere, and join Ol, Ol'. Suppose a polygon circumscribed about the sector $Olal'$ such that its sides, excluding the two radii, are $2n$ in number

and all equal, as $LK, \ldots BA, AB', \ldots K'L'$; and let OA be that
radius of the great circle which bisects the segment lal'.

The circle circumscribing the polygon will then have the
same centre O as the given great circle.

Now suppose the polygon and the two circles to revolve
together about OA. The two circles will describe spheres, the

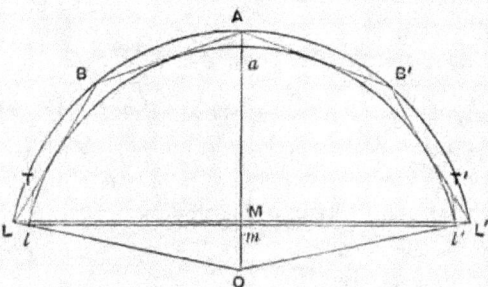

angular points except A will describe circles on the outer
sphere, with diameters BB' etc., the points of contact of the
sides with the inner segment will describe circles on the inner
sphere, the sides themselves will describe the surfaces of cones
or frusta of cones, and the whole figure circumscribed to the
segment of the inner sphere by the revolution of the equal
sides of the polygon will have for its base the circle on LL'
as diameter.

*The surface of the solid figure so circumscribed about the
sector of the sphere [excluding its base] will be greater than that
of the segment of the sphere whose base is the circle on ll' as
diameter.*

For draw the tangents lT, $l'T''$ to the inner segment at l, l'.
These with the sides of the polygon will describe by their
revolution a solid whose surface is greater than that of the
segment [*Assumptions*, 4].

But the surface described by the revolution of lT is less
than that described by the revolution of LT, since the angle TlL
is a right angle, and therefore $LT > lT$.

Hence, *a fortiori*, the surface described by $LK \ldots A \ldots K'L'$
is greater than that of the segment.

COR. *The surface of the figure so described about the sector of the sphere is equal to a circle the square on whose radius is equal to the rectangle*

$$AB(BB' + CC' + \ldots + KK' + \tfrac{1}{2}LL').$$

For the circumscribed figure is inscribed in the outer sphere, and the proof of Prop. 35 therefore applies.

Proposition 40.

The surface of the figure circumscribed to the sector as before is greater than a circle whose radius is equal to al.

Let the diameter AaO meet the great circle and the circle circumscribing the revolving polygon again in a', A'. Join $A'B$, and let ON be drawn to N, the point of contact of AB with the inner circle.

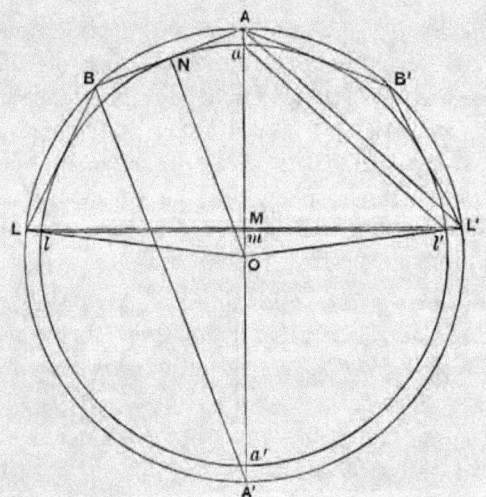

Now, by Prop. 39, Cor., the surface of the solid figure circumscribed to the sector $OlAl'$ is equal to a circle the square on whose radius is equal to the rectangle

$$AB\left(BB' + CC' + \ldots + KK' + \frac{LL'}{2}\right).$$

But this rectangle is equal to $A'B \cdot AM$ [as in Prop. 22].

Next, since AL', al' are parallel, the triangles AML', aml' are similar. And $AL' > al'$; therefore $AM > am$.

Also $\qquad\qquad A'B = 2ON = aa'$.

Therefore $\qquad\quad A'B . AM > am . aa'$

$$> al'^2.$$

Hence the surface of the solid figure circumscribed to the sector is greater than a circle whose radius is equal to al', or al.

COR. 1. *The volume of the figure circumscribed about the sector together with the cone whose apex is O and base the circle on LL' as diameter, is equal to the volume of a cone whose base is equal to the surface of the circumscribed figure and whose height is ON.*

For the figure is inscribed in the outer sphere which has the same centre as the inner. Hence the proof of Prop. 38 applies.

COR. 2. *The volume of the circumscribed figure with the cone OLL' is greater than the cone whose base is a circle with radius equal to al and whose height is equal to the radius (Oa) of the inner sphere.*

For the volume of the figure with the cone OLL' is equal to a cone whose base is equal to the surface of the figure and whose height is equal to ON.

And the surface of the figure is greater than a circle with radius equal to al [Prop. 40], while the heights Oa, ON are equal.

Proposition 41.

Let lal' be a segment of a great circle of a sphere which is less than a semicircle.

Suppose a polygon inscribed in the sector $Olal'$ such that the sides $lk, \ldots ba$, $ab', \ldots k'l'$ are $2n$ in number and all equal. Let a similar polygon be circumscribed about the sector so that its sides are parallel to those of the first polygon; and draw the circle circumscribing the outer polygon.

Now let the polygons and circles revolve together about OaA, the radius bisecting the segment lal'.

Then (1) *the surfaces of the outer and inner solids of revolution so described are in the ratio of AB^2 to ab^2, and (2) their volumes together with the corresponding cones with the same base and with apex O in each case are as AB^3 to ab^3.*

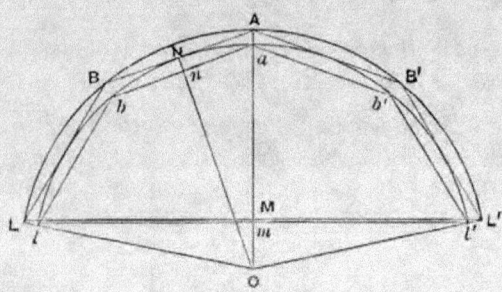

(1) For the surfaces are equal to circles the squares on whose radii are equal respectively to

$$AB\left(BB' + CC' + \ldots + KK' + \frac{LL'}{2}\right),$$

[Prop. 39, Cor.]

and $ab\left(bb' + cc' + \ldots + kk' + \dfrac{ll'}{2}\right).$ [Prop. 35]

But these rectangles are in the ratio of AB^2 to ab^2. Therefore so are the surfaces.

(2) Let OnN be drawn perpendicular to ab and AB; and suppose the circles which are equal to the surfaces of the outer and inner solids of revolution to be denoted by S, s respectively.

Now the volume of the circumscribed solid together with the cone OLL' is equal to a cone whose base is S and whose height is ON [Prop. 40, Cor. 1].

And the volume of the inscribed figure with the cone Oll' is equal to a cone with base s and height On [Prop. 38].

But $S : s = AB^2 : ab^2,$

and $ON : On = AB : ab.$

Therefore the volume of the circumscribed solid together with the cone OLL' is to the volume of the inscribed solid together with the cone Oll' as AB^3 is to ab^3 [*Lemma* 5].

Proposition 42.

If lal' be a segment of a sphere less than a hemisphere and Oa the radius perpendicular to the base of the segment, the surface of the segment is equal to a circle whose radius is equal to al.

Let R be a circle whose radius is equal to al. Then the surface of the segment, which we will call S, must, if it be not equal to R, be either greater or less than R.

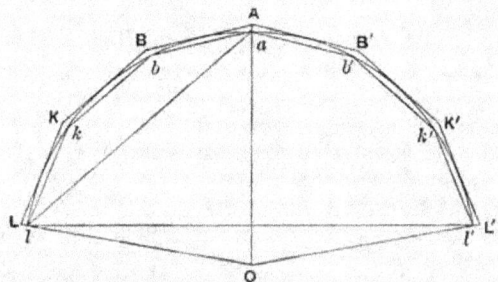

I. Suppose, if possible, $S > R$.

Let *lal'* be a segment of a great circle which is less than a semicircle. Join Ol, Ol', and let similar polygons with $2n$ equal sides be circumscribed and inscribed to the sector, as in the previous propositions, but such that

(circumscribed polygon) : (inscribed polygon) $< S : R$.

[Prop. 6]

Let the polygons now revolve with the segment about OaA, generating solids of revolution circumscribed and inscribed to the segment of the sphere.

Then

(surface of outer solid) : (surface of inner solid)

$= AB^2 : ab^2$ [Prop. 41]

$=$ (circumscribed polygon) : (inscribed polygon)

$< S : R$, by hypothesis.

But the surface of the outer solid is greater than S [Prop. 39].

Therefore the surface of the inner solid is greater than R; which is impossible, by Prop. 37.

II. Suppose, if possible, $S < R$.

In this case we circumscribe and inscribe polygons such that their ratio is less than $R : S$; and we arrive at the result that

(surface of outer solid) : (surface of inner solid)

$$< R : S.$$

But the surface of the outer solid is greater than R [Prop. 40]. Therefore the surface of the inner solid is greater than S: which is impossible [Prop. 36].

Hence, since S is neither greater nor less than R,

$$S = R.$$

Proposition 43.

Even if the segment of the sphere is greater than a hemisphere, its surface is still equal to a circle whose radius is equal to al.

For let $lal'a'$ be a great circle of the sphere, aa' being the diameter perpendicular to ll'; and let $la'l'$ be a segment less than a semi-circle.

Then, by Prop. 42, the surface of the segment $la'l'$ of the sphere is equal to a circle with radius equal to $a'l$.

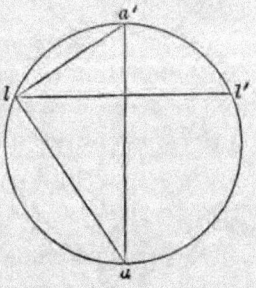

Also the surface of the whole sphere is equal to a circle with radius equal to aa' [Prop. 33].

But $aa'^2 - a'l^2 = al^2$, and circles are to one another as the squares on their radii.

Therefore the surface of the segment lal', being the difference between the surfaces of the sphere and of $la'l'$, is equal to a circle with radius equal to al.

Proposition 44.

The volume of any sector of a sphere is equal to a cone whose base is equal to the surface of the segment of the sphere included in the sector, and whose height is equal to the radius of the sphere.

Let R be a cone whose base is equal to the surface of the segment lal' of a sphere and whose height is equal to the radius of the sphere: and let S be the volume of the sector $Olal'$.

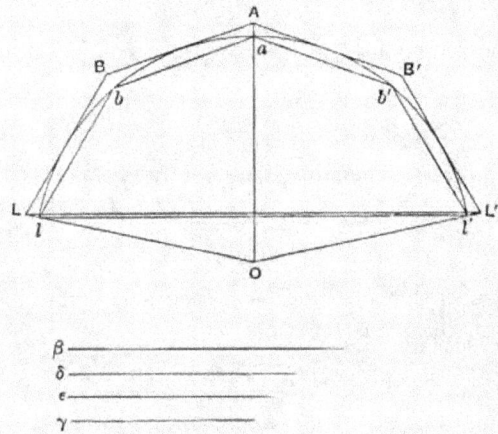

Then, if S is not equal to R, it must be either greater or less.

I.   Suppose, if possible, that $S > R$.

Find two straight lines β, γ, of which β is the greater, such that

$$\beta : \gamma < S : R;$$

and let δ, ϵ be two arithmetic means between β, γ.

Let lal' be a segment of a great circle of the sphere. Join Ol, Ol', and let similar polygons with $2n$ equal sides be circumscribed and inscribed to the sector of the circle as before, but such that their sides are in a ratio less than $\beta : \delta$. [Prop. 4].

Then let the two polygons revolve with the segment about OaA, generating two solids of revolution.

Denoting the volumes of these solids by V, v respectively, we have

$$(V + \text{cone } OLL') : (v + \text{cone } Oll') = AB^3 : ab^3 \qquad [\text{Prop. 41}]$$
$$< \beta^3 : \delta^3$$
$$< \beta : \gamma, \text{ a fortiori*},$$
$$< S : R, \text{ by hypothesis.}$$

Now $\qquad\qquad (V + \text{cone } OLL') > S.$

Therefore also $\qquad (v + \text{cone } Oll') > R.$

But this is impossible, by Prop. 38, Cor. combined with Props. 42, 43.

Hence $\qquad\qquad\qquad S \not> R.$

II. Suppose, if possible, that $S < R$.

In this case we take β, γ such that

$$\beta : \gamma < R : S,$$

and the rest of the construction proceeds as before.

We thus obtain the relation

$$(V + \text{cone } OLL') : (v + \text{cone } Oll') < R : S.$$

Now $\qquad\qquad (v + \text{cone } Oll') < S.$

Therefore $\qquad (V + \text{cone } OLL') < R;$

which is impossible, by Prop. 40, Cor. 2 combined with Props. 42, 43.

Since then S is neither greater nor less than R,

$$S = R.$$

---

* Cf. note on Prop. 34, p. 42..

ON THE SPHERE AND CYLINDER.

BOOK II.

"ARCHIMEDES to Dositheus greeting.

On a former occasion you asked me to write out the proofs of the problems the enunciations of which I had myself sent to Conon. In point of fact they depend for the most part on the theorems of which I have already sent you the demonstrations, namely (1) that the surface of any sphere is four times the greatest circle in the sphere, (2) that the surface of any segment of a sphere is equal to a circle whose radius is equal to the straight line drawn from the vertex of the segment to the circumference of its base, (3) that the cylinder whose base is the greatest circle in any sphere and whose height is equal to the diameter of the sphere is itself in magnitude half as large again as the sphere, while its surface [including the two bases] is half as large again as the surface of the sphere, and (4) that any solid sector is equal to a cone whose base is the circle which is equal to the surface of the segment of the sphere included in the sector, and whose height is equal to the radius of the sphere. Such then of the theorems and problems as depend on these theorems I have written out in the book which I send herewith; those which are discovered by means of a different sort of investigation, those namely which relate to spirals and the conoids, I will endeavour to send you soon.

The first of the problems was as follows: *Given a sphere, to find a plane area equal to the surface of the sphere.*

The solution of this is obvious from the theorems aforesaid. For four times the greatest circle in the sphere is both a plane area and equal to the surface of the sphere.

The second problem was the following."

Proposition 1. (Problem.)

Given a cone or a cylinder, to find a sphere equal to the cone or to the cylinder.

If V be the given cone or cylinder, we can make a cylinder equal to $\frac{3}{2}V$. Let this cylinder be the cylinder whose base is the circle on AB as diameter and whose height is OD.

Now, if we could make another cylinder, equal to the cylinder (OD) but such that its height is equal to the diameter of its base, the problem would be solved, because this latter cylinder would be equal to $\frac{3}{2}V$, and the sphere whose diameter is equal to the height (or to the diameter of the base) of the same cylinder would then be the sphere required [I. 34, Cor.].

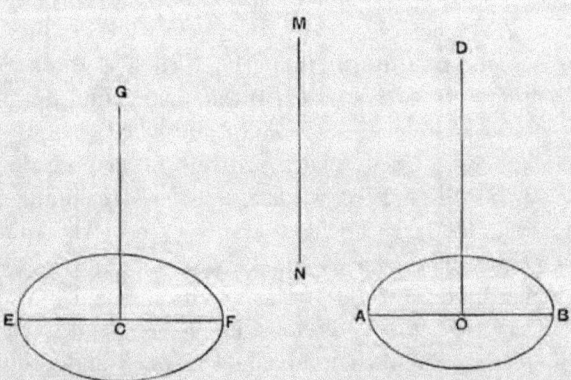

Suppose the problem solved, and let the cylinder (CG) be equal to the cylinder (OD), while EF, the diameter of the base, is equal to the height CG.

Then, since in equal cylinders the heights and bases are reciprocally proportional,

$$AB^2 : EF^2 = CG : OD$$
$$= EF : OD \dots\dots\dots\dots(1).$$

Suppose MN to be such a line that

$$EF^2 = AB . MN \dots\dots\dots\dots(2).$$

Hence $\qquad AB : EF = EF : MN,$

and, combining (1) and (2), we have

$$AB : MN = EF : OD,$$

or $\qquad AB : EF = MN : OD.$

Therefore $\qquad AB : EF = EF : MN = MN : OD,$

and EF, MN are two mean proportionals between AB, OD.

The synthesis of the problem is therefore as follows. Take two mean proportionals EF, MN between AB and OD, and describe a cylinder whose base is a circle on EF as diameter and whose height CG is equal to EF.

Then, since

$$AB : EF = EF : MN = MN : OD,$$
$$EF^2 = AB . MN,$$

and therefore $\qquad AB^2 : EF^2 = AB : MN$

$$= EF : OD$$
$$= CG : OD;$$

whence the bases of the two cylinders (OD), (CG) are reciprocally proportional to their heights.

Therefore the cylinders are equal, and it follows that

$$\text{cylinder } (CG) = \tfrac{3}{2} V.$$

The sphere on EF as diameter is therefore the sphere required, being equal to V.

Proposition 2.

If BAB' be a segment of a sphere, BB' a diameter of the base of the segment, and O the centre of the sphere, and if AA' be the diameter of the sphere bisecting BB' in M, then the volume of the segment is equal to that of a cone whose base is the same as that of the segment and whose height is h, where

$$h : AM = OA' + A'M : A'M.$$

Measure MH along MA equal to h, and MH' along MA' equal to h', where

$$h' : A'M = OA + AM : AM.$$

Suppose the three cones constructed which have O, H H' for their apices and the base (BB') of the segment for their common base. Join AB, $A'B$.

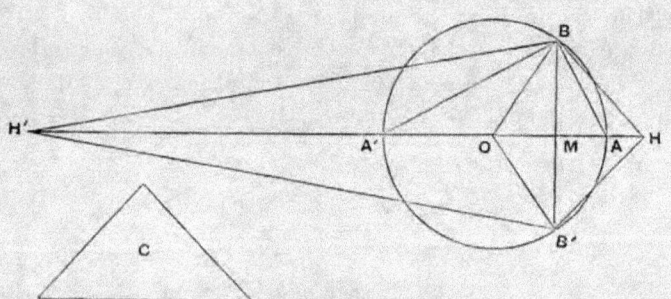

Let C be a cone whose base is equal to the surface of the segment BAB' of the sphere, i.e. to a circle with radius equal to AB [I. 42], and whose height is equal to OA.

Then the cone C is equal to the solid sector $OBAB'$ [I. 44].

Now, since $HM : MA = OA' + A'M : A'M$,

dividendo, $\qquad HA : AM = OA : A'M$,

and, alternately, $HA : AO = AM : MA'$,

so that

$$HO : OA = AA' : A'M$$
$$= AB^2 : BM^2$$
$$= \text{(base of cone } C) : \text{(circle on } BB' \text{ as diameter)}.$$

But OA is equal to the height of the cone C; therefore, since cones are equal if their bases and heights are reciprocally proportional, it follows that the cone C (or the solid sector $OBAB'$) is equal to a cone whose base is the circle on BB' as diameter and whose height is equal to OH.

And this latter cone is equal to the sum of two others having the same base and with heights OM, MH, i.e. to the solid rhombus $OBHB'$.

Hence the sector $OBAB'$ is equal to the rhombus $OBHB'$.

Taking away the common part, the cone OBB',

the segment $BAB' =$ the cone HBB'.

Similarly, by the same method, we can prove that

the segment $BA'B' =$ the cone $H'BB'$.

Alternative proof of the latter property.

Suppose D to be a cone whose base is equal to the surface of the whole sphere and whose height is equal to OA.

Thus D is equal to the volume of the sphere.   [I. 33, 34]

Now, since $OA' + A'M : A'M = HM : MA$,

dividendo and *alternando*, as before,

$$OA : AH = A'M : MA.$$

Again, since $H'M : MA' = OA + AM : AM$,

$$H'A' : OA = A'M : MA$$
$$= OA : AH, \text{ from above.}$$

Componendo, $H'O : OA = OH : HA$ ....(1).

Alternately, $H'O : OH = OA : AH$ ....(2),

and, *componendo,* $HH' : HO = OH : HA$,

$$= H'O : OA, \text{ from (1),}$$

whence $HH' . OA = H'O . OH$ ....(3).

Next, since $H'O : OH = OA : AH$, by (2),

$$= A'M : MA,$$

$(H'O + OH)^2 : H'O . OH = (A'M + MA)^2 : A'M . MA,$

whence, by means of (3),

$$HH'^2 : HH'.OA = AA'^2 : A'M.MA,$$

or $\qquad HH' : OA = AA'^2 : BM^2.$

Now the cone D, which is equal to the sphere, has for its base a circle whose radius is equal to AA', and for its height a line equal to OA.

Hence this cone D is equal to a cone whose base is the circle on BB' as diameter and whose height is equal to HH';

therefore $\qquad$ the cone $D =$ the rhombus $HBH'B'$,

or $\qquad$ the rhombus $HBH'B' =$ the sphere.

But $\qquad$ the segment $BAB' =$ the cone HBB';

therefore the remaining segment $BA'B' =$ the cone $H'BB'$.

COR. *The segment BAB' is to a cone with the same base and equal height in the ratio of $OA' + A'M$ to $A'M$.*

Proposition 3. (Problem.)

To cut a given sphere by a plane so that the surfaces of the segments may have to one another a given ratio.

Suppose the problem solved. Let AA' be a diameter of a great circle of the sphere, and suppose that a plane perpendicular to AA' cuts the plane of the great circle in the straight

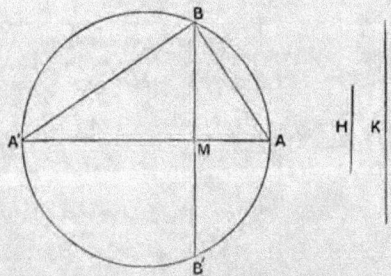

line BB', and AA' in M, and that it divides the sphere so that the surface of the segment BAB' has to the surface of the segment $BA'B'$ the given ratio.

Now these surfaces are respectively equal to circles with radii equal to AB, $A'B$ [I. 42, 43].

Hence the ratio $AB^2 : A'B^2$ is equal to the given ratio, i.e. AM is to MA' in the given ratio.

Accordingly the synthesis proceeds as follows.

If $H : K$ be the given ratio, divide AA' in M so that

$$AM : MA' = H : K.$$

Then $AM : MA' = AB^2 : A'B^2$

     $= $ (circle with radius AB) : (circle with radius $A'B$)

     $= $ (surface of segment BAB') : (surface of segment $BA'B'$).

Thus the ratio of the surfaces of the segments is equal to the ratio $H : K$.

Proposition 4. (Problem.)

To cut a given sphere by a plane so that the volumes of the segments are to one another in a given ratio.

Suppose the problem solved, and let the required plane cut the great circle ABA' at right angles in the line BB'. Let AA' be that diameter of the great circle which bisects BB' at right angles (in M), and let O be the centre of the sphere.

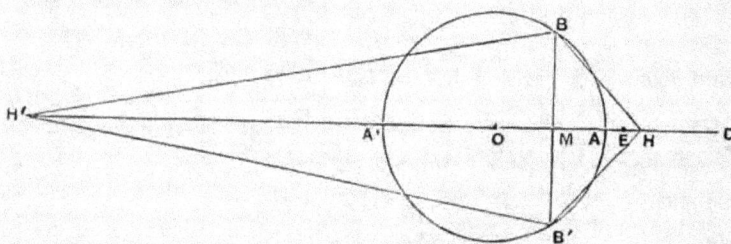

Take H on OA produced, and H' on OA' produced, such that

$$OA' + A'M : A'M = HM : MA, \dots\dots\dots\dots(1),$$

and $\quad\quad\quad OA + AM : AM = H'M : MA' \dots\dots\dots\dots(2).$

Join BH, $B'H$, BH', $B'H'$.

Then the cones HBB', $H'BB'$ are respectively equal to the segments BAB', $BA'B'$ of the sphere [Prop. 2].

Hence the ratio of the cones, and therefore of their altitudes, is given, i.e.

$$HM : H'M = \text{the given ratio}\ldots\ldots\ldots\ldots(3).$$

We have now three equations (1), (2), (3), in which there appear three as yet undetermined points M, H, H'; and it is first necessary to find, by means of them, another equation in which only one of these points (M) appears, i.e. we have, so to speak, to *eliminate* H, H'.

Now, from (3), it is clear that $HH' : H'M$ is also a given ratio; and Archimedes' method of elimination is, *first*, to find values for each of the ratios $A'H' : H'M$ and $HH' : H'A'$ which are alike independent of H, H', and then, *secondly*, to equate the ratio compounded of these two ratios to the known value of the ratio $HH' : H'M$.

(*a*) To find such a value for $A'H' : H'M$.

It is at once clear from equation (2) above that

$$A'H' : H'M = OA : OA + AM \ldots\ldots\ldots\ldots(4).$$

(*b*) To find such a value for $HH' : A'H'$.

From (1) we derive

$$A'M : MA = OA' + A'M : HM$$
$$= OA' : AH \ldots\ldots\ldots\ldots\ldots(5);$$

and, from (2), $\quad A'M : MA = H'M : OA + AM$
$$= A'H' : OA \ldots\ldots\ldots\ldots\ldots(6).$$

Thus $\qquad HA : AO = OA' : A'H'$,

whence $\qquad OH : OA' = OH' : A'H'$,

or $\qquad OH : OH' = OA' : A'H'$.

It follows that

$$HH' : OH' = OH' : A'H',$$

or $\qquad HH' \cdot H'A' = OH'^2.$

Therefore $\quad HH' : H'A' = OH'^2 : H'A'^2$

$$= AA'^2 : A'M^2, \text{ by means of (6)}$$

(c) To express the ratios $A'H' : H'M$ and $HH' : H'M$ more simply we make the following construction. Produce OA to D so that $OA = AD$. (D will lie beyond H, for $A'M > MA$, and therefore, by (5), $OA > AH$.)

Then $\qquad A'H' : H'M = OA : OA + AM$

$$= AD : DM \dotfill (7).$$

Now divide AD at E so that

$$HH' : H'M = AD : DE \dotfill (8).$$

Thus, using equations (8), (7) and the value of $HH' : H'A'$ above found, we have

$$AD : DE = HH' : H'M$$

$$= (HH' : H'A') . (A'H' : H'M)$$

$$= (AA'^2 : A'M^2) . (AD : DM).$$

But $\qquad AD : DE = (DM : DE) . (AD : DM).$

Therefore $\qquad MD : DE = AA'^2 : A'M^2 \dotfill (9).$

And D is given, since $AD = OA$. Also $AD : DE$ (being equal to $HH' : H'M$) is a given ratio. Therefore DE is given.

Hence the problem reduces itself to the problem of dividing $A'D$ into two parts at M so that

$$MD : \text{(a given length)} = \text{(a given area)} : A'M^2.$$

Archimedes adds: "If the problem is propounded in this general form, it requires a διορισμός [i.e. it is necessary to investigate the limits of possibility], but, if there be added the conditions subsisting in the present case, it does not require a διορισμός."

In the present case the problem is:

Given a straight line $A'A$ produced to D so that $A'A = 2AD$, and given a point E on AD, to cut AA' in a point M so that

$$AA'^2 : A'M^2 = MD : DE.$$

"And the analysis and synthesis of both problems will be given at the end*."

The synthesis of the main problem will be as follows. Let $R : S$ be the given ratio, R being less than S. AA' being a

---

* See the note following this proposition.

diameter of a great circle, and O the centre, produce OA to D so that $OA = AD$, and divide AD in E so that

$$AE : ED = R : S.$$

Then cut AA' in M so that

$$MD : DE = AA'^2 : A'M^2.$$

Through M erect a plane perpendicular to AA'; this plane will then divide the sphere into segments which will be to one another as R to S.

Take H on $A'A$ produced, and H' on AA' produced, so that

$$OA' + A'M : A'M = HM : MA, \dots\dots\dots\dots(1),$$
$$OA + AM : AM = H'M : MA' \dots\dots\dots\dots(2).$$

We have then to show that

$$HM : MH' = R : S, \text{ or } AE : ED.$$

(a) We first find the value of $HH' : H'A'$ as follows.

As was shown in the analysis (b),

$$HH' . H'A' = OH'^2,$$

or $\qquad HH' : H'A' = OH'^2 : H'A'^2$

$$= AA'^2 : A'M^2$$

$$= MD : DE, \text{ by construction.}$$

(β) Next we have

$$H'A' : H'M = OA : OA + AM$$

$$= AD : DM.$$

Therefore $\quad HH' : H'M = (HH' : H'A') . (H'A' : H'M)$

$$= (MD : DE) . (AD : DM)$$

$$= AD : DE,$$

whence $\qquad HM : MH' = AE : ED$

$$= R : S. \qquad \text{Q. E. D.}$$

Note. The solution of the subsidiary problem to which the original problem of Prop. 4 is reduced, and of which Archimedes promises a discussion, is given in a highly interesting and important note by Eutocius, who introduces the subject with the following explanation.

" He [Archimedes] promised to give a solution of this problem at the end, but we do not find the promise kept in any of the copies. Hence we find that Dionysodorus too failed to light upon the promised discussion and, being unable to grapple with the omitted lemma, approached the original problem in a different way, which I shall describe later. Diocles also expressed in his work περὶ πυρίων the opinion that Archimedes made the promise but did not perform it, and tried to supply the omission himself. His attempt I shall also give in its order. It will however be seen to have no relation to the omitted discussion but to give, like Dionysodorus, a construction arrived at by a different method of proof. On the other hand, as the result of unremitting and extensive research, I found in a certain old book some theorems discussed which, although the reverse of clear owing to errors and in many ways faulty as regards the figures, nevertheless gave the substance of what I sought, and moreover to some extent kept to the Doric dialect affected by Archimedes, while they retained the names familiar in old usage, the parabola being called a section of a right-angled cone, and the hyperbola a section of an obtuse-angled cone; whence I was led to consider whether these theorems might not in fact be what he promised he would give at the end. For this reason I paid them the closer attention, and, after finding great difficulty with the actual text owing to the multitude of the mistakes above referred to, I made out the sense gradually and now proceed to set it out, as well as I can, in more familiar and clearer language. And first the theorem will be treated generally, in order that what Archimedes says about the limits of possibility may be made clear; after which there will follow the special application to the conditions stated in his analysis of the problem."

The investigation which follows may be thus reproduced. The general problem is:

Given two straight lines AB, AC and an area D, to divide AB at M so that

$$AM : AC = D : MB^2.$$

Analysis.

Suppose M found, and suppose AC placed at right angles to AB. Join CM and produce it. Draw EBN through B parallel to AC meeting CM in N, and through C draw CHE parallel to AB meeting EBN in E. Complete the parallelogram $CENF$, and through M draw PMH parallel to AC meeting FN in P.

Measure EL along EN so that

$$CE . EL \text{ (or } AB . EL) = D.$$

Then, by hypothesis,

$$AM : AC = CE . EL : MB^2.$$

And

$$AM : AC = CE : EN,$$

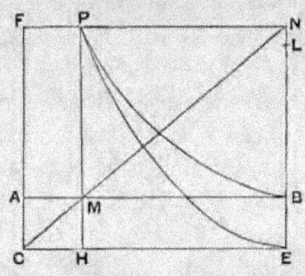

by similar triangles,

$$= CE . EL : EL . EN.$$

It follows that $PN^2 = MB^2 = EL . EN.$

Hence, if a parabola be described with vertex E, axis EN, and parameter equal to EL, it will pass through P; and it will be given in position, since EL is given.

Therefore P lies on a given parabola.

Next, since the rectangles FH, AE are equal,

$$FP . PH = AB . BE.$$

Hence, if a rectangular hyperbola be described with CE, CF as asymptotes and passing through B, it will pass through P. And the hyperbola is given in position.

Therefore P lies on a given hyperbola.

Thus P is determined as the intersection of the parabola and hyperbola. And since P is thus given, M is also given.

διορισμός.

Now, since $AM : AC = D : MB^2,$

$$AM . MB^2 = AC . D.$$

But $AC . D$ is given, and *it will be proved later that the maximum value of $AM . MB^2$ is that which it assumes when $BM = 2AM$.*

5—2

Hence *it is a necessary condition of the possibility of a solution that* $AC.D$ *must not be greater than* $\frac{1}{3}AB.(\frac{2}{3}AB)^2$, *or* $\frac{4}{27}AB^2$.

Synthesis.

If O be such a point on AB that $BO = 2AO$, we have seen that, in order that the solution may be possible,

$$AC.D \not> AO.OB^2.$$

Thus $AC.D$ is either equal to, or less than, $AO.OB^2$.

(1) If $AC.D = AO.OB^2$, then the point O itself solves the problem.

(2) Let $AC.D$ be less than $AO.OB^2$.

Place AC at right angles to AB. Join CO, and produce it to R. Draw EBR through B parallel to AC meeting CO in R, and through C draw CE parallel to AB in E. Com-plete the parallelogram $CERF$, and through O draw QOK parallel to AC meeting FR in Q and CE in K.

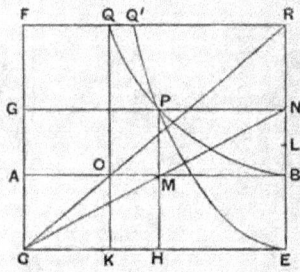

Then, since

$$AC.D < AO.OB^2,$$

measure RQ' along RQ so that

$$AC.D = AO.Q'R^2,$$

or $\qquad\qquad AO : AC = D : Q'R^2.$

Measure EL along ER so that

$$D = CE.EL \text{ (or } AB.EL).$$

Now, since $\quad AO : AC = D : Q'R^2$, by hypothesis,

$$= CE.EL : Q'R^2,$$

and $\qquad\qquad AO : AC = CE : ER$, by similar triangles,

$$= CE.EL : EL.ER,$$

it follows that

$$Q'R^2 = EL.ER.$$

Describe a parabola with vertex E, axis ER, and parameter equal to EL. This parabola will then pass through Q'.

Again, rect. $FK =$ rect. AE,

or $FQ \cdot QK = AB \cdot BE$;

and, if we describe a rectangular hyperbola with asymptotes CE, CF and passing through B, it will also pass through Q.

Let the parabola and hyperbola intersect at P, and through P draw PMH parallel to AC meeting AB in M and CE in H, and GPN parallel to AB meeting CF in G and ER in N.

Then shall M be the required point of division.

Since $PG \cdot PH = AB \cdot BE$,

rect. $GM =$ rect. ME,

and therefore CMN is a straight line.

Thus $AB \cdot BE = PG \cdot PH = AM \cdot EN \ldots\ldots\ldots\ldots(1)$.

Again, by the property of the parabola,

$$PN^2 = EL \cdot EN,$$

or $MB^2 = EL \cdot EN \ldots\ldots\ldots\ldots\ldots\ldots\ldots(2)$.

From (1) and (2)

$$AM : EL = AB \cdot BE : MB^2,$$

or $AM \cdot AB : AB \cdot EL = AB \cdot AC : MB^2$.

Alternately,

$$AM \cdot AB : AB \cdot AC = AB \cdot EL : MB^2,$$

or $AM : AC = D : MB^2$.

Proof of διορισμός.

It remains to be proved that, *if AB be divided at O so that $BO = 2AO$, then $AO \cdot OB^2$ is the maximum value of $AM \cdot MB^2$,*

or $AO \cdot OB^2 > AM \cdot MB^2$,

where M is any point on AB other than O.

Suppose that $\qquad AO : AC = CE . EL' : OB^2$,

so that $\qquad\qquad AO . OB^2 = CE . EL' . AC$.

Join CO, and produce it to N; draw EBN through B parallel to AC, and complete the parallelogram $CENF$.

Through O draw POH parallel to AC meeting FN in P and CE in H.

With vertex E, axis EN, and parameter EL', describe a parabola. This will pass through P, as shown in the analysis above, and beyond P will meet the diameter CF of the parabola in some point.

Next draw a rectangular hyperbola with asymptotes CE, CF and passing through B. This hyperbola will also pass through P, as shown in the analysis.

Produce NE to T so that $TE = EN$. Join TP meeting CE in Y, and produce it to meet CF in W. Thus TP will touch the parabola at P.

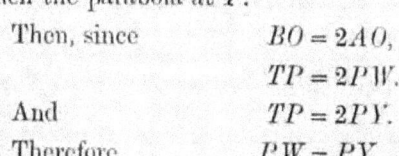

Then, since $\qquad\qquad BO = 2AO$,

$$TP = 2PW.$$

And $\qquad\qquad\qquad TP = 2PY.$

Therefore $\qquad\qquad PW = PY.$

Since, then, WY between the asymptotes is bisected at P, the point where it meets the hyperbola,

$$WY \text{ is a tangent to the hyperbola.}$$

Hence the hyperbola and parabola, having a common tangent at P, touch one another at P.

Now take any point M on AB, and through M draw QMK parallel to AC meeting the hyperbola in Q and CE in K. Lastly, draw $GqQR$ through Q parallel to AB meeting CF in G, the parabola in q, and EN in R.

Then, since, by the property of the hyperbola, the rectangles GK, AE are equal, CMR is a straight line.

By the property of the parabola,

$$qR^2 = EL' . ER,$$

so that $$QR^2 < EL' . ER.$$

Suppose $$QR^2 = EL . ER,$$

and we have $AM : AC = CE : ER$

$$= CE . EL : EL . ER$$

$$= CE . EL : QR^2$$

$$= CE . EL : MB^2,$$

or $$AM . MB^2 = CE . EL . AC.$$

Therefore $AM . MB^2 < CE . EL' . AC$

$$< AO . OB^2.$$

If $AC . D < AO . OB^2$, there are two solutions because there will be two points of intersection between the parabola and the hyperbola.

For, if we draw with vertex E and axis EN a parabola whose parameter is equal to EL, the parabola will pass through the point Q (see the last figure); and, since the parabola meets the diameter CF beyond Q, it must meet the hyperbola again (which has CF for its asymptote).

[If we put $AB = a$, $BM = x$, $AC = c$, and $D = b^2$, the proportion

$$AM : AC = D : MB^2$$

is seen to be equivalent to the equation

$$x^2 (a - x) = b^2 c,$$

being a *cubic equation* with the term containing x omitted.

Now suppose EN, EC to be axes of coordinates, EN being the axis of y.

Then the parabola used in the above solution is the parabola

$$x^2 = \frac{b^2}{a} \cdot y,$$

and the rectangular hyperbola is

$$y(a - x) = ac.$$

Thus the solution of the cubic equation and the conditions under which there are no positive solutions, or one, or two positive solutions are obtained by the use of the two conics.]

[For the sake of completeness, and for their intrinsic interest, the solutions of the original problem in Prop. 4 given by Dionysodorus and Diocles are here appended.

Dionysodorus' solution.

Let AA' be a diameter of the given sphere. It is required to find a plane cutting AA' at right angles (in a point M, suppose) so that the segments into which the sphere is divided are in a given ratio, as $CD : DE$.

Produce $A'A$ to F so that $AF = OA$, where O is the centre of the sphere.

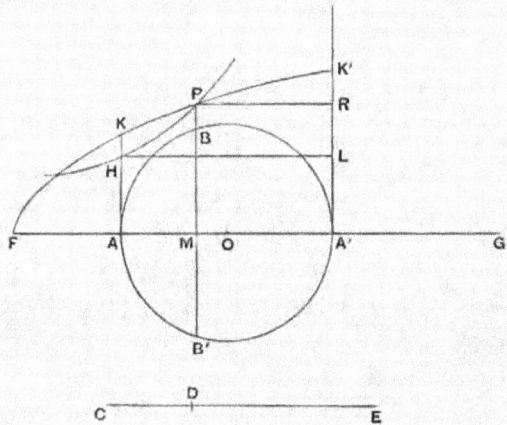

Draw AH perpendicular to AA' and of such length that

$$FA : AH = CE : ED,$$

and produce AH to K so that

$$AK^2 = FA . AH \quad\dots\dots\dots\dots\dots\dots(\alpha).$$

With vertex F, axis FA, and parameter equal to AH describe a parabola. This will pass through K, by the equation (α).

Draw $A'K'$ parallel to AK and meeting the parabola in K'; and with $A'F$, $A'K'$ as asymptotes describe a rectangular hyperbola passing through H. This hyperbola will meet the parabola at some point, as P, between K and K'.

Draw PM perpendicular to AA' meeting the great circle in B, B', and from H, P draw HL, PR both parallel to AA' and meeting $A'K'$ in L, R respectively.

Then, by the property of the hyperbola,

$$PR . PM = AH . HL,$$

i.e. $\qquad\qquad PM . MA' = HA . AA',$

or $\qquad\qquad PM : AH = AA' : A'M,$

and $\qquad\qquad PM^2 : AH^2 = AA'^2 : A'M^2.$

Also, by the property of the parabola,

$$PM^2 = FM . AH,$$

i.e. $\qquad\qquad FM : PM = PM : AH,$

or $\qquad\qquad FM : AH = PM^2 : AH^2$

$$= AA'^2 : A'M^2, \text{ from above.}$$

Thus, since circles are to one another as the squares of their radii, the cone whose base is the circle with $A'M$ as radius and whose height is equal to FM, and the cone whose base is the circle with AA' as radius and whose height is equal to AH, have their bases and heights reciprocally proportional.

Hence the cones are equal; i.e., if we denote the first cone by the symbol $c(A'M)$, FM, and so on,

$$c(A'M),\ FM = c(AA'),\ AH.$$

Now $c(AA')$, $FA : c(AA')$, $AH = FA : AH$

$$= CE : ED, \text{ by construction.}$$

Therefore

$$c(AA'), FA : c(A'M), FM = CE : ED \ldots\ldots(\beta).$$

But (1) $c(AA'), FA = \text{the sphere.}$ [I. 34]

(2) $c(A'M), FM$ can be proved equal to the segment of the sphere whose vertex is A' and height $A'M$.

For take G on AA' produced such that

$$GM : MA' = FM : MA$$
$$= OA + AM : AM.$$

Then the cone GBB' is equal to the segment $A'BB'$ [Prop. 2].

And $FM : MG = AM : MA'$, by hypothesis,
$$= BM^2 : A'M^2.$$

Therefore

$$(\text{circle with rad. } BM) : (\text{circle with rad. } A'M)$$
$$= FM : MG,$$

so that $c(A'M), FM = c(BM), MG$
$$= \text{the segment } A'BB'.$$

We have therefore, from the equation (β) above,

$$(\text{the sphere}) : (\text{segmt. } A'BB') = CE : ED,$$

whence $(\text{segmt. } ABB') : (\text{segmt. } A'BB') = CD : DE.$

Diocles' solution.

Diocles starts, like Archimedes, from the property, proved in Prop. 2, that, if the plane of section cut a diameter AA' of the sphere at right angles in M, and if H, H' be taken on OA, OA' produced respectively so that

$$OA' + A'M : A'M = HM : MA,$$
$$OA + AM : AM = H'M : MA',$$

then the cones $HBB', H'BB'$ are respectively equal to the segments $ABB', A'BB'$.

Then, drawing the inference that
$$HA : AM = OA' : A'M,$$
$$H'A' : A'M = OA : AM,$$

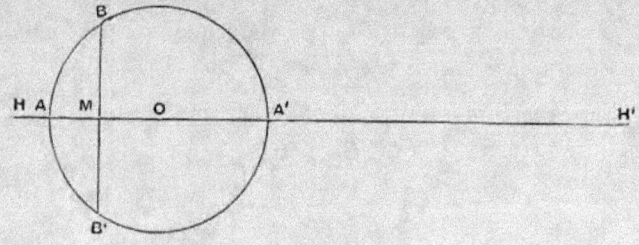

he proceeds to state the problem in the following form, slightly generalising it by the substitution of *any* given straight line for *OA* or *OA'*:

Given a straight line AA', its extremities A, A', a ratio C : D, and another straight line as AK, to divide AA' at M and to find two points H, H' on A'A and AA' produced respectively so that the following relations may hold simultaneously,

$$C : D = HM : MH' \quad \text{............(α),}$$
$$HA : AM = AK : A'M \quad \text{............(β),}$$
$$H'A' : A'M = AK : AM \quad \text{............(γ).}$$

Analysis.

Suppose the problem solved and the points M, H, H' all found.

Place AK at right angles to AA', and draw $A'K'$ parallel and equal to AK. Join $KM, K'M$, and produce them to meet $K'A', KA$ respectively in E, F. Join KK', draw EG through E parallel to $A'A$ meeting KF in G, and through M draw QMN parallel to AK meeting EG in Q and KK' in N.

Now $\quad HA : AM = A'K' : A'M$, by (β),

$\qquad\qquad\qquad = FA : AM$, by similar triangles,

whence $\qquad HA = FA.$

Similarly $\qquad H'A' = A'E.$

Next,

$$FA + AM : A'K' + A'M = AM : A'M$$
$$= AK + AM : EA' + A'M, \text{ by similar triangles.}$$

Therefore

$$(FA + AM).(EA' + A'M) = (KA + AM).(K'A' + A'M).$$

Take AR along AH and $A'R'$ along $A'H'$ such that

$$AR = A'R' = AK.$$

Then, since $FA + AM = HM$, $EA' + A'M = MH'$, we have

$$HM.MH' = RM.MR' \quad ....(\delta).$$

(Thus, if R falls between A and H, R' falls on the side of H' remote from A', and *vice versa*.)

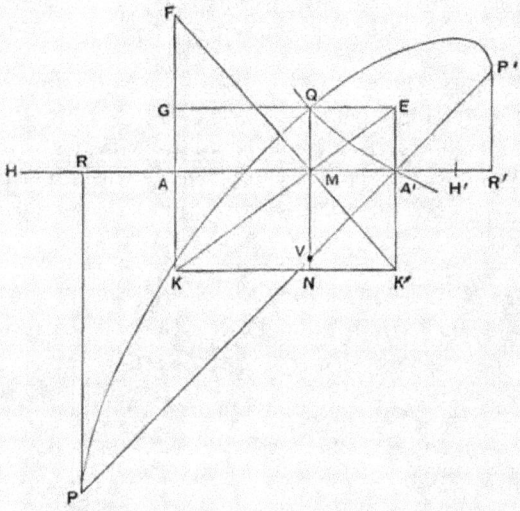

Now $C : D = HM : MH'$, by hypothesis,

$$= HM.MH' : MH'^2$$

$$= RM.MR' : MH'^2, \text{ by } (\delta).$$

Measure MV along MN so that $MV = A'M$. Join $A'V$ and produce it both ways. Draw RP, $R'P'$ perpendicular to RR' meeting $A'V$ produced in P, P' respectively. Then, the angle $MA'V$ being half a right angle, PP' is given in position, and, since R, R' are given, so are P, P'.

And, by parallels,

$$P'V : PV = R'M : MR.$$

Therefore $PV . P'V : PV^2 = RM . MR' : RM^2$.

But $PV^2 = 2RM^2$.

Therefore $PV . P'V = 2RM . MR'$.

And it was shown that

$$RM . MR' : MH'^2 = C : D.$$

Hence $PV . P'V : MH'^2 = 2C : D$.

But $MH' = A'M + A'E = VM + MQ = QV$.

Therefore $QV^2 : PV . P'V = D : 2C$, a given ratio.

Thus, if we take a line p such that

$$D : 2C = p : PP'^*,$$

and if we describe an ellipse with PP' as a diameter and p as the corresponding parameter $[= DD'^2/PP'$ in the ordinary notation of geometrical conics], and such that the ordinates to PP' are inclined to it at an angle equal to half a right angle, i.e. are parallel to QV or AK, then the ellipse will pass through Q.

Hence Q lies on an ellipse given in position.

Again, since EK is a diagonal of the parallelogram GK',

$$GQ . QN = AA' . A'K'.$$

If therefore a rectangular hyperbola be described with KG, KK' as asymptotes and passing through A', it will also pass through Q.

Hence Q lies on a given rectangular hyperbola.

Thus Q is determined as the intersection of a given ellipse

---

* There is a mistake in the Greek text here which seems to have escaped the notice of all the editors up to the present. The words are ἐὰν ἄρα ποιήσωμεν, ὡς τὴν Δ πρὸς τὴν διπλασίαν τῆς Γ, οὕτως τὴν ΤΥ πρὸς ἄλλην τινά ὡς τὴν Φ, i.e. (with the lettering above) "If we take a length p such that $D : 2C = PP' : p$." This cannot be right, because we should then have

$$QV^2 : PV . P'V = PP' : p,$$

whereas the two latter terms should be reversed, the correct property of the ellipse being

$$QV^2 : PV . P'V = p : PP'. \qquad \text{[Apollonius I. 21]}$$

The mistake would appear to have originated as far back as Eutocius, but I think that Eutocius is more likely to have made the slip than Diocles himself, because any intelligent mathematician would be more likely to make such a slip in writing out another man's work than to overlook it if made by another.

and a given hyperbola, and is therefore given. Thus M is given, and H, H' can at once be found.

Synthesis.

Place AA', AK at right angles, draw $A'K'$ parallel and equal to AK, and join KK'.

Make AR (measured along $A'A$ produced) and $A'R'$ (measured along AA' produced) each equal to AK, and through R, R' draw perpendiculars to RR'.

Then through A' draw PP' making an angle $(AA'P)$ with AA' equal to half a right angle and meeting the perpendiculars just drawn in P, P' respectively.

Take a length p such that

$$D : 2C = p : PP'^{*},$$

and with PP' as diameter and p as the corresponding parameter describe an ellipse such that the ordinates to PP' are inclined to it at an angle equal to $AA'P$, i.e. are parallel to AK.

With asymptotes KA, KK' draw a rectangular hyperbola passing through A'.

Let the hyperbola and ellipse meet in Q, and from Q draw $QMVN$ perpendicular to AA' meeting AA' in M, PP' in V and KK' in N. Also draw GQE parallel to AA' meeting AK, $A'K'$ respectively in G, E.

Produce KA, $K'M$ to meet in F.

Then, from the property of the hyperbola,

$$GQ.QN = AA'.A'K',$$

and, since these rectangles are equal, KME is a straight line.

Measure AH along AR equal to AF, and $A'H'$ along $A'R'$ equal to $A'E$.

From the property of the ellipse,

$$QV^{2} : PV.P'V = p : PP'$$
$$= D : 2C.$$

---

* Here too the Greek text repeats the same error as that noted on p. 77.

And, by parallels,

$$PV : P'V = RM : R'M,$$

or $$PV . P'V : P'V^2 = RM . MR' : R'M^2,$$

while $P'V^2 = 2R'M^2$, since the angle $RA'P$ is half a right angle.

Therefore $$PV . P'V = 2RM . MR',$$

whence $$QV^2 : 2RM . MR' = D : 2C.$$

But $$QV = EA' + A'M = MH'.$$

Therefore $$RM . MR' : MH'^2 = C : D.$$

Again, by similar triangles,

$$FA + AM : K'A' + A'M = AM : A'M$$
$$= KA + AM : EA' + A'M.$$

Therefore

$$(FA + AM) . (EA' + A'M) = (KA + AM) . (K'A' + A'M)$$

or $$HM . MH' = RM . MR'.$$

It follows that

$$HM . MH' : MH'^2 = C : D,$$

or $$HM : MH' = C : D \dots\dots\dots\dots\dots\dots\dots (\alpha).$$

Also $$HA : AM = FA : AM,$$
$$= A'K' : A'M, \text{ by similar}$$
$$\text{triangles} \dots (\beta),$$

and $$H'A' : A'M = EA' : A'M$$
$$= AK : AM \dots\dots\dots\dots\dots (\gamma).$$

Hence the points M, H, H' satisfy the three given relations.]

Proposition 5. (Problem.)

To construct a segment of a sphere similar to one segment and equal in volume to another.

Let ABB' be one segment whose vertex is A and whose base is the circle on BB' as diameter; and let DEF be another segment whose vertex is D and whose base is the circle on EF

as diameter.   Let AA', DD' be diameters of the great circles passing through BB', EF respectively, and let O, C be the respective centres of the spheres.

Suppose it required to draw a segment similar to DEF and equal in volume to ABB'.

Analysis.   Suppose the problem solved, and let def be the required segment, d being the vertex and ef the diameter of the base.   Let dd' be the diameter of the sphere which bisects ef at right angles, c the centre of the sphere.

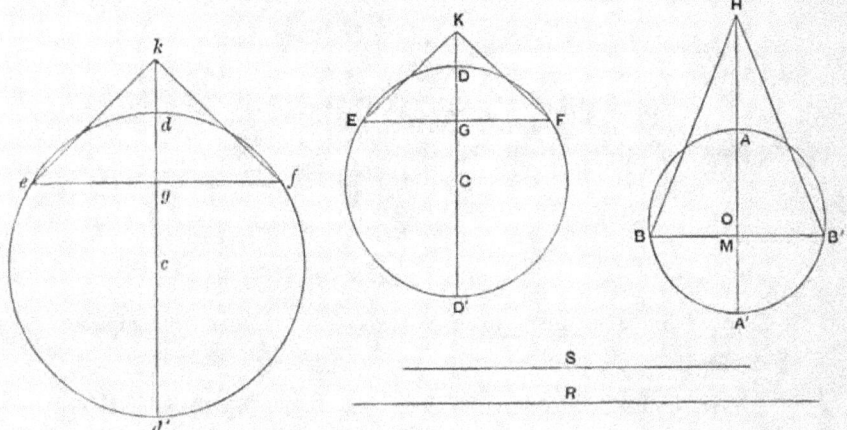

Let M, G, g be the points where BB', EF, ef are bisected at right angles by AA', DD', dd' respectively, and produce OA, CD, cd respectively to H, K, k, so that

$$OA' + A'M : A'M = HM : MA$$
$$CD' + D'G : D'G = KG : GD$$
$$cd' + d'g : d'g = kg : gd$$

and suppose cones formed with vertices H, K, k and with the same bases as the respective segments.   The cones will then be equal to the segments respectively [Prop. 2].

Therefore, by hypothesis,

$$\text{the cone } HBB' = \text{the cone } kef.$$

Hence

(circle on diameter BB') : (circle on diameter ef) $= kg : HM$,

so that $\qquad BB'^2 : ef^2 = kg : HM$ .... (1).

But, since the segments DEF, def are similar, so are the cones KEF, kef.

Therefore $\qquad KG : EF = kg : ef$.

And the ratio $KG : EF$ is given. Therefore the ratio $kg : ef$ is given.

Suppose a length R taken such that

$$kg : ef = HM : R \qquad\qquad (2).$$

Thus R is given.

Again, since $kg : HM = BB'^2 : ef^2 = ef : R$, by (1) and (2), suppose a length S taken such that

$$ef^2 = BB' . S,$$

or $\qquad\qquad BB'^2 : ef^2 = BB' : S.$

Thus $\qquad\qquad BB' : ef = ef : S = S : R,$

and ef, S are *two mean proportionals in continued proportion between BB', R*.

Synthesis. Let ABB', DEF be great circles, AA', DD' the diameters bisecting BB', EF at right angles in M, G respectively, and O, C the centres.

Take H, K in the same way as before, and construct the cones HBB', KEF, which are therefore equal to the respective segments ABB', DEF.

Let R be a straight line such that

$$KG : EF = HM : R,$$

and between BB', R take two mean proportionals ef, S.

On ef as base describe a segment of a circle with vertex d and similar to the segment of a circle DEF. Complete the circle, and let dd' be the diameter through d, and c the centre. Conceive a sphere constructed of which def is a great circle, and through ef draw a plane at right angles to dd'.

Then shall *def* be the required segment of a sphere.

For the segments *DEF*, *def* of the spheres are similar, like the circular segments *DEF*, *def*.

Produce *cd* to *k* so that

$$cd' + d'g : d'g = kg : gd.$$

The cones *KEF*, *kef* are then similar.

Therefore $kg : ef = KG : EF = HM : R,$

whence $kg : HM = ef : R.$

But, since *BB'*, *ef*, *S*, *R* are in continued proportion,

$$BB'^2 : ef^2 = BB' : S$$

$$= ef : R$$

$$= kg : HM.$$

Thus the bases of the cones *HBB'*, *kef* are reciprocally proportional to their heights. The cones are therefore equal, and *def* is the segment required, being equal in volume to the cone *kef*.   [Prop. 2]

Proposition 6. (Problem.)

Given two segments of spheres, to find a third segment of a sphere similar to one of the given segments and having its surface equal to that of the other.

Let *ABB'* be the segment to whose surface the surface of the required segment is to be equal, *ABA'B'* the great circle whose plane cuts the plane of the base of the segment *ABB'* at right angles in *BB'*. Let *AA'* be the diameter which bisects *BB'* at right angles.

Let *DEF* be the segment to which the required segment is to be similar, *DED'F* the great circle cutting the base of the segment at right angles in *EF*. Let *DD'* be the diameter bisecting *EF* at right angles in *G*.

Suppose the problem solved, *def* being a segment similar to *DEF* and having its surface equal to that of *ABB'*; and complete the figure for *def* as for *DEF*, corresponding points being denoted by small and capital letters respectively.

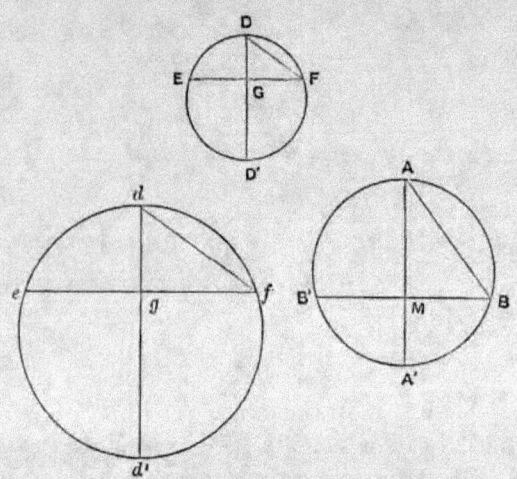

Join *AB*, *DF*, *df*.

Now, since the surfaces of the segments *def*, *ABB′* are equal, so are the circles on *df*, *AB* as diameters;　　　　　[I. 42, 43]

that is,　　　　　　　　　　$df = AB.$

From the similarity of the segments *DEF*, *def* we obtain

$$d'd : dg = D'D : DG,$$

and　　　　　　　　$dg : df = DG : DF;$

whence　　　　　　$d'd : df = D'D : DF,$

or　　　　　　　　$d'd : AB = D'D : DF.$

But *AB*, *D′D*, *DF* are all given;

therefore *d′d* is given.

Accordingly the synthesis is as follows.

Take *d′d* such that

$$d'd : AB = D'D : DF \dots\dots\dots\dots(1).$$

Describe a circle on *d′d* as diameter, and conceive a sphere constructed of which this circle is a great circle.

Divide $d'd$ at g so that
$$d'g : gd = D'G : GD,$$
and draw through g a plane perpendicular to $d'd$ cutting off the segment def of the sphere and intersecting the plane of the great circle in ef. The segments def, DEF are thus similar, and
$$dg : df = DG : DF.$$

But from above, *componendo*,
$$d'd : dg = D'D : DG.$$

Therefore, *ex aequali*, $d'd : df = D'D : DF$, whence, by (1), $df = AB$.

Therefore the segment def has its surface equal to the surface of the segment ABB' [1. 42, 43], while it is also similar to the segment DEF.

Proposition 7.   (Problem.)

From a given sphere to cut off a segment by a plane so that the segment may have a given ratio to the cone which has the same base as the segment and equal height.

Let AA' be the diameter of a great circle of the sphere. It is required to draw a plane at right angles to AA' cutting off a segment, as ABB', such that the segment ABB' has to the cone ABB' a given ratio.

Analysis.

Suppose the problem solved, and let the plane of section cut the plane of the great circle in BB', and the diameter AA' in M. Let O be the centre of the sphere.

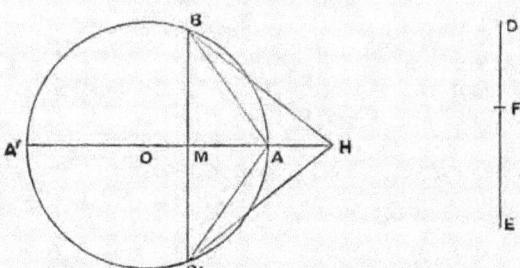

Produce OA to H so that
$$OA' + A'M : A'M = HM : MA \ldots \ldots \ldots (1).$$

Thus the cone HBB' is equal to the segment ABB'.   [Prop. 2]

Therefore the given ratio must be equal to the ratio of the cone HBB' to the cone ABB', *i.e.* to the ratio $HM : MA$.

Hence the ratio $OA' + A'M : A'M$ is given; and therefore $A'M$ is given.

διορισμός.

Now $\qquad OA' : A'M > OA' : A'A,$

so that $\qquad OA' + A'M : A'M > OA' + A'A : A'A$

$$> 3 : 2.$$

Thus, *in order that a solution may be possible, it is a necessary condition that the given ratio must be greater than* $3 : 2$.

The **synthesis** proceeds thus.

Let AA' be a diameter of a great circle of the sphere, O the centre.

Take a line DE, and a point F on it, such that $DE : EF$ is equal to the given ratio, being greater than $3 : 2$.

Now, since $\qquad OA' + A'A : A'A = 3 : 2,$

$$DE : EF > OA' + A'A : A'A,$$

so that $\qquad DF : FE > OA' : A'A.$

Hence a point M can be found on AA' such that

$$DF : FE = OA' : A'M. \dots\dots\dots\dots\dots(2).$$

Through M draw a plane at right angles to AA' intersecting the plane of the great circle in BB', and cutting off from the sphere the segment ABB'.

As before, take H on OA produced such that

$$OA' + A'M : A'M = HM : MA.$$

Therefore $HM : MA = DE : EF$, by means of (2).

It follows that the cone HBB', or the segment ABB', is to the cone ABB' in the given ratio $DE : EF$.

Proposition 8.

If a sphere be cut by a plane not passing through the centre into two segments A'BB', ABB', of which A'BB' is the greater, then the ratio

(segmt. $A'BB'$) : (segmt. ABB')

$$< (surface\ of\ A'BB')^2 : (surface\ of\ ABB')^2$$

$$but > (surface\ of\ A'BB')^{\frac{3}{2}} : (surface\ of\ ABB')^{\frac{3}{2}*}.$$

Let the plane of section cut a great circle $A'BAB'$ at right angles in BB', and let AA' be the diameter bisecting BB' at right angles in M.

Let O be the centre of the sphere.

Join $A'B, AB$.

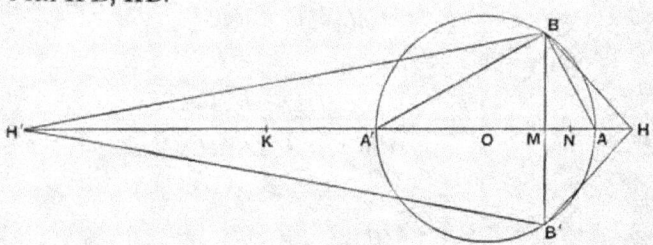

As usual, take H on OA produced, and H' on OA' produced, so that

$$OA' + A'M : A'M = HM : MA \dots \dots \dots \dots (1),$$

$$OA + AM : AM = H'M : MA' \dots \dots \dots (2),$$

and conceive cones drawn each with the same base as the two segments and with apices H, H' respectively. The cones are then respectively equal to the segments [Prop. 2], and they are in the ratio of their heights $HM, H'M$.

Also

(surface of $A'BB'$) : (surface of ABB') $= A'B^2 : AB^2$ [I. 42, 43]

$$= A'M : AM.$$

---

* This is expressed in Archimedes' phrase by saying that the greater segment has to the lesser a ratio "less than the duplicate (διπλάσιον) of that which the surface of the greater segment has to the surface of the lesser, but greater than the sesquialterate (ἡμιόλιον) [of that ratio]."

We have therefore to prove

(a) that $\qquad H'M : MH < A'M^2 : MA^2$,

(b) that $\qquad H'M : MH > A'M^3 : MA^3$.

(a) From (2) above,

$$A'M : AM = H'M : OA + AM$$
$$= H'A' : OA', \text{ since } OA = OA'.$$

Since $A'M > AM$, $H'A' > OA'$; therefore, if we take K on $H'A'$ so that $OA' = A'K$, K will fall between H' and A'.

And, by (1), $\qquad A'M : AM = KM : MH$.

Thus $\qquad KM : MH = H'A' : A'K$, since $A'K = OA'$,
$$> H'M : MK.$$

Therefore $\qquad H'M . MH < KM^2$.

It follows that

$$H'M . MH : MH^2 < KM^2 : MH^2,$$
or $\qquad H'M : MH < KM^2 : MH^2$
$$< A'M^2 : AM^2, \text{ by (1)}.$$

(b) Since $OA' = OA$,
$$A'M . MA < A'O . OA,$$
or $\qquad A'M : OA' < OA : AM$
$$< H'A' : A'M, \text{ by means of (2)}.$$

Therefore $\qquad A'M^2 < H'A' . OA'$
$$< H'A' . A'K.$$

Take a point N on $A'A$ such that
$$A'N^2 = H'A' . A'K.$$

Thus $\qquad H'A' : A'K = A'N^2 : A'K^2$ ....(3).

Also $\qquad H'A' : A'N = A'N : A'K$,

and, componendo,

$$H'N : A'N = NK : A'K,$$
whence $\qquad A'N^2 : A'K^2 = H'N^2 : NK^2$.

Therefore, by (3),

$$H'A' : A'K = H'N^2 : NK^2.$$

Now $\qquad H'M : MK > H'N : NK.$

Therefore $\quad H'M^2 : MK^2 > H'A' : A'K$

$$> H'A' : OA'$$

$$> A'M : MA, \text{ by (2), as above,}$$

$$> OA' + A'M : MH, \text{ by (1),}$$

$$> KM : MH.$$

Hence $\qquad H'M^2 : MH^2 = (H'M^2 : MK^2).(KM^2 : MH^2)$

$$> (KM : MH).(KM^2 : MH^2).$$

It follows that

$$H'M : MH > KM^{\frac{3}{2}} : MH^{\frac{3}{2}}$$

$$> A'M^{\frac{3}{2}} : AM^{\frac{3}{2}}, \text{ by (1).}$$

[The text of Archimedes adds an alternative proof of this proposition, which is here omitted because it is in fact neither clearer nor shorter than the above.]

Proposition 9.

Of all segments of spheres which have equal surfaces the hemisphere is the greatest in volume.

Let $ABA'B'$ be a great circle of a sphere, AA' being a diameter, and O the centre. Let the sphere be cut by a plane, not passing through O, perpendicular to AA' (at M), and intersecting the plane of the great circle in BB'. The segment ABB' may then be either less than a hemisphere as in Fig. 1, or greater than a hemisphere as in Fig. 2.

Let $DED'E'$ be a great circle of another sphere, DD' being a diameter and C the centre. Let the sphere be cut by a plane through C perpendicular to DD' and intersecting the plane of the great circle in the diameter EE'.

Suppose the surfaces of the segment ABB' and of the hemisphere DEE' to be equal.

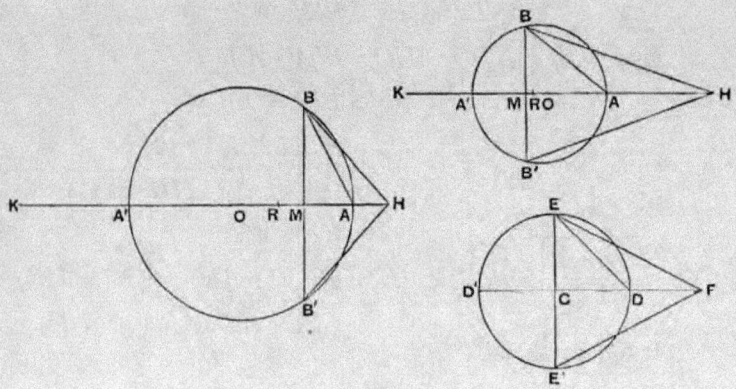

Since the surfaces are equal, $AB = DE$.   [I. 42, 43]

Now, in Fig. 1, $AB^2 > 2AM^2$ and $< 2AO^2$,

and, in Fig. 2, $AB^2 < 2AM^2$ and $> 2AO^2$.

Hence, if R be taken on AA' such that

$$AR^2 = \tfrac{1}{2}AB^2,$$

R will fall between O and M.

Also, since $AB^2 = DE^2$, $AR = CD$.

Produce OA' to K so that $OA' = A'K$, and produce $A'A$ to H so that

$$A'K : A'M = HA : AM,$$

or, *componendo*, $A'K + A'M : A'M = HM : MA$...........(1).

Thus the cone HBB' is equal to the segment ABB'.

[Prop. 2]

Again, produce CD to F so that $CD = DF$, and the cone FEE' will be equal to the hemisphere DEE'.   [Prop. 2]

Now $AR \cdot RA' > AM \cdot MA'$,

and $AR^2 = \tfrac{1}{2}AB^2 = \tfrac{1}{2}AM \cdot AA' = AM \cdot A'K$.

Hence

$$AR \cdot RA' + RA^2 > AM \cdot MA' + AM \cdot A'K,$$

or $\qquad AA' \cdot AR > AM \cdot MK$

$$> HM \cdot A'M, \text{ by (1)}.$$

Therefore $\quad AA' : A'M > HM : AR,$

or $\qquad\qquad AB^2 : BM^2 > HM : AR,$

i.e. $\qquad\qquad AR^2 : BM^2 > HM : 2AR, \text{ since } AB^2 = 2AR^2,$

$$> HM : CF.$$

Thus, since $AR = CD$, or CE,

(circle on diam. EE') : (circle on diam. BB') $> HM : CF$.

It follows that

(the cone FEE') $>$ (the cone HBB'),

and therefore the hemisphere DEE' is greater in volume than the segment ABB'.

MEASUREMENT OF A CIRCLE.

Proposition 1.

The area of any circle is equal to a right-angled triangle in which one of the sides about the right angle is equal to the radius, and the other to the circumference, of the circle.

Let $ABCD$ be the given circle, K the triangle described.

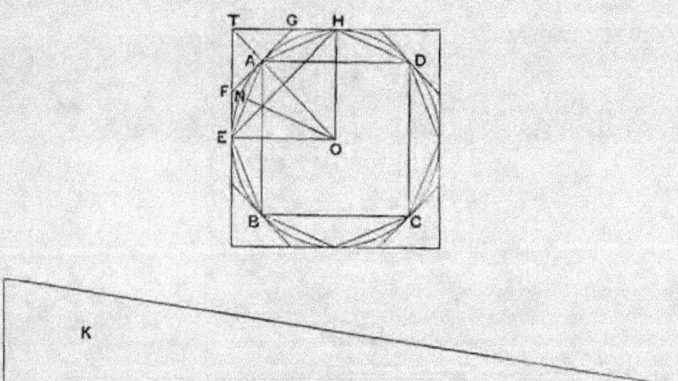

Then, if the circle is not equal to K, it must be either greater or less.

I. If possible, let the circle be greater than K.

Inscribe a square $ABCD$, bisect the arcs AB, BC, CD, DA, then bisect (if necessary) the halves, and so on, until the sides of the inscribed polygon whose angular points are the points of division subtend segments whose sum is less than the excess of the area of the circle over K.

Thus the area of the polygon is greater than K.

Let AE be any side of it, and ON the perpendicular on AE from the centre O.

Then ON is less than the radius of the circle and therefore less than one of the sides about the right angle in K. Also the perimeter of the polygon is less than the circumference of the circle, i.e. less than the other side about the right angle in K.

Therefore the area of the polygon is less than K; which is inconsistent with the hypothesis.

Thus the area of the circle is not greater than K.

II. If possible, let the circle be less than K.

Circumscribe a square, and let two adjacent sides, touching the circle in $E, H,$ meet in T. Bisect the arcs between adjacent points of contact and draw the tangents at the points of bisection. Let A be the middle point of the arc EH, and FAG the tangent at A.

Then the angle TAG is a right angle.

Therefore $$TG > GA$$
$$> GH.$$

It follows that the triangle FTG is greater than half the area $TEAH$.

Similarly, if the arc AH be bisected and the tangent at the point of bisection be drawn, it will cut off from the area GAH more than one-half.

Thus, by continuing the process, we shall ultimately arrive at a circumscribed polygon such that the spaces intercepted between it and the circle are together less than the excess of K over the area of the circle.

Thus the area of the polygon will be less than K.

Now, since the perpendicular from O on any side of the polygon is equal to the radius of the circle, while the perimeter of the polygon is greater than the circumference of the circle, it follows that the area of the polygon is greater than the triangle K; which is impossible.

Therefore the area of the circle is not less than K.

Since then the area of the circle is neither greater nor less than K, it is equal to it.

Proposition 2.

The area of a circle is to the square on its diameter as 11 *to* 14.

[The text of this proposition is not satisfactory, and Archimedes cannot have placed it before Proposition 3, as the approximation depends upon the result of that proposition.]

Proposition 3.

The ratio of the circumference of any circle to its diameter is less than $3\frac{1}{7}$ *but greater than* $3\frac{10}{71}$.

[In view of the interesting questions arising out of the arithmetical content of this proposition of Archimedes, it is necessary, in reproducing it, to distinguish carefully the actual steps set out in the text as we have it from the intermediate steps (mostly supplied by Eutocius) which it is convenient to put in for the purpose of making the proof easier to follow. Accordingly all the steps not actually appearing in the text have been enclosed in square brackets, in order that it may be clearly seen how far Archimedes omits actual calculations and only gives results. It will be observed that he gives two fractional approximations to $\sqrt{3}$ (one being less and the other greater than the real value) without any explanation as to how he arrived at them; and in like manner approximations to the square roots of several large numbers which are not complete squares are merely stated. These various approximations and the machinery of Greek arithmetic in general will be found discussed in the Introduction, Chapter IV.]

I. Let AB be the diameter of any circle, O its centre, AC the tangent at A; and let the angle AOC be one-third of a right angle.

Then $OA : AC [= \sqrt{3} : 1] > 265 : 153$............ (1),
and $OC : CA [= 2 : 1] = 306 : 153$............... (2).

First, draw OD bisecting the angle AOC and meeting AC in D.

Now $CO : OA = CD : DA,$ [Eucl. VI. 3]

so that $[CO + OA : OA = CA : DA,$ or]

$CO + OA : CA = OA : AD.$

Therefore [by (1) and (2)]

$OA : AD > 571 : 153$ .... (3).

Hence $OD^2 : AD^2 [= (OA^2 + AD^2) : AD^2$

$> (571^2 + 153^2) : 153^2]$

$> 349450 : 23409,$

so that $OD : DA > 591\frac{1}{8} : 153$ .... (4).

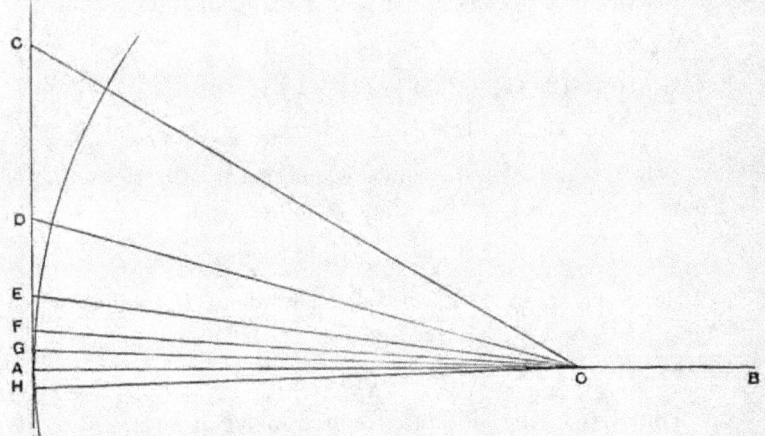

Secondly, let OE bisect the angle AOD, meeting AD in E.

[Then $DO : OA = DE : EA,$

so that $DO + OA : DA = OA : AE.$]

Therefore $OA : AE [> (591\frac{1}{8} + 571) : 153,$ by (3) and (4)]

$> 1162\frac{1}{8} : 153$.................... (5).

[It follows that

$$OE^2 : EA^2 > \{(1162\tfrac{1}{8})^2 + 153^2\} : 153^2$$

$$> (1350534\tfrac{33}{64} + 23409) : 23409$$

$$> 1373943\tfrac{33}{64} : 23409.]$$

Thus $OE : EA > 1172\tfrac{1}{8} : 153$......................(6).

Thirdly, let OF bisect the angle AOE and meet AE in F.

We thus obtain the result [corresponding to (3) and (5) above] that

$$OA : AF \,[> (1162\tfrac{1}{8} + 1172\tfrac{1}{8}) : 153]$$

$$> 2334\tfrac{1}{4} : 153$$......................(7).

[Therefore $OF^2 : FA^2 > \{(2334\tfrac{1}{4})^2 + 153^2\} : 153^2$

$$> 5472132\tfrac{1}{16} : 23409.]$$

Thus $OF : FA > 2339\tfrac{1}{4} : 153$......................(8).

Fourthly, let OG bisect the angle AOF, meeting AF in G.

We have then

$$OA : AG \,[> (2334\tfrac{1}{4} + 2339\tfrac{1}{4}) : 153, \text{ by means of (7) and (8)]}$$

$$> 4673\tfrac{1}{2} : 153.$$

Now the angle AOC, which is one-third of a right angle, has been bisected four times, and it follows that

$$\angle AOG = \tfrac{1}{48} \text{ (a right angle).}$$

Make the angle AOH on the other side of OA equal to the angle AOG, and let GA produced meet OH in H.

Then $\angle GOH = \tfrac{1}{24}$ (a right angle).

Thus GH is one side of a regular polygon of 96 sides circumscribed to the given circle.

And, since $OA : AG > 4673\tfrac{1}{2} : 153$,

while $AB = 2OA,\quad GH = 2AG$,

it follows that

AB : (perimeter of polygon of 96 sides) $[> 4673\tfrac{1}{2} : 153 \times 96]$

$$> 4673\tfrac{1}{2} : 14688,$$

But $\qquad \dfrac{14688}{4673\frac{1}{2}} = 3 + \dfrac{667\frac{1}{2}}{4673\frac{1}{2}}$

$$\left[< 3 + \dfrac{667\frac{1}{2}}{4672\frac{1}{2}} \right]$$

$$< 3\tfrac{1}{7}.$$

Therefore the circumference of the circle (being less than the perimeter of the polygon) is *a fortiori* less than $3\frac{1}{7}$ times the diameter AB.

II.   Next let AB be the diameter of a circle, and let AC, meeting the circle in C, make the angle CAB equal to one-third of a right angle.   Join BC.

Then $\qquad AC : CB\,[= \sqrt{3} : 1] < 1351 : 780.$

First, let AD bisect the angle BAC and meet BC in d and the circle in D.   Join BD.

Then $\qquad\qquad\qquad \angle BAD = \angle dAC$

$$= \angle dBD,$$

and the angles at D, C are both right angles.

It follows that the triangles ADB, $[ACd]$, BDd are similar.

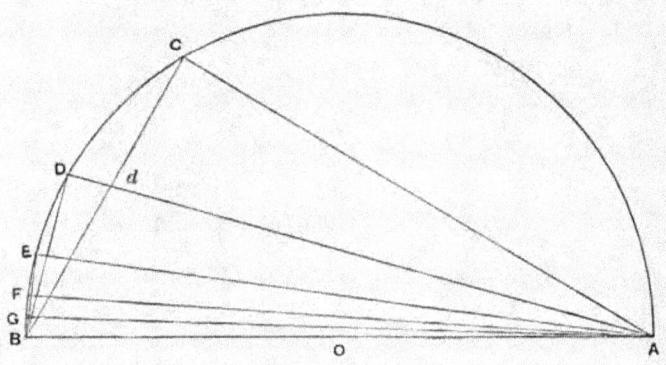

Therefore $\qquad AD : DB = BD : Dd$

$$[= AC : Cd]$$

$$= AB : Bd \qquad\qquad \text{[Eucl. VI. 3]}$$

$$= AB + AC : Bd + Cd$$

$$= AB + AC : BC$$

or $\qquad\qquad BA + AC : BC = AD : DB.$

[But $\quad AC : CB < 1351 : 780$, from above,

while $\qquad\qquad BA : BC = 2 : 1$

$\qquad\qquad\qquad = 1560 : 780.$]

Therefore $\qquad AD : DB < 2911 : 780$ ....(1).

[Hence $\qquad AB^2 : BD^2 < (2911^2 + 780^2) : 780^2$

$\qquad\qquad\qquad < 9082321 : 608400.$]

Thus $\qquad AB : BD < 3013\frac{3}{4} : 780$ ....(2).

Secondly, let AE bisect the angle BAD, meeting the circle in E; and let BE be joined.

Then we prove, in the same way as before, that

$\qquad AE : EB [= BA + AD : BD$

$\qquad\qquad < (3013\frac{3}{4} + 2911) : 780$, by (1) and (2)]

$\qquad\qquad < 5924\frac{3}{4} : 780$

$\qquad\qquad < 5924\frac{3}{4} \times \frac{4}{13} : 780 \times \frac{4}{13}$

$\qquad\qquad < 1823 : 240$ ....(3).

[Hence $\quad AB^2 : BE^2 < (1823^2 + 240^2) : 240^2$

$\qquad\qquad\qquad < 3380929 : 57600.$]

Therefore $\qquad AB : BE < 1838\frac{9}{11} : 240$ ....(4).

Thirdly, let AF bisect the angle BAE, meeting the circle in F.

Thus $\qquad AF : FB [= BA + AE : BE$

$\qquad\qquad < 3661\frac{9}{11} : 240$, by (3) and (4)]

$\qquad\qquad < 3661\frac{9}{11} \times \frac{11}{40} : 240 \times \frac{11}{40}$

$\qquad\qquad < 1007 : 66$ ....(5).

[It follows that

$\qquad\qquad AB^2 : BF^2 < (1007^2 + 66^2) : 66^2$

$\qquad\qquad\qquad < 1018405 : 4356.$]

Therefore $\qquad AB : BF < 1009\frac{1}{6} : 66$ ....(6).

Fourthly, let the angle BAF be bisected by AG meeting the circle in G.

Then $\qquad AG : GB [= BA + AF : BF]$

$\qquad\qquad < 2016\frac{1}{6} : 66$, by (5) and (6).

H. A.

[And $AB^2 : BG^2 < \{(2016\frac{1}{6})^2 + 66^2\} : 66^2$

                    $< 4069284\frac{1}{36} : 4356.$]

Therefore $AB : BG < 2017\frac{1}{4} : 66,$

whence $BG : AB > 66 : 2017\frac{1}{4}$.....................(7).

[Now the angle BAG which is the result of the fourth bisection of the angle BAC, or of one-third of a right angle, is equal to one-fortyeighth of a right angle.

Thus the angle subtended by BG at the centre is

$$\tfrac{1}{24} \text{ (a right angle).]}$$

Therefore BG is a side of a regular inscribed polygon of 96 sides.

It follows from (7) that

(perimeter of polygon) : $AB \, [> 96 \times 66 : 2017\frac{1}{4}]$

                    $> 6336 : 2017\frac{1}{4}.$

And $\dfrac{6336}{2017\frac{1}{4}} > 3\frac{10}{71}.$

Much more then is the circumference of the circle greater than $3\frac{10}{71}$ times the diameter.

Thus the ratio of the circumference to the diameter

$$< 3\tfrac{1}{7} \text{ but } > 3\tfrac{10}{71}.$$

ON CONOIDS AND SPHEROIDS.

Introduction*.

"ARCHIMEDES to Dositheus greeting.

In this book I have set forth and send you the proofs of the remaining theorems not included in what I sent you before, and also of some others discovered later which, though I had often tried to investigate them previously, I had failed to arrive at because I found their discovery attended with some difficulty. And this is why even the propositions themselves were not published with the rest. But afterwards, when I had studied them with greater care, I discovered what I had failed in before.

Now the remainder of the earlier theorems were propositions concerning the right-angled conoid [paraboloid of revolution]; but the discoveries which I have now added relate to an obtuse-angled conoid [hyperboloid of revolution] and to spheroidal figures, some of which I call *oblong* (παραμάκεα) and others *flat* (ἐπιπλατέα).

I. Concerning the *right-angled conoid* it was laid down that, if a section of a right-angled cone [a parabola] be made to revolve about the diameter [axis] which remains fixed and

---

* The whole of this introductory matter, including the definitions, is translated literally from the Greek text in order that the terminology of Archimedes may be faithfully represented. When this has once been set out, nothing will be lost by returning to modern phraseology and notation. These will accordingly be employed, as usual, when we come to the actual propositions of the treatise.

return to the position from which it started, the figure comprehended by the section of the right-angled cone is called a **right-angled conoid,** and the diameter which has remained fixed is called its **axis,** while its **vertex** is the point in which the axis meets (ἅπτεται) the surface of the conoid. And if a plane touch the right-angled conoid, and another plane drawn parallel to the tangent plane cut off a segment of the conoid, the **base** of the segment cut off is defined as the portion intercepted by the section of the conoid on the cutting plane, the **vertex** [of the segment] as the point in which the first plane touches the conoid, and the **axis** [of the segment] as the portion cut off within the segment from the line drawn through the vertex of the segment parallel to the axis of the conoid.

The questions propounded for consideration were

(1) why, if a segment of the right-angled conoid be cut off by a plane at right angles to the axis, will the segment so cut off be half as large again as the cone which has the same base as the segment and the same axis, and

(2) why, if two segments be cut off from the right-angled conoid by planes drawn in any manner, will the segments so cut off have to one another the duplicate ratio of their axes.

II. Respecting the *obtuse-angled conoid* we lay down the following premisses. If there be in a plane a section of an obtuse-angled cone [a hyperbola], its diameter [axis], and the nearest lines to the section of the obtuse-angled cone [*i.e.* the asymptotes of the hyperbola], and if, the diameter [axis] remaining fixed, the plane containing the aforesaid lines be made to revolve about it and return to the position from which it started, the nearest lines to the section of the obtuse-angled cone [the asymptotes] will clearly comprehend an isosceles cone whose vertex will be the point of concourse of the nearest lines and whose axis will be the diameter [axis] which has remained fixed. The figure comprehended by the section of the obtuse-angled cone is called an **obtuse-angled conoid** [hyperboloid of revolution], its **axis** is the diameter which has remained fixed, and its **vertex** the point in which the axis meets the surface

of the conoid. The cone comprehended by the nearest lines to
the section of the obtuse-angled cone is called [the cone]
enveloping the conoid (περιέχων τὸ κωνοειδές), and the
straight line between the vertex of the conoid and the vertex
of the cone enveloping the conoid is called [the line] **adjacent
to the axis** (ποτεοῦσα τῷ ἄξονι). And if a plane touch the
obtuse-angled conoid, and another plane drawn parallel to the
tangent plane cut off a segment of the conoid, the **base** of
the segment so cut off is defined as the portion intercepted by
the section of the conoid on the cutting plane, the **vertex** [of
the segment] as the point of contact of the plane which touches
the conoid, the **axis** [of the segment] as the portion cut off
within the segment from the line drawn through the vertex of
the segment and the vertex of the cone enveloping the conoid;
and the straight line between the said vertices is called
adjacent to the axis.

Right-angled conoids are all similar; but of obtuse-angled
conoids let those be called similar in which the cones enveloping
the conoids are similar.

The following questions are propounded for consideration,

(1) why, if a segment be cut off from the obtuse-angled
conoid by a plane at right angles to the axis, the segment so
cut off has to the cone which has the same base as the segment
and the same axis the ratio which the line equal to the sum
of the axis of the segment and three times the line adjacent
to the axis bears to the line equal to the sum of the axis of
the segment and twice the line adjacent to the axis, and

(2) why, if a segment of the obtuse-angled conoid be cut
off by a plane not at right angles to the axis, the segment so
cut off will bear to the figure which has the same base as
the segment and the same axis, being a segment of a cone*
(ἀπότμαμα κώνου), the ratio which the line equal to the sum
of the axis of the segment and three times the line adjacent
to the axis bears to the line equal to the sum of the axis of the
segment and twice the line adjacent to the axis.

---

* A *segment of a cone* is defined later (p. 104).

III.   Concerning spheroidal figures we lay down the following premisses.   If a section of an acute-angled cone [ellipse] be
made to revolve about the greater diameter [major axis] which
remains fixed and return to the position from which it started,
the figure comprehended by the section of the acute-angled
cone is called an **oblong spheroid** (παραμᾶκες σφαιροειδές).
But if the section of the acute-angled cone revolve about the
lesser diameter [minor axis] which remains fixed and return
to the position from which it started, the figure comprehended
by the section of the acute-angled cone is called a **flat spheroid**
(ἐπιπλατὺ σφαιροειδές).   In either of the spheroids the **axis**
is defined as the diameter [axis] which has remained fixed, the
vertex as the point in which the axis meets the surface of the
spheroid, the **centre** as the middle point of the axis, and the
diameter as the line drawn through the centre at right angles
to the axis.   And, if parallel planes touch, without cutting,
either of the spheroidal figures, and if another plane be drawn
parallel to the tangent planes and cutting the spheroid, the
base of the resulting segments is defined as the portion intercepted by the section of the spheroid on the cutting plane, their
vertices as the points in which the parallel planes touch the
spheroid, and their **axes** as the portions cut off within the
segments from the straight line joining their vertices.   And
that the planes touching the spheroid meet its surface at one
point only, and that the straight line joining the points of
contact passes through the centre of the spheroid, we shall
prove.   Those spheroidal figures are called **similar** in which
the axes have the same ratio to the 'diameters.'   And let
segments of spheroidal figures and conoids be called **similar** if
they are cut off from similar figures and have their bases
similar, while their axes, being either at right angles to the
planes of the bases or making equal angles with the corresponding diameters [axes] of the bases, have the same ratio
to one another as the corresponding diameters [axes] of the
bases.

The following questions about spheroids are propounded for
consideration,

(1)   why, if one of the spheroidal figures be cut by a plane

through the centre at right angles to the axis, each of the
resulting segments will be double of the cone having the same
base as the segment and the same axis; while, if the plane of
section be at right angles to the axis without passing through
the centre, (*a*) the greater of the resulting segments will bear
to the cone which has the same base as the segment and the
same axis the ratio which the line equal to the sum of half the
straight line which is the axis of the spheroid and the axis of
the lesser segment bears to the axis of the lesser segment, and
(*b*) the lesser segment bears to the cone which has the same
base as the segment and the same axis the ratio which the line
equal to the sum of half the straight line which is the axis
of the spheroid and the axis of the greater segment bears to the
axis of the greater segment;

(2) why, if one of the spheroids be cut by a plane passing
through the centre but not at right angles to the axis, each of
the resulting segments will be double of the figure having the
same base as the segment and the same axis and consisting of a
segment of a cone*.

(3) But, if the plane cutting the spheroid be neither
through the centre nor at right angles to the axis, (*a*) the
greater of the resulting segments will have to the figure
which has the same base as the segment and the same axis
the ratio which the line equal to the sum of half the line
joining the vertices of the segments and the axis of the lesser
segment bears to the axis of the lesser segment, and (*b*) the
lesser segment will have to the figure with the same base
as the segment and the same axis the ratio which the line
equal to the sum of half the line joining the vertices of the
segments and the axis of the greater segment bears to the axis
of the greater segment. And the figure referred to is in these
cases also a segment of a cone*.

When the aforesaid theorems are proved, there are discovered by means of them many theorems and problems.

Such, for example, are the theorems

(1) that similar spheroids and similar segments both of

---

* See the definition of a *segment of a cone* (ἀπότμαμα κώνου) on p. 104.

spheroidal figures and conoids have to one another the triplicate ratio of their axes, and

(2)   that in equal spheroidal figures the squares on the 'diameters' are reciprocally proportional to the axes, and, if in spheroidal figures the squares on the 'diameters' are reciprocally proportional to the axes, the spheroids are equal.

Such also is the problem, From a given spheroidal figure or conoid to cut off a segment by a plane drawn parallel to a given plane so that the segment cut off is equal to a given cone or cylinder or to a given sphere.

After prefixing therefore the theorems and directions ($\dot{\epsilon}\pi\iota$-$\tau\dot{\alpha}\gamma\mu\alpha\tau\alpha$) which are necessary for the proof of them, I will then proceed to expound the propositions themselves to you. Farewell.

DEFINITIONS.

If a cone be cut by a plane meeting all the sides [generators] of the cone, the section will be either a circle or a section of an acute-angled cone [an ellipse]. If then the section be a circle, it is clear that the segment cut off from the cone towards the same parts as the vertex of the cone will be a cone. But, if the section be a section of an acute-angled cone [an ellipse], let the figure cut off from the cone towards the same parts as the vertex of the cone be called a **segment of a cone.** Let the **base** of the segment be defined as the plane comprehended by the section of the acute-angled cone, its **vertex** as the point which is also the vertex of the cone, and its **axis** as the straight line joining the vertex of the cone to the centre of the section of the acute-angled cone.

And if a cylinder be cut by two parallel planes meeting all the sides [generators] of the cylinder, the sections will be either circles or sections of acute-angled cones [ellipses] equal and similar to one another. If then the sections be circles, it is clear that the figure cut off from the cylinder between the parallel planes will be a cylinder. But, if the sections be sections of acute-angled cones [ellipses], let the figure cut off from the cylinder between the parallel planes be called a **frustum** ($\tau\dot{o}\mu\sigma$) **of a cylinder.** And let the **bases** of the frustum be defined as the planes comprehended by the sections of the acute-angled cones [ellipses], and the **axis** as the straight line joining the centres of the sections of the acute-angled cones, so that the axis will be in the same straight line with the axis of the cylinder."

Lemma.

If in an ascending arithmetical progression consisting of the magnitudes A_1, A_2, ... A_n the common difference be equal to the least term A_1, then

$$n \cdot A_n < 2(A_1 + A_2 + ... + A_n),$$

and $\qquad > 2(A_1 + A_2 + ... + A_{n-1}).$

[The proof of this is given incidentally in the treatise *On Spirals*, Prop. 11. By placing lines side by side to represent the terms of the progression and then producing each so as to make it equal to the greatest term, Archimedes gives the equivalent of the following proof.

If $\qquad S_n = A_1 + A_2 + ... + A_{n-1} + A_n,$

we have also $\quad S_n = A_n + A_{n-1} + A_{n-2} + ... + A_1.$

And $\qquad A_1 + A_{n-1} = A_2 + A_{n-2} = ... = A_n.$

Therefore $\qquad 2S_n = (n+1) A_n,$

whence $\qquad n \cdot A_n < 2S_n,$

and $\qquad n \cdot A_n > 2S_{n-1}.$

Thus, if the progression is $a, 2a, ... na,$

$$S_n = \frac{n(n+1)}{2} a,$$

and $\qquad n^2 a < 2S_n,$

but $\qquad > 2S_{n-1}.]$

Proposition 1.

If A_1, B_1, C_1, ...K_1 and A_2, B_2, C_2, ...K_2 be two series of magnitudes such that

$$A_1 : B_1 = A_2 : B_2,$$
$$B_1 : C_1 = B_2 : C_2, \text{ and so on} \left.\right\} \quad ....(\alpha),$$

and if A_3, B_3, C_3, ...K_3 and A_4, B_4, C_4, ...K_4 be two other series such that

$$A_1 : A_3 = A_2 : A_4,$$
$$\left. B_1 : B_3 = B_2 : B_4, \text{ and so on} \right\} \ldots\ldots\ldots(\beta),$$

then $\quad (A_1 + B_1 + C_1 + \ldots + K_1) : (A_3 + B_3 + C_3 + \ldots + K_3)$
$$= (A_2 + B_2 + C_2 + \ldots + K_2) : (A_4 + B_4 + \ldots + K_4).$$

The proof is as follows.

Since $\qquad\qquad A_3 : A_1 = A_4 : A_2,$

and $\qquad\qquad A_1 : B_1 = A_2 : B_2,$

while $\qquad\qquad B_1 : B_3 = B_2 : B_4,$

we have, *ex aequali*, $\quad A_3 : B_3 = A_4 : B_4.$
$$\left. \text{Similarly} \qquad B_3 : C_3 = B_4 : C_4, \text{and so on} \right\} \ldots\ldots(\gamma).$$

Again, it follows from equations (α) that

$$A_1 : A_2 = B_1 : B_2 = C_1 : C_2 = \ldots$$

Therefore

$$A_1 : A_2 = (A_1 + B_1 + C_1 + \ldots + K_1) : (A_2 + B_2 + \ldots + K_2),$$

or $(A_1 + B_1 + C_1 + \ldots + K_1) : A_1 = (A_2 + B_2 + C_2 + \ldots + K_2) : A_2;$

and $\qquad\qquad\qquad A_1 : A_3 = A_2 : A_4,$

while from equations (γ) it follows in like manner that

$$A_3 : (A_3 + B_3 + C_3 + \ldots + K_3) = A_4 : (A_4 + B_4 + C_4 + \ldots + K_4).$$

By the last three equations, *ex aequali*,

$$(A_1 + B_1 + C_1 + \ldots + K_1) : (A_3 + B_3 + C_3 + \ldots + K_3)$$
$$= (A_2 + B_2 + C_2 + \ldots + K_2) : (A_4 + B_4 + C_4 + \ldots + K_4).$$

COR. If any terms in the third and fourth series corresponding to terms in the first and second be left out, the result is the same. For example, if the last terms K_3, K_4 are absent,

$$(A_1 + B_1 + C_1 + \ldots + K_1) : (A_3 + B_3 + C_3 + \ldots + I_3)$$
$$= (A_2 + B_2 + C_2 + \ldots + K_2) : (A_4 + B_4 + C_4 + \ldots + I_4),$$

where I immediately precedes K in each series.

Lemma to Proposition 2.

[On Spirals, Prop. 10.]

If A_1, A_2, A_3, ...A_n be n lines forming an ascending arithmetical progression in which the common difference is equal to the least term A_1, then

$$(n+1) A_n^2 + A_1(A_1 + A_2 + A_3 + ... + A_n)$$
$$= 3 (A_1^2 + A_2^2 + A_3^2 + ... + A_n^2).$$

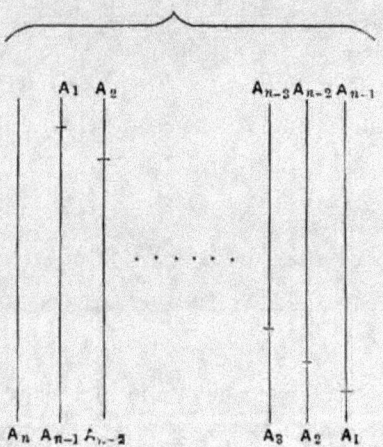

Let the lines A_n, A_{n-1}, A_{n-2}, ...A_1 be placed in a row from left to right. Produce A_{n-1}, A_{n-2}, ...A_1 until they are each equal to A_n, so that the parts produced are respectively equal to A_1, A_2, ...A_{n-1}.

Taking each line successively, we have

$$2A_n^2 = 2A_n^2,$$
$$(A_1 + A_{n-1})^2 = A_1^2 + A_{n-1}^2 + 2A_1 . A_{n-1},$$
$$(A_2 + A_{n-2})^2 = A_2^2 + A_{n-2}^2 + 2A_2 . A_{n-2},$$
$$.................................$$
$$(A_{n-1} + A_1)^2 = A_{n-1}^2 + A_1^2 + 2A_{n-1} . A_1.$$

And, by addition,

$$(n+1)A_n^2 = 2(A_1^2 + A_2^2 + \ldots + A_n^2)$$
$$+ 2A_1 . A_{n-1} + 2A_2 . A_{n-2} + \ldots + 2A_{n-1} . A_1.$$

Therefore, in order to obtain the required result, we have to prove that

$$2(A_1.A_{n-1} + A_2 . A_{n-2} + \ldots + A_{n-1}.A_1) + A_1(A_1 + A_2 + A_3 + \ldots + A_n)$$
$$= A_1^2 + A_2^y + \ldots + A_n^2 \ldots\ldots\ldots (\alpha).$$

Now $2A_2 . A_{n-2} = A_1 . 4A_{n-2}$, because $A_2 = 2A_1$,

$2A_3 . A_{n-3} = A_1 . 6A_{n-3}$, because $A_3 = 3A_1$,

....

$2A_{n-1} . A_1 = A_1 . 2(n-1)A_1.$

It follows that

$$2(A_1.A_{n-1} + A_2.A_{n-2} + \ldots + A_{n-1}.A_1) + A_1(A_1 + A_2 + \ldots + A_n)$$
$$= A_1\{A_n + 3A_{n-1} + 5A_{n-2} + \ldots + (2n-1)A_1\}.$$

And this last expression can be proved to be equal to

$$A_1^2 + A_2^2 + \ldots + A_n^2.$$

For $A_n^2 = A_1(n . A_n)$

$$= A_1\{A_n + (n-1)A_n\}$$

$$= A_1\{A_n + 2(A_{n-1} + A_{n-2} + \ldots + A_1)\},$$

because $(n-1)A_n = A_{n-1} + A_1$

$$+ A_{n-2} + A_2$$

$$+ \ldots\ldots\ldots$$

$$+ A_1 + A_{n-1}.$$

Similarly $A_{n-1}^2 = A_1\{A_{n-1} + 2(A_{n-2} + A_{n-3} + \ldots + A_1)\},$

....

$$A_2^2 = A_1(A_2 + 2A_1),$$
$$A_1^2 = A_1 . A_1;$$

whence, by addition,

$$A_1^2 + A_2^2 + A_3^2 + \ldots + A_n^2$$
$$= A_1\{A_n + 3A_{n-1} + 5A_{n-2} + \ldots + (2n-1)A_1\}.$$

Thus the equation marked (a) above is true; and it follows that

$$(n+1)A_n^2 + A_1(A_1 + A_2 + A_3 + \ldots + A_n) = 3(A_1^2 + A_2^2 + \ldots + A_n^2).$$

COR. 1. *From this it is evident that*

$$n \cdot A_n^2 < 3(A_1^2 + A_2^2 + \ldots + A_n^2) \ldots\ldots\ldots\ldots (1).$$

Also $\quad A_n^2 = A_1\{A_n + 2(A_{n-1} + A_{n-2} + \ldots + A_1)\}$, as above,

so that $\quad A_n^2 > A_1(A_n + A_{n-1} + \ldots + A_1),$

and therefore

$$A_n^2 + A_1(A_1 + A_2 + \ldots + A_n) < 2A_n^2.$$

It follows from the proposition that

$$n \cdot A_n^2 > 3(A_1^2 + A_2^2 + \ldots + A_{n-1}^2) \ldots\ldots\ldots\ldots (2).$$

COR. 2. All these results will hold if we substitute *similar figures* for squares on all the lines; for similar figures are in the duplicate ratio of their sides.

[In the above proposition the symbols A_1, A_2, ...A_n have been used instead of a, $2a$, $3a$, ...na in order to exhibit the geometrical character of the proof; but, if we now substitute the latter terms in the various results, we have (1)

$$(n+1)\, n^2 a^2 + a(a + 2a + \ldots + na)$$
$$= 3\{a^2 + (2a)^2 + (3a)^2 + \ldots + (na)^2\}.$$

Therefore $\quad a^2 + (2a)^2 + (3a)^2 + \ldots + (na)^2$

$$= \frac{a^2}{3}\left\{(n+1)\, n^2 + \frac{n(n+1)}{2}\right\}$$

$$= a^2 \cdot \frac{n(n+1)(2n+1)}{6}.$$

Also (2) $\quad n^3 < 3(1^2 + 2^2 + 3^2 + \ldots + n^2),$

and (3) $\quad n^3 > 3(1^2 + 2^2 + 3^2 + \ldots + \overline{n-1}|^2).]$

Proposition 2.

If $A_1, A_2 \ldots A_n$ be any number of areas such that

$$A_1 = ax + x^2,$$
$$A_2 = a \cdot 2x + (2x)^2,$$
$$A_3 = a \cdot 3x + (3x)^2,$$
$$\ldots\ldots\ldots\ldots\ldots\ldots$$
$$A_n = a \cdot nx + (nx)^2,$$

then $\quad n \cdot A_n : (A_1 + A_2 + \ldots + A_n) < (a + nx) : \left(\dfrac{a}{2} + \dfrac{nx}{3}\right),$

and $\quad n \cdot A_n : (A_1 + A_2 + \ldots + A_{n-1}) > (a + nx) : \left(\dfrac{a}{2} + \dfrac{nx}{3}\right).$

For, by the Lemma immediately preceding Prop. 1,

$$n \cdot anx < 2\,(ax + a \cdot 2x + \ldots + a \cdot nx),$$

and $\qquad\qquad\qquad > 2\,(ax + a \cdot 2x + \ldots + a \cdot \overline{n-1}\,x).$

Also, by the Lemma preceding this proposition,

$$n \cdot (nx)^2 < 3\,\{x^2 + (2x)^2 + (3x)^2 + \ldots + (nx)^2\}$$

and $\qquad\qquad\qquad > 3\,\{x^2 + (2x)^2 + \ldots + (\overline{n-1}\,x)^2\}.$

Hence

$$\frac{an^2x}{2} + \frac{n\,(nx)^2}{3} < [(ax + x^2) + \{a \cdot 2x + (2x)^2\} + \ldots + \{a \cdot nx + (nx)^2\}],$$

and

$$> [(ax + x^2) + \{a \cdot 2x + (2x)^2\} + \ldots + \{a \cdot \overline{n-1}\,x + (\overline{n-1}\,x)^2\}],$$

or $\qquad\qquad \dfrac{an^2x}{2} + \dfrac{n\,(nx)^2}{3} < A_1 + A_2 + \ldots + A_n,$

and $\qquad\qquad\qquad\qquad\qquad > A_1 + A_2 + \ldots + A_{n-1}.$

It follows that

$$n \cdot A_n : (A_1 + A_2 + \ldots + A_n) < n\,\{a \cdot nx + (nx)^2\} : \left\{\frac{an^2x}{2} + \frac{n\,(nx)^2}{3}\right\},$$

or $\qquad n \cdot A_n : (A_1 + A_2 + \ldots + A_n) < (a + nx) : \left(\dfrac{a}{2} + \dfrac{nx}{3}\right);$

also $\qquad n \cdot A_n : (A_1 + A_2 + \ldots + A_{n-1}) > (a + nx) : \left(\dfrac{a}{2} + \dfrac{nx}{3}\right).$

---

* The phraseology of Archimedes here is that associated with the traditional method of application of areas: εἰ κα...παρ' ἑκάσταν αὐτᾶν παραπέσῃ τι χωρίον ὑπερβάλλον εἴδει τετραγώνῳ, "if to each of the lines there be applied a space [rectangle] exceeding by a square figure." Thus A_1 is a rectangle of height x applied to a line a but overlapping it so that the base extends a distance x beyond a.

Proposition 3.

(1) *If TP, TP′ be two tangents to any conic meeting in T, and if Qq, Q′q′ be any two chords parallel respectively to TP, TP′ and meeting in O, then*

$$QO.Oq : Q'O.Oq' = TP^2 : TP'^2.$$

"And this is proved in the elements of conics*."

(2) *If QQ′ be a chord of a parabola bisected in V by the diameter PV, and if PV be of constant length, then the areas of the triangle PQQ′ and of the segment PQQ′ are both constant whatever be the direction of QQ′.*

Let *ABB′* be the particular segment of the parabola whose vertex is *A*, so that *BB′* is bisected perpendicularly by the axis at the point *H*, where *AH = PV*.

Draw *QD* perpendicular to *PV*.

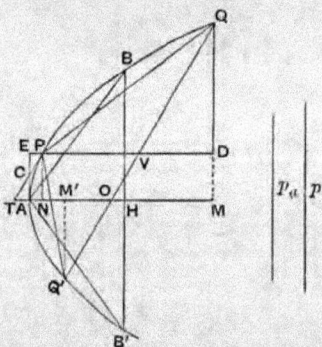

Let p_a be the parameter of the principal ordinates, and let *p* be another line of such length that

$$QV^2 : QD^2 = p : p_a;$$

it will then follow that *p* is equal to the parameter of the ordinates to the diameter *PV*, i.e. those which are parallel to *QV*.

---

* i.e. in the treatises on conics by Aristaeus and Euclid.

"For this is proved in the conics*."

Thus $\qquad\qquad QV^2 = p . PV.$

And $BH^2 = p_a . AH$, while $AH = PV$.

Therefore $\qquad QV^2 : BH^2 = p : p_a.$

But $\qquad\qquad QV^2 : QD^2 = p : p_a;$

hence $\qquad\qquad\qquad BH = QD.$

Thus $\qquad\qquad BH . AH = QD . PV,$

and therefore $\qquad \triangle ABB' = \triangle PQQ';$

that is, the area of the triangle PQQ' is constant so long as PV is of constant length.

Hence also the area of the segment PQQ' is constant under the same conditions; for the segment is equal to $\frac{4}{3}\triangle PQQ'$. [*Quadrature of the Parabola*, Prop. 17 or 24.]

* The theorem which is here assumed by Archimedes as known can be proved in various ways.

(1) It is easily deduced from Apollonius I. 49 (cf. *Apollonius of Perga*, pp. liii, 39). If in the figure the tangents at A and P be drawn, the former meeting PV in E, and the latter meeting the axis in T, and if AE, PT meet at C, the proposition of Apollonius is to the effect that

$$CP : PE = p : 2PT,$$

where p is the parameter of the ordinates to PV.

(2) It may be proved independently as follows.

Let QQ' meet the axis in O, and let QM, $Q'M'$, PN be ordinates to the axis.

Then $\qquad AM : AM' = QM^2 : Q'M'^2 = OM^2 : OM'^2,$

whence $\qquad AM : MM' = OM^2 : OM^2 - OM'^2$

$$= OM^2 : (OM - OM') . MM',$$

so that $\qquad OM^2 = AM . (OM - OM').$

That is to say, $\quad (AM - AO)^2 = AM . (AM + AM' - 2AO),$

or $\qquad\qquad AO^2 = AM . AM'.$

And, since $QM^2 = p_a . AM$, and $Q'M'^2 = p_a . AM'$,

it follows that $\qquad QM . Q'M' = p_a . AO$ .... (a).

Now $\qquad QV^2 : QD^2 = QV^2 : \left(\dfrac{QM + Q'M'}{2} \right)^2$

$$= QV^2 : \left(\dfrac{QM - Q'M'}{2} \right)^2 + QM . Q'M'$$

$$= QV^2 : (PN^2 + QM . Q'M')$$

$$= p . PV : p_a . (AN + AO), \text{ by (a)}.$$

But $\qquad\qquad PV = TO = AN + AO.$

Therefore $\qquad QV^2 : QD^2 = p : p_a.$

Proposition 4.

The area of any ellipse is to that of the auxiliary circle as the minor axis to the major.

Let AA' be the major and BB' the minor axis of the ellipse, and let BB' meet the auxiliary circle in b, b'.

Suppose O to be such a circle that

$$(\text{circle } AbA'b') : O = CA : CB.$$

Then shall O be equal to the area of the ellipse.

For, if not, O must be either greater or less than the ellipse.

I. If possible, let O be greater than the ellipse.

We can then inscribe in the circle O an equilateral polygon of $4n$ sides such that its area is greater than that of the ellipse. [cf. *On the Sphere and Cylinder*, I. 6.]

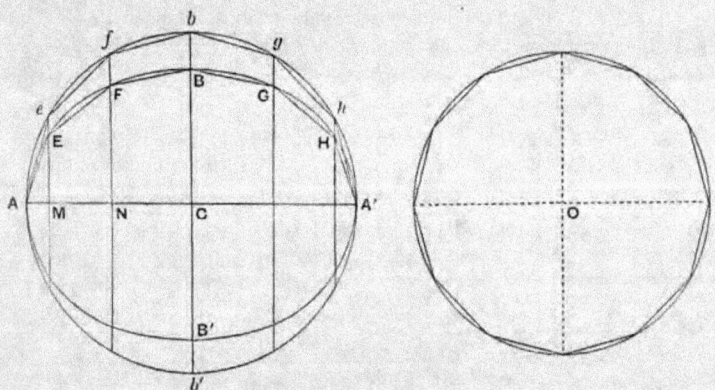

Let this be done, and inscribe in the auxiliary circle of the ellipse the polygon $AefbghA'$... similar to that inscribed in O. Let the perpendiculars eM, fN,... on AA' meet the ellipse in $E, F,$... respectively. Join $AE, EF, FB,$....

Suppose that P' denotes the area of the polygon inscribed in the auxiliary circle, and P that of the polygon inscribed in the ellipse.

Then, since all the lines eM, fN,... are cut in the same proportions at E, F,...,

i.e. $eM : EM = fN : FN = ... = bC : BC$,

the pairs of triangles, as eAM, EAM, and the pairs of trapeziums, as $eMNf$, $EMNF$, are all in the same ratio to one another as bC to BC, or as CA to CB.

Therefore, by addition,

$$P' : P = CA : CB.$$

Now P' : (polygon inscribed in O)

$$= (\text{circle } AbA'b') : O$$

$$= CA : CB, \text{ by hypothesis.}$$

Therefore P is equal to the polygon inscribed in O.

But this is impossible, because the latter polygon is by hypothesis greater than the ellipse, and *a fortiori* greater than P.

Hence O is not greater than the ellipse.

II.   If possible, let O be less than the ellipse.

In this case we inscribe in the *ellipse* a polygon P with $4n$ equal sides such that $P > O$.

Let the perpendiculars from the angular points on the axis AA' be produced to meet the auxiliary circle, and let the corresponding polygon (P') in the circle be formed.

Inscribe in O a polygon similar to P'.

Then $P' : P = CA : CB$

$$= (\text{circle } AbA'b') : O, \text{ by hypothesis,}$$

$$= P' : (\text{polygon inscribed in } O).$$

Therefore the polygon inscribed in O is equal to the polygon P; which is impossible, because $P > O$.

Hence O, being neither greater nor less than the ellipse, is equal to it; and the required result follows.

Proposition 5.

If AA′, BB′ be the major and minor axis of an ellipse respectively, and if d be the diameter of any circle, then

$$(\text{area of ellipse}) : (\text{area of circle}) = AA' . BB' : d^2.$$

For

$$(\text{area of ellipse}) : (\text{area of auxiliary circle}) = BB' : AA' \quad [\text{Prop. 4}]$$
$$= AA' . BB' : AA'^2.$$

And

$$(\text{area of aux. circle}) : (\text{area of circle with diam. } d) = AA'^2 : d^2.$$

Therefore the required result follows *ex aequali.*

Proposition 6.

The areas of ellipses are as the rectangles under their axes.

This follows at once from Props. 4, 5.

COR. *The areas of similar ellipses are as the squares of corresponding axes.*

Proposition 7.

Given an ellipse with centre C, and a line CO drawn perpendicular to its plane, it is possible to find a circular cone with vertex O and such that the given ellipse is a section of it [or, in other words, to find the circular sections of the cone with vertex O passing through the circumference of the ellipse].

Conceive an ellipse with BB' as its minor axis and lying in a plane perpendicular to that of the paper. Let CO be drawn perpendicular to the plane of the ellipse, and let O be the vertex of the required cone. Produce OB, OC, OB', and in the same plane with them draw BED meeting OC, OB' produced in E, D respectively and in such a direction that

$$BE . ED : EO^2 = CA^2 : CO^2,$$

where CA is half the major axis of the ellipse.

8—2

"And this is possible, since

$$BE.ED : EO^2 > BC.CB' : CO^2."$$

[Both the construction and this proposition are assumed as known.]

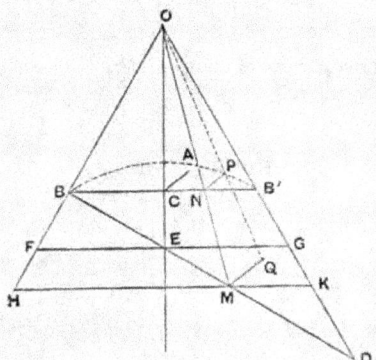

Now conceive a circle with BD as diameter lying in a plane at right angles to that of the paper, and describe a cone with this circle for its base and with vertex O.

We have therefore to prove that the given ellipse is a section of the cone, or, if P be any point on the ellipse, that P lies on the surface of the cone.

Draw PN perpendicular to BB'. Join ON and produce it to meet BD in M, and let MQ be drawn in the plane of the circle on BD as diameter perpendicular to BD and meeting the circle in Q. Also let FG, HK be drawn through E, M respectively parallel to BB'.

We have then

$$QM^2 : HM.MK = BM.MD : HM.MK$$
$$= BE.ED : FE.EG$$
$$= (BE.ED : EO^2).(EO^2 : FE.EG)$$
$$= (CA^2 : CO^2).(CO^2 : BC.CB')$$
$$= CA^2 : CB^2$$
$$= PN^2 : BN.NB'.$$

Therefore $QM^2 : PN^2 = HM . MK : BN . NB'$
$$= OM^2 : ON^2;$$

whence, since PN, QM are parallel, OPQ is a straight line.

But Q is on the circumference of the circle on BD as diameter; therefore OQ is a generator of the cone, and hence P lies on the cone.

Thus the cone passes through all points on the ellipse.

Proposition 8.

Given an ellipse, a plane through one of its axes AA' and perpendicular to the plane of the ellipse, and a line CO drawn from C, the centre, in the given plane through AA' but not perpendicular to AA', it is possible to find a cone with vertex O such that the given ellipse is a section of it [or, in other words, to find the circular sections of the cone with vertex O whose surface passes through the circumference of the ellipse].

By hypothesis, OA, OA' are unequal. Produce OA' to D so that $OA = OD$. Join AD, and draw FG through C parallel to it.

The given ellipse is to be supposed to lie in a plane perpendicular to the plane of the paper. Let BB' be the other axis of the ellipse.

Conceive a plane through AD perpendicular to the plane of the paper, and in it describe either (a), if $CB^2 = FC . CG$, a circle with diameter AD, or (b), if not, an ellipse on AD as axis such that, if d be the other axis,

$$d^2 : AD^2 = CB^2 : FC . CG.$$

Take a cone with vertex O whose surface passes through the circle or ellipse just drawn. .This is possible even when the curve is an ellipse, because the line from O to the middle point of AD is perpendicular to the plane of the ellipse, and the construction is effected by means of Prop. 7.

Let P be any point on the given ellipse, and we have only to prove that P lies on the surface of the cone so described.

Draw PN perpendicular to AA'. Join ON, and produce it to meet AD in M. Through M draw HK parallel to $A'A$.

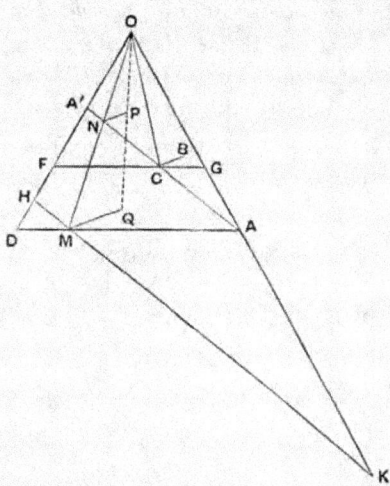

Lastly, draw MQ perpendicular to the plane of the paper (and therefore perpendicular to both HK and AD) meeting the ellipse or circle about AD (and therefore the surface of the cone) in Q.

Then

$$QM^2 : HM.MK = (QM^2 : DM.MA).(DM.MA : HM.MK)$$
$$= (d^2 : AD^2).(FC.CG : A'C.CA)$$
$$= (CB^2 : FC.CG).(FC.CG : A'C.CA)$$
$$= CB^2 : CA^2$$
$$= PN^2 : A'N.NA.$$

Therefore, alternately,

$$QM^2 : PN^2 = HM.MK : A'N.NA$$
$$= OM^2 : ON^2.$$

Thus, since PN, QM are parallel, OPQ is a straight line; and, Q being on the surface of the cone, it follows that P is also on the surface of the cone.

Similarly all points on the ellipse are also on the cone, and the ellipse is therefore a section of the cone.

Proposition 9.

Given an ellipse, a plane through one of its axes and perpendicular to that of the ellipse, and a straight line CO drawn from the centre C of the ellipse in the given plane through the axis but not perpendicular to that axis, it is possible to find a cylinder with axis OC such that the ellipse is a section of it [or, in other words, to find the circular sections of the cylinder with axis OC whose surface passes through the circumference of the given ellipse].

Let AA' be an axis of the ellipse, and suppose the plane of the ellipse to be perpendicular to that of the paper, so that OC lies in the plane of the paper.

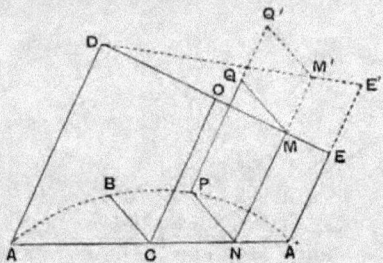

Draw AD, $A'E$ parallel to CO, and let DE be the line through O perpendicular to both AD and $A'E$.

We have now three different cases according as the other axis BB' of the ellipse is (1) equal to, (2) greater than, or (3) less than, DE.

(1) Suppose $BB' = DE$.

Draw a plane through DE at right angles to OC, and in this plane describe a circle on DE as diameter. Through this circle describe a cylinder with axis OC.

This cylinder shall be the cylinder required, or its surface shall pass through every point P of the ellipse.

For, if P be any point on the ellipse, draw PN perpendicular to AA'; through N draw NM parallel to CO meeting DE in M, and through M, in the plane of the circle on DE as diameter, draw MQ perpendicular to DE, meeting the circle in Q.

Then, since $\qquad DE = BB'$,

$$PN^2 : AN.NA' = DO^2 : AC.CA'.$$

And $\qquad DM.ME : AN.NA' = DO^2 : AC'^2,$

since AD, NM, CO, $A'E$ are parallel.

Therefore $\qquad PN^2 = DM.ME$

$$= QM^2,$$

by the property of the circle.

Hence, since PN, QM are equal as well as parallel, PQ is parallel to MN and therefore to CO. It follows that PQ is a generator of the cylinder, whose surface accordingly passes through P.

(2) If $BB' > DE$, we take E' on $A'E$ such that $DE' = BB'$ and describe a circle on DE' as diameter in a plane perpendicular to that of the paper; and the rest of the construction and proof is exactly similar to those given for case (1).

(3) Suppose $BB' < DE$.

Take a point K on CO produced such that

$$DO^2 - CB^2 = OK^2.$$

From K draw KR perpendicular to the plane of the paper and equal to CB.

Thus $\qquad OR^2 = OK^2 + CB^2 = OD^2.$

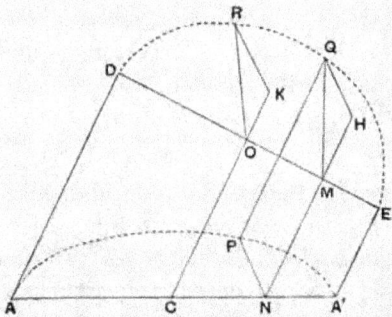

In the plane containing DE, OR describe a circle on DE as diameter. Through this circle (which must pass through R) draw a cylinder with axis OC.

We have then to prove that, if P be any point on the given ellipse, P lies on the cylinder so described.

Draw PN perpendicular to AA', and through N draw NM parallel to CO meeting DE in M. In the plane of the circle on DE as diameter draw MQ perpendicular to DE and meeting the circle in Q.

Lastly, draw QH perpendicular to NM produced. QH will then be perpendicular to the plane containing AC, DE, i.e. the plane of the paper.

Now $\qquad QH^2 : QM^2 = KR^2 : OR^2$, by similar triangles.

And $\quad QM^2 : AN . NA' = DM . ME : AN . NA'$
$$= OD^2 : CA^2.$$

Hence, *ex aequali*, since $OR = OD$,
$$QH^2 : AN . NA' = KR^2 : CA^2$$
$$= CB^2 : CA^2$$
$$= PN^2 : AN . NA'.$$

Thus $QH = PN$. And QH, PN are also parallel. Accordingly PQ is parallel to MN, and therefore to CO, so that PQ is a generator, and the cylinder passes through P.

Proposition 10.

It was proved by the earlier geometers that *any two cones have to one another the ratio compounded of the ratios of their bases and of their heights*[*]. The same method of proof will show that *any segments of cones have to one another the ratio compounded of the ratios of their bases and of their heights*.

The proposition that *any 'frustum' of a cylinder is triple of the conical segment which has the same base as the frustum and equal height* is also proved in the same manner as the proposition that *the cylinder is triple of the cone which has the same base as the cylinder and equal height*[†].

---

[*] This follows from Eucl. xii. 11 and 14 taken together. Cf. *On the Sphere and Cylinder* i, Lemma 1.

[†] This proposition was proved by Endoxus, as stated in the preface to *On the Sphere and Cylinder* i. Cf. Eucl. xii. 10.

Proposition 11.

(1) *If a paraboloid of revolution be cut by a plane through, or parallel to, the axis, the section will be a parabola equal to the original parabola which by its revolution generates the paraboloid. And the axis of the section will be the intersection between the cutting plane and the plane through the axis of the paraboloid at right angles to the cutting plane.*

If the paraboloid be cut by a plane at right angles to its axis, the section will be a circle whose centre is on the axis.

(2) *If a hyperboloid of revolution be cut by a plane through the axis, parallel to the axis, or through the centre, the section will be a hyperbola, (a) if the section be through the axis, equal, (b) if parallel to the axis, similar, (c) if through the centre, not similar, to the original hyperbola which by its revolution generates the hyperboloid. And the axis of the section will be the intersection of the cutting plane and the plane through the axis of the hyperboloid at right angles to the cutting plane.*

Any section of the hyperboloid by a plane at right angles to the axis will be a circle whose centre is on the axis.

(3) *If any of the spheroidal figures be cut by a plane through the axis or parallel to the axis, the section will be an ellipse, (a) if the section be through the axis, equal, (b) if parallel to the axis, similar, to the ellipse which by its revolution generates the figure. And the axis of the section will be the intersection of the cutting plane and the plane through the axis of the spheroid at right angles to the cutting plane.*

If the section be by a plane at right angles to the axis of the spheroid, it will be a circle whose centre is on the axis.

(4) *If any of the said figures be cut by a plane through the axis, and if a perpendicular be drawn to the plane of section from any point on the surface of the figure but not on the section, that perpendicular will fall within the section.*

"And the proofs of all these propositions are evident."[*]

[*] Cf. the Introduction, chapter III. § 4.

Proposition 12.

If a paraboloid of revolution be cut by a plane neither parallel nor perpendicular to the axis, and if the plane through the axis perpendicular to the cutting plane intersect it in a straight line of which the portion intercepted within the paraboloid is RR', the section of the paraboloid will be an ellipse whose major axis is RR' and whose minor axis is equal to the perpendicular distance between the lines through R, R' parallel to the axis of the paraboloid.

Suppose the cutting plane to be perpendicular to the plane of the paper, and let the latter be the plane through the axis *ANF* of the paraboloid which intersects the cutting plane at right angles in *RR'*. Let *RH* be parallel to the axis of the paraboloid, and *R'H* perpendicular to *RH*.

Let *Q* be any point on the section made by the cutting plane, and from *Q* draw *QM* perpendicular to *RR'*. *QM* will therefore be perpendicular to the plane of the paper.

Through *M* draw *DMFE* perpendicular to the axis *ANF* meeting the parabolic section made by the plane of the paper in *D, E.* Then *QM* is perpendicular to *DE*, and, if a plane be drawn through *DE*, *QM*, it will be perpendicular to the axis and will cut the paraboloid in a circular section.

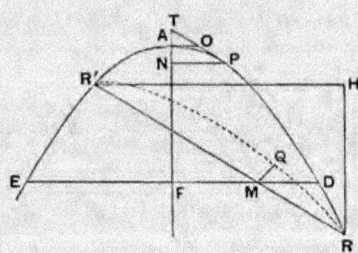

Since *Q* is on this circle,

$$QM^2 = DM \cdot ME.$$

Again, if *PT* be that tangent to the parabolic section in the plane of the paper which is parallel to RR', and if the tangent at A meet PT in O, then, from the property of the parabola,

$$DM.ME:RM.MR'=AO^2:OP^2 \qquad \text{[Prop. 3 (1)]}$$
$$= AO^2:OT^2, \text{ since } AN=AT.$$

Therefore $\quad QM^2:RM.MR'=AO^2:OT^2$
$$= R'H^2:RR'^2,$$

by similar triangles.

Hence Q lies on an ellipse whose major axis is RR' and whose minor axis is equal to $R'H$.

Propositions 13, 14.

If a hyperboloid of revolution be cut by a plane meeting all the generators of the enveloping cone, or if an 'oblong' spheroid be cut by a plane not perpendicular to the axis*, and if a plane through the axis intersect the cutting plane at right angles in a straight line on which the hyperboloid or spheroid intercepts a length RR', then the section by the cutting plane will be an ellipse whose major axis is RR'.

Suppose the cutting plane to be at right angles to the plane of the paper, and suppose the latter plane to be that

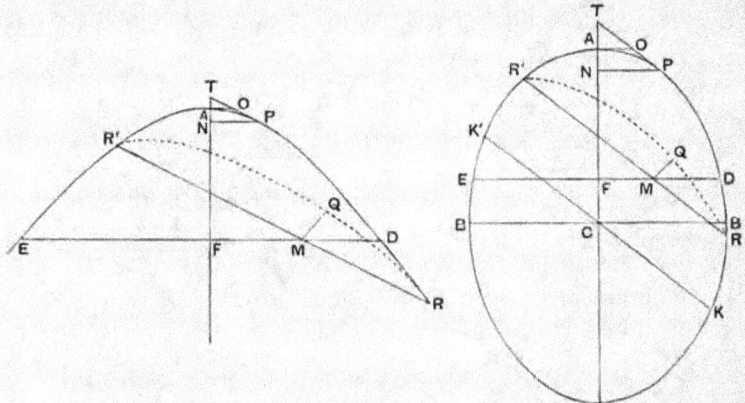

* Archimedes begins Prop. 14 for the *spheroid* with the remark that, when the cutting plane passes through or is parallel to the axis, the case is clear ($\delta \tilde{\eta} \lambda o \nu$). Cf. Prop. 11 (3).

through the axis ANF which intersects the cutting plane at right angles in RR'. The section of the hyperboloid or spheroid by the plane of the paper is thus a hyperbola or ellipse having ANF for its transverse or major axis.

Take any point on the section made by the cutting plane, as Q, and draw QM perpendicular to RR'. QM will then be perpendicular to the plane of the paper.

Through M draw DFE at right angles to the axis ANF meeting the hyperbola or ellipse in D, E; and through QM, DE let a plane be described. This plane will accordingly be perpendicular to the axis and will cut the hyperboloid or spheroid in a circular section.

Thus $$QM^2 = DM \cdot ME.$$

Let PT be that tangent to the hyperbola or ellipse which is parallel to RR', and let the tangent at A meet PT in O.

Then, by the property of the hyperbola or ellipse,

$$DM \cdot ME : RM \cdot MR' = OA^2 : OP^2,$$

or $$QM^2 : RM \cdot MR' = OA^2 : OP^2.$$

Now (1) in the hyperbola $OA < OP$, because $AT < AN^*$, and accordingly $OT < OP$, while $OA < OT$,

(2) in the ellipse, if KK' be the diameter parallel to RR', and BB' the minor axis,

$$BC \cdot CB' : KC \cdot CK' = OA^2 : OP^2;$$

and $BC \cdot CB' < KC \cdot CK'$, so that $OA < OP$.

Hence in both cases the locus of Q is an ellipse whose major axis is RR'.

Cor. 1. If the spheroid be a 'flat' spheroid, the section will be an ellipse, and everything will proceed as before except that RR' will in this case be the *minor* axis.

Cor. 2. In all conoids or spheroids parallel sections will be similar, since the ratio $OA^2 : OP^2$ is the same for all the parallel sections.

---

* With reference to this assumption cf. the Introduction, chapter III. § 3.

Proposition 15.

(1) *If from any point on the surface of a conoid a line be drawn, in the case of the paraboloid, parallel to the axis, and, in the case of the hyperboloid, parallel to any line passing through the vertex of the enveloping cone, the part of the straight line which is in the same direction as the convexity of the surface will fall without it, and the part which is in the other direction within it.*

For, if a plane be drawn, in the case of the paraboloid, through the axis and the point, and, in the case of the hyperboloid, through the given point and through the given straight line drawn through the vertex of the enveloping cone, the section by the plane will be (*a*) in the paraboloid a parabola whose axis is the axis of the paraboloid, (*b*) in the hyperboloid a hyperbola in which the given line through the vertex of the enveloping cone is a diameter*.   [Prop. 11]

Hence the property follows from the plane properties of the conics.

(2) *If a plane touch a conoid without cutting it, it will touch it at one point only, and the plane drawn through the point of contact and the axis of the conoid will be at right angles to the plane which touches it.*

For, if possible, let the plane touch at two points. Draw through each point a parallel to the axis. The plane passing through both parallels will therefore either pass through, or be parallel to, the axis. Hence the section of the conoid made by this plane will be a conic [Prop. 11 (1), (2)], the two points will lie on this conic, and the line joining them will lie within the conic and therefore within the conoid. But this line will be in the tangent plane, since the two points are in it. Therefore some portion of the tangent plane will be within the conoid; which is impossible, since the plane does not cut it.

---

* There seems to be some error in the text here, which says that "the *diameter*" (i.e. axis) of the hyperbola is "the straight line drawn in the conoid from the vertex of the cone." But this straight line is not, in general, the *axis* of the section.

Therefore the tangent plane touches in one point only.

That the plane through the point of contact and the axis is perpendicular to the tangent plane is evident in the particular case where the point of contact is the vertex of the conoid. For, if two planes through the axis cut it in two conics, the tangents at the vertex in both conics will be perpendicular to the axis of the conoid. And all such tangents will be in the tangent plane, which must therefore be perpendicular to the axis and to any plane through the axis.

If the point of contact P is not the vertex, draw the plane passing through the axis AN and the point P. It will cut the conoid in a conic whose axis is AN and the tangent plane in a line DPE touching the conic at P. Draw PNP' perpendicular to the axis, and draw a plane through it also perpendicular to the axis. This plane will make a circular section and meet the tangent plane in a tangent to the circle, which will therefore be at right angles to PN. Hence the tangent to the circle will be at right angles to the plane containing PN, AN; and it follows that this last plane is perpendicular to the tangent plane.

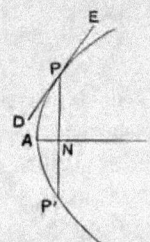

Proposition 16.

(1) *If a plane touch any of the spheroidal figures without cutting it, it will touch at one point only, and the plane through the point of contact and the axis will be at right angles to the tangent plane.*

This is proved by the same method as the last proposition.

(2) *If any conoid or spheroid be cut by a plane through the axis, and if through any tangent to the resulting conic a plane be erected at right angles to the plane of section, the plane so erected will touch the conoid or spheroid in the same point as that in which the line touches the conic.*

For it cannot meet the surface at any other point. If it did, the perpendicular from the second point on the cutting plane would be perpendicular also to the tangent to the conic and would therefore fall outside the surface. But it must fall within it.   [Prop. 11 (4)]

(3) *If two parallel planes touch any of the spheroidal figures, the line joining the points of contact will pass through the centre of the spheroid.*

If the planes are at right angles to the axis, the proposition is obvious. If not, the plane through the axis and one point of contact is at right angles to the tangent plane at that point. It is therefore at right angles to the parallel tangent plane, and therefore passes through the second point of contact. Hence both points of contact lie on one plane through the axis, and the proposition is reduced to a plane one.

Proposition 17.

If two parallel planes touch any of the spheroidal figures, and another plane be drawn parallel to the tangent planes and passing through the centre, the line drawn through any point of the circumference of the resulting section parallel to the chord of contact of the tangent planes will fall outside the spheroid.

This is proved at once by reduction to a plane proposition.

Archimedes adds that it is evident that, if the plane parallel to the tangent planes does not pass through the centre, a straight line drawn in the manner described will fall without the spheroid in the direction of the smaller segment but within it in the other direction.

Proposition 18.

Any spheroidal figure which is cut by a plane through the centre is divided, both as regards its surface and its volume, into two equal parts by that plane.

To prove this, Archimedes takes another equal and similar spheroid, divides it similarly by a plane through the centre, and then uses the method of application.

Propositions 19, 20.

Given a segment cut off by a plane from a paraboloid or hyperboloid of revolution, or a segment of a spheroid less than half the spheroid also cut off by a plane, it is possible to inscribe in the segment one solid figure and to circumscribe about it another solid figure, each made up of cylinders or 'frusta' of cylinders of equal height, and such that the circumscribed figure exceeds the inscribed figure by a volume less than that of any given solid.

Let the plane base of the segment be perpendicular to the plane of the paper, and let the plane of the paper be the plane through the axis of the conoid or spheroid which cuts the base of the segment at right angles in BC. The section in the plane of the paper is then a conic BAC.   [Prop. 11]

Let EAF be that tangent to the conic which is parallel to BC, and let A be the point of contact. Through EAF draw a plane parallel to the plane through BC bounding the segment. The plane so drawn will then touch the conoid or spheroid at A.   [Prop. 16]

(1) If the base of the segment is at right angles to the axis of the conoid or spheroid, A will be the vertex of the conoid or spheroid, and its axis AD will bisect BC at right angles.

(2) If the base of the segment is not at right angles to the axis of the conoid or spheroid, we draw AD

(*a*) in the paraboloid, parallel to the axis,

(*b*) in the hyperboloid, through the centre (or the vertex of the enveloping cone),

(*c*) in the spheroid, through the centre,

and in all the cases it will follow that AD bisects BC in D.

Then A will be the vertex of the segment, and AD will be its axis.

Further, the base of the segment will be a circle or an ellipse with BC as diameter or as an axis respectively, and with centre D. We can therefore describe through this circle or ellipse a cylinder or a 'frustum' of a cylinder whose axis is AD. [Prop. 9]

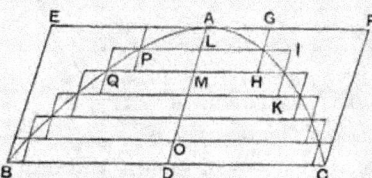

Dividing this cylinder or frustum continually into equal parts by planes parallel to the base, we shall at length arrive at a cylinder or frustum less in volume than any given solid.

Let this cylinder or frustum be that whose axis is OD, and let AD be divided into parts equal to OD, at $L, M, \ldots$. Through $L, M, \ldots$ draw lines parallel to BC meeting the conic in $P, Q, \ldots$, and through these lines draw planes parallel to the base of the segment. These will cut the conoid or spheroid in circles or similar ellipses. On each of these circles or ellipses describe two cylinders or frusta of cylinders each with axis equal to OD, one of them lying in the direction of A and the other in the direction of D, as shown in the figure.

Then the cylinders or frusta of cylinders drawn in the direction of A make up a circumscribed figure, and those in the direction of D an inscribed figure, in relation to the segment.

Also the cylinder or frustum PG in the circumscribed figure is equal to the cylinder or frustum PH in the inscribed figure, QI in the circumscribed figure is equal to QK in the inscribed figure, and so on.

Therefore, by addition,

(circumscribed fig.) = (inscr. fig.)

+ (cylinder or frustum whose axis is OD).

But the cylinder or frustum whose axis is OD is less than the given solid figure; whence the proposition follows.

___

"Having set out these preliminary propositions, let us proceed to demonstrate the theorems propounded with reference to the figures."

Propositions 21, 22.

Any segment of a paraboloid of revolution is half as large again as the cone or segment of a cone which has the same base and the same axis.

Let the base of the segment be perpendicular to the plane of the paper, and let the plane of the paper be the plane through the axis of the paraboloid which cuts the base of the segment at right angles in BC and makes the parabolic section BAC.

Let EF be that tangent to the parabola which is parallel to BC, and let A be the point of contact.

Then (1), if the plane of the base of the segment is perpendicular to the axis of the paraboloid, that axis is the line AD bisecting BC at right angles in D.

(2) If the plane of the base is not perpendicular to the axis of the paraboloid, draw AD parallel to the axis of the paraboloid. AD will then bisect BC, but not at right angles.

Draw through EF a plane parallel to the base of the segment. This will touch the paraboloid at A, and A will be the vertex of the segment, AD its axis.

The base of the segment will be a circle with diameter BC or an ellipse with BC as major axis.

Accordingly a cylinder or a frustum of a cylinder can be found passing through the circle or ellipse and having AD for its axis [Prop. 9]; and likewise a cone or a segment of a cone can be drawn passing through the circle or ellipse and having A for vertex and AD for axis. [Prop. 8]

Suppose X to be a cone equal to $\frac{3}{2}$ (cone or segment of cone ABC). The cone X is therefore equal to half the cylinder or frustum of a cylinder EC. [Cf. Prop. 10]

We shall prove that the volume of the segment of the paraboloid is equal to X.

If not, the segment must be either greater or less than X.

I. If possible, let the segment be greater than X.

We can then inscribe and circumscribe, as in the last

9—2

proposition, figures made up of cylinders or frusta of cylinders with equal height and such that

(circumscribed fig.) − (inscribed fig.) < (segment) − X.

Let the greatest of the cylinders or frusta forming the circumscribed figure be that whose base is the circle or ellipse about BC and whose axis is OD, and let the smallest of them be that whose base is the circle or ellipse about PP' and whose axis is AL.

Let the greatest of the cylinders forming the inscribed figure be that whose base is the circle or ellipse about RR' and whose axis is OD, and let the smallest be that whose base is the circle or ellipse about PP' and whose axis is LM.

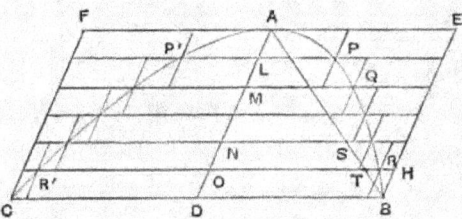

Produce all the plane bases of the cylinders or frusta to meet the surface of the complete cylinder or frustum EC.

Now, since

(circumscribed fig.) − (inscr. fig.) < (segment) − X,

it follows that   (inscribed figure) > X ....(α).

Next, comparing successively the cylinders or frusta with heights equal to OD and respectively forming parts of the complete cylinder or frustum EC and of the inscribed figure, we have

(first cylinder or frustum in EC) : (first in inscr. fig.)

$$= BD^2 : RO^2$$
$$= AD : AO$$
$$= BD : TO, \text{ where } AB \text{ meets } OR \text{ in } T.$$

And (second cylinder or frustum in EC) : (second in inscr. fig.)

$$= HO : SN, \text{ in like manner,}$$

and so on.

Hence [Prop. 1] (cylinder or frustum EC) : (inscribed figure)

$$= (BD + HO + ...) : (TO + SN + ...),$$

where $BD, HO,...$ are all equal, and $BD, TO, SN,...$ diminish in arithmetical progression.

But [Lemma preceding Prop. 1]

$$BD + HO + ... > 2 (TO + SN + ...).$$

Therefore   (cylinder or frustum EC) > 2 (inscribed fig.),

or $X >$ (inscribed fig.);

which is impossible, by (α) above.

II.   If possible, let the segment be less than X.

In this case we inscribe and circumscribe figures as before, but such that

(circumscr. fig.) − (inscr. fig.) < X − (segment),

whence it follows that

(circumscribed figure) < X ....(β).

And, comparing the cylinders or frusta making up the complete cylinder or frustum CE and the *circumscribed* figure respectively, we have

(first cylinder or frustum in CE) : (first in circumscr. fig.)

$$= BD^2 : BD^2$$
$$= BD : BD.$$

(second in CE) : (second in circumscr. fig.)

$$= HO^2 : RO^2$$
$$= AD : AO$$
$$= HO : TO,$$

and so on.

Hence [Prop. 1]

(cylinder or frustum CE) : (circumscribed fig.)

$$= (BD + HO + ...) : (BD + TO + ...),$$
$$< 2 : 1,   \text{[Lemma preceding Prop. 1]}$$

and it follows that

$$X < \text{(circumscribed fig.)};$$

which is impossible, by (β).

Thus the segment, being neither greater nor less than X, is equal to it, and therefore to $\frac{3}{2}$ (cone or segment of cone ABC).

Proposition 23.

*If from a paraboloid of revolution two segments be cut off,
one by a plane perpendicular to the axis, the other by a plane not
perpendicular to the axis, and if the axes of the segments are
equal, the segments will be equal in volume.*

Let the two planes be supposed perpendicular to the plane
of the paper, and let the latter plane be the plane through the
axis of the paraboloid cutting the other two planes at right
angles in BB', QQ' respectively and the paraboloid itself in the
parabola $QPQ'B'$.

Let AN, PV be the equal axes of the segments, and A, P
their respective vertices.

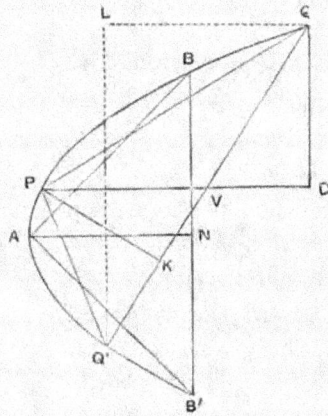

Draw QL parallel to AN or PV and $Q'L$ perpendicular
to QL.

Now, since the segments of the parabolic section cut off by
BB', QQ' have equal axes, the triangles ABB', PQQ' are equal
[Prop. 3]. Also, if QD be perpendicular to PV, $QD = BN$ (as
in the same Prop. 3).

Conceive two cones drawn with the same bases as the
segments and with A, P as vertices respectively. The height
of the cone PQQ' is then PK, where PK is perpendicular to
QQ'.

Now the cones are in the ratio compounded of the ratios of their bases and of their heights, i.e. the ratio compounded of (1) the ratio of the circle about BB' to the ellipse about QQ', and (2) the ratio of AN to PK.

That is to say, we have, by means of Props. 5, 12,

(cone ABB') : (cone PQQ') = $(BB'^2 : QQ' . Q'L).(AN : PK)$.

And $BB' = 2BN = 2QD = Q'L$, while $QQ' = 2QV$.

Therefore

$$\text{(cone } ABB') : \text{(cone } PQQ') = (QD : QV).(AN : PK)$$
$$= (PK : PV).(AN : PK)$$
$$= AN : PV.$$

Since $AN = PV$, the ratio of the cones is a ratio of equality : and it follows that the segments, being each half as large again as the respective cones [Prop. 22], are equal.

Proposition 24.

If from a paraboloid of revolution two segments be cut off by planes drawn in any manner, the segments will be to one another as the squares on their axes.

For let the paraboloid be cut by a plane through the axis in the parabolic section $P'PApp'$, and let the axis of the parabola and paraboloid be ANN'.

Measure along ANN' the lengths AN, AN' equal to the respective axes of the given segments, and through N, N' draw planes perpendicular to the axis, making circular sections on Pp, $P'p'$ as diameters respectively. With these circles as bases and with the common vertex A let two cones be described.

Now the segments of the paraboloid whose bases are the circles about Pp, $P'p'$ are equal to the given segments respectively, since their respective axes are equal [Prop. 23]; and, since the segments APp, $AP'p'$ are half as large again as the cones APp, $AP'p'$ respectively, we have only to show that the cones are in the ratio of AN^2 to AN'^2.

But

$$(\text{cone } APp) : (\text{cone } AP'p') = (PN^2 : P'N'^2).(AN : AN')$$
$$= (AN : AN').(AN : AN')$$
$$= AN^2 : AN'^2;$$

thus the proposition is proved.

Propositions 25, 26.

In any hyperboloid of revolution, if A be the vertex and AD the axis of any segment cut off by a plane, and if CA be the semidiameter of the hyperboloid through A (CA being of course in the same straight line with AD), then

(segment) : (cone with same base and axis)
$$= (AD + 3CA) : (AD + 2CA).$$

Let the plane cutting off the segment be perpendicular to the plane of the paper, and let the latter plane be the plane through the axis of the hyperboloid which intersects the cutting plane at right angles in BB', and makes the hyperbolic segment BAB'. Let C be the centre of the hyperboloid (or the vertex of the enveloping cone).

Let EF be that tangent to the hyperbolic section which is parallel to BB'. Let EF touch at A, and join CA. Then CA produced will bisect BB' at D, CA will be a semi-diameter of the hyperboloid, A will be the vertex of the segment, and AD its axis. Produce AC to A' and H, so that $AC = CA' = A'H$.

Through EF draw a plane parallel to the base of the segment. This plane will touch the hyperboloid at A.

Then (1), if the base of the segment is at right angles to the axis of the hyperboloid, A will be the vertex, and AD the axis, of the hyperboloid as well as of the segment, and the base of the segment will be a circle on BB' as diameter.

(2) If the base of the segment is not perpendicular to the axis of the hyperboloid, the base will be an ellipse on BB' as major axis. [Prop. 13]

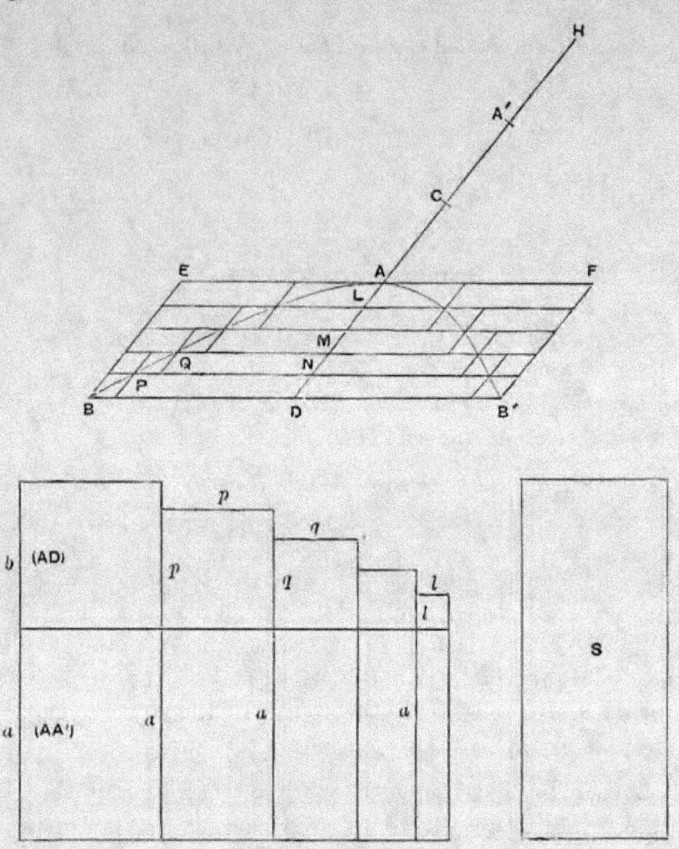

Then we can draw a cylinder or a frustum of a cylinder $EBB'F$ passing through the circle or ellipse about BB' and having AD for its axis; also we can describe a cone or a segment of a cone through the circle or ellipse and having A for its vertex.

We have to prove that

(segment ABB') : (cone or segment of cone ABB') = $HD : A'D$.

Let V be a cone such that

$$V : (\text{cone or segment of cone } ABB') = HD : A'D, \ldots\ldots(\alpha)$$

and we have to prove that V is equal to the segment.

Now

(cylinder or frustum EB') : (cone or segmt. of cone ABB') $= 3 : 1$.

Therefore, by means of (α),

$$(\text{cylinder or frustum } EB') : V = A'D : \frac{HD}{3} \ldots\ldots(\beta).$$

If the segment is not equal to V, it must either be greater or less.

I.   If possible, let the segment be greater than V.

Inscribe and circumscribe to the segment figures made up of cylinders or frusta of cylinders, with axes along AD and all equal to one another, such that

$$(\text{circumscribed fig.}) - (\text{inscr. fig.}) < (\text{segmt.}) - V,$$

whence $(\text{inscribed figure}) > V \ldots\ldots\ldots\ldots\ldots\ldots(\gamma).$

Produce all the planes forming the bases of the cylinders or frusta of cylinders to meet the surface of the complete cylinder or frustum EB'.

Then, if ND be the axis of the greatest cylinder or frustum in the circumscribed figure, the complete cylinder will be divided into cylinders or frusta each equal to this greatest cylinder or frustum.

Let there be a number of straight lines a equal to AA' and as many in number as the parts into which AD is divided by the bases of the cylinders or frusta. To each line a apply a rectangle which shall overlap it by a square, and let the greatest of the rectangles be equal to the rectangle $AD . A'D$ and the least equal to the rectangle $AL . A'L$; also let the sides of the overlapping squares b, p, $q, \ldots l$ be in descending arithmetical progression. Thus b, p, $q, \ldots l$ will be respectively equal to AD, AN, $AM, \ldots AL$, and the rectangles $(ab + b^2)$, $(ap + p^2), \ldots (al + l^2)$ will be respectively equal to $AD . A'D$, $AN . A'N, \ldots AL . A'L$.

Suppose, further, that we have a series of spaces S each equal to the largest rectangle $AD.A'D$ and as many in number as the diminishing rectangles.

Comparing now the successive cylinders or frusta (1) in the complete cylinder or frustum EB' and (2) in the inscribed figure, beginning from the base of the segment, we have

(first cylinder or frustum in EB') : (first in inscr. figure)

$$= BD^2 : PN^2$$

$$= AD.A'D : AN.A'N, \text{ from the hyperbola,}$$

$$= S : (ap + p^2).$$

Again

(second cylinder or frustum in EB') : (second in inscr. fig.)

$$= BD^2 : QM^2$$

$$= AD.A'D : AM.A'M$$

$$= S : (aq + q^2),$$

and so on.

The last cylinder or frustum in the complete cylinder or frustum EB' has no cylinder or frustum corresponding to it in the inscribed figure.

Combining the proportions, we have   [Prop. 1]

(cylinder or frustum EB') : (inscribed figure)

$$= (\text{sum of all the spaces } S) : (ap + p^2) + (aq + q^2) + \dots$$

$$> (a + b) : \left(\frac{a}{2} + \frac{b}{3}\right) \qquad\qquad\qquad [\text{Prop. 2}]$$

$$> A'D : \frac{HD}{3}, \quad \text{since } a = AA', \ b = AD,$$

$$> (EB') : V, \quad \text{by } (\beta) \text{ above.}$$

Hence   (inscribed figure) $< V$.

But this is impossible, because, by (γ) above, the inscribed figure is greater than V.

II. Next suppose, if possible, that the segment is less than V.

In this case we circumscribe and inscribe figures such that

(circumscribed fig.) — (inscribed fig.) $< V -$ (segment),

whence we derive

$$V > \text{(circumscribed figure)} \ldots\ldots\ldots\ldots(\delta).$$

We now compare successive cylinders or frusta in the complete cylinder or frustum and in the *circumscribed* figure; and we have

(first cylinder or frustum in EB') : (first in circumscribed fig.)

$$= S : S$$
$$= S : (ab + b^2),$$

(second in EB') : (second in circumscribed fig.)

$$= S : (ap + p^2),$$

and so on.

Hence [Prop. 1]

(cylinder or frustum EB') : (circumscribed fig.)

$$= \text{(sum of all spaces } S) : (ab + b^2) + (ap + p^2) + \ldots$$

$$< (a + b) : \left(\frac{a}{2} + \frac{b}{3}\right) \qquad\qquad \text{[Prop. 2]}$$

$$< A'D : \frac{HD}{3}$$

$$< (EB') : V, \text{ by } (\beta) \text{ above.}$$

Hence the circumscribed figure is greater than V; which is impossible, by (δ) above.

Thus the segment is neither greater nor less than V, and is therefore equal to it.

Therefore, by (α),

(segment ABB') : (cone or segment of cone ABB')

$$= (AD + 3CA) : (AD + 2CA).$$

Propositions 27, 28, 29, 30.

(1) *In any spheroid whose centre is C, if a plane meeting the axis cut off a segment not greater than half the spheroid and having A for its vertex and AD for its axis, and if A'D be the axis of the remaining segment of the spheroid, then*

(first segmt.) : (cone or segmt. of cone with same base and axis)

$$= CA + A'D : A'D$$

$$[= 3CA - AD : 2CA - AD].$$

(2) *As a particular case, if the plane passes through the centre, so that the segment is half the spheroid, half the spheroid is double of the cone or segment of a cone which has the same vertex and axis.*

Let the plane cutting off the segment be at right angles to the plane of the paper, and let the latter plane be the plane through the axis of the spheroid which intersects the cutting plane in BB' and makes the elliptic section $ABA'B'$.

Let $EF, E'F'$ be the two tangents to the ellipse which are parallel to BB', let them touch it in A, A', and through the tangents draw planes parallel to the base of the segment. These planes will touch the spheroid at A, A', which will be the vertices of the two segments into which it is divided. Also AA' will pass through the centre C and bisect BB' in D.

Then (1) if the base of the segments be perpendicular to the axis of the spheroid, A, A' will be the vertices of the spheroid as well as of the segments, AA' will be the axis of the spheroid, and the base of the segments will be a circle on BB' as diameter;

(2) if the base of the segments be not perpendicular to the axis of the spheroid, the base of the segments will be an ellipse of which BB' is one axis, and $AD, A'D$ will be the axes of the segments respectively.

We can now draw a cylinder or a frustum of a cylinder $EBB'F$ through the circle or ellipse about BB' and having AD for its axis; and we can also draw a cone or a segment of a cone passing through the circle or ellipse about BB' and having A for its vertex.

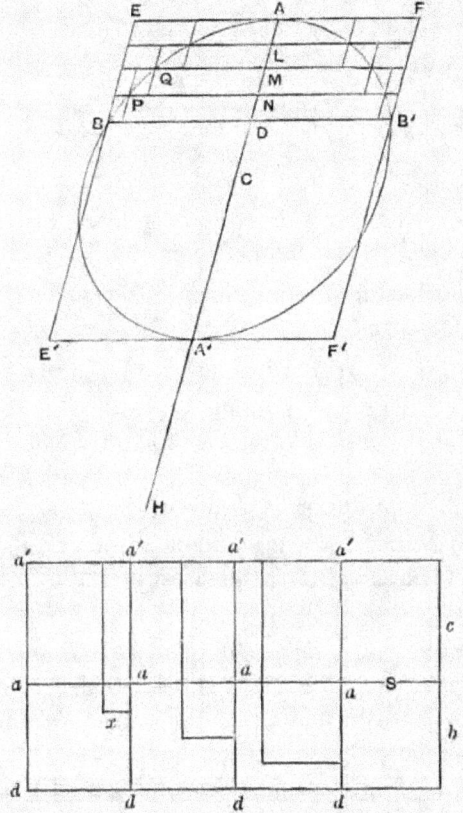

We have then to show that, if CA' be produced to H so that $CA' = A'H$,

(segment ABB') : (cone or segment of cone ABB') $= HD : A'D$.

Let V be such a cone that

$$V : (\text{cone or segment of cone } ABB') = HD : A'D \;\dots (\alpha);$$

and we have to show that the segment ABB' is equal to V.

But, since

(cylinder or frustum EB') : (cone or segment of cone ABB')
$$= 3 : 1,$$

we have, by the aid of (α),

$$\text{(cylinder or frustum } EB') : V = A'D : \frac{HD}{3}\ldots\ldots(\beta).$$

Now, if the segment ABB' is not equal to V, it must be either greater or less.

I. Suppose, if possible, that the segment is greater than V.

Let figures be inscribed and circumscribed to the segment consisting of cylinders or frusta of cylinders, with axes along AD and all equal to one another, such that

(circumscribed fig.) − (inscribed fig.) < (segment) − V,

whence it follows that

$$\text{(inscribed fig.)} > V \quad \ldots\ldots\ldots\ldots\ldots\ldots (\gamma).$$

Produce all the planes forming the bases of the cylinders or frusta to meet the surface of the complete cylinder or frustum EB'. Thus, if ND be the axis of the greatest cylinder or frustum of a cylinder in the circumscribed figure, the complete cylinder or frustum EB' will be divided into cylinders or frusta of cylinders each equal to the greatest of those in the circumscribed figure.

Take straight lines da' each equal to $A'D$ and as many in number as the parts into which AD is divided by the bases of the cylinders or frusta, and measure da along da' equal to AD. It follows that $aa' = 2CD$.

Apply to each of the lines $a'd$ rectangles with height equal to ad, and draw the squares on each of the lines ad as in the figure. Let S denote the area of each complete rectangle.

From the first rectangle take away a gnomon with breadth equal to AN (i.e. with each end of a length equal to AN); take away from the second rectangle a gnomon with breadth equal to AM, and so on, the last rectangle having no gnomon taken from it.

Then

the first gnomon $= A'D . AD - ND . (A'D - AN)$

$\qquad = A'D . AN + ND . AN$

$\qquad = AN . A'N.$

Similarly,

the second gnomon $= AM . A'M,$

and so on.

And the last gnomon (that in the last rectangle but one) is equal to $AL . A'L.$

Also, after the gnomons are taken away from the successive rectangles, the remainders (which we will call $R_1, R_2,... R_n$, where n is the number of rectangles and accordingly $R_n = S$) are rectangles applied to straight lines each of length aa' and "exceeding by squares" whose sides are respectively equal to $DN, DM,... DA.$

For brevity, let DN be denoted by x, and aa' or $2CD$ by c, so that $R_1 = cx + x^2, R_2 = c . 2x + (2x)^2,...$

Then, comparing successively the cylinders or frusta of cylinders (1) in the complete cylinder or frustum EB' and (2) in the inscribed figure, we have

(first cylinder or frustum in EB') : (first in inscribed fig.)

$\qquad = BD^2 : PN^2$

$\qquad = AD . A'D : AN . A'N$

$\qquad = S : (\text{first gnomon}) ;$

(second cylinder or frustum in EB') : (second in inscribed fig.)

$\qquad = S : (\text{second gnomon}),$

and so on.

The last of the cylinders or frusta in the cylinder or frustum EB' has none corresponding to it in the inscribed figure, and there is no corresponding gnomon.

Combining the proportions, we have [by Prop. 1]

(cylinder or frustum EB') : (inscribed fig.)

$\qquad = (\text{sum of all spaces } S) : (\text{sum of gnomons}).$

Now the differences between S and the successive gnomons are $R_1, R_2, \ldots R_n$, while

$$R_1 = cx + x^2,$$
$$R_2 = c \cdot 2x + (2x)^2,$$
$$\ldots\ldots\ldots\ldots\ldots\ldots$$
$$R_n = cb + b^2 = S,$$

where $b = nx = AD$.

Hence [Prop. 2]

(sum of all spaces S) : $(R_1 + R_2 + \ldots + R_n) < (c + b) : \left(\dfrac{c}{2} + \dfrac{b}{3}\right)$.

It follows that

(sum of all spaces S) : (sum of gnomons) $> (c + b) : \left(\dfrac{c}{2} + \dfrac{2b}{3}\right)$

$$> A'D : \frac{HD}{3}.$$

Thus (cylinder or frustum EB') : (inscribed fig.)

$$> A'D : \frac{HD}{3}$$

$$> \text{(cylinder or frustum } EB') : V,$$

from (β) above.

Therefore　　　　　(inscribed fig.) $< V$;

which is impossible, by (γ) above.

Hence the segment ABB' is not greater than V.

II.　If possible, let the segment ABB' be less than V.

We then inscribe and circumscribe figures such that

(circumscribed fig.) $-$ (inscribed fig.) $< V -$ (segment),

whence　　　　　$V > $ (circumscribed fig.)...................(δ).

In this case we compare the cylinders or frusta in (EB') with those in the *circumscribed* figure.

Thus

(first cylinder or frustum in EB') : (first in circumscribed fig.)

$$= S : S;$$

(second in EB') : (second in circumscribed fig.)

$$= S : \text{(first gnomon)},$$

and so on.

Lastly   (last in EB') : (last in circumscribed fig.)

$$= S : \text{(last gnomon)}.$$

Now

$$[S + \text{(all the gnomons)}] = nS - (R_1 + R_2 + \ldots + R_{n-1}).$$

And $nS : R_1 + R_2 + \ldots + R_{n-1} > (c+b) : \left(\dfrac{c}{2} + \dfrac{b}{3}\right),$ [Prop. 2]

so that

$$nS : [S + \text{(all the gnomons)}] < (c+b) : \left(\dfrac{c}{2} + \dfrac{2b}{3}\right).$$

It follows that, if we combine the above proportions as in Prop. 1, we obtain

(cylinder or frustum EB') : (circumscribed fig.)

$$< (c+b) : \left(\dfrac{c}{2} + \dfrac{2b}{3}\right)$$

$$< A'D : \dfrac{HD}{3}$$

$$< (EB') : V, \text{ by } (\beta) \text{ above.}$$

Hence the circumscribed figure is greater than V; which is impossible, by (δ) above.

Thus, since the segment ABB' is neither greater nor less than V, it is equal to it; and the proposition is proved.

(2) The particular case [Props. 27, 28] where the segment is half the spheroid differs from the above in that the distance CD or $c/2$ vanishes, and the rectangles $cb + b^2$ are simply squares (b^2), so that the gnomons are simply the differences between b^2 and x^2, b^2 and $(2x)^2$, and so on.

Instead therefore of Prop. 2 we use the *Lemma to Prop. 2, Cor. 1*, given above [*On Spirals*, Prop. 10], and instead of the ratio $(c+b) : \left(\dfrac{c}{2} + \dfrac{2b}{3}\right)$ we obtain the ratio $3 : 2$, whence

(segment ABB') : (cone or segment of cone ABB') $= 2 : 1$.

[This result can also be obtained by simply substituting CA for AD in the ratio $(3CA - AD) : (2CA - AD)$.]

Propositions 31, 32.

If a plane divide a spheroid into two unequal segments, and if AN, A'N be the axes of the lesser and greater segments respectively, while C is the centre of the spheroid, then

(greater segmt.) : (cone or segmt. of cone with same base and axis)
$$= CA + AN : AN.$$

Let the plane dividing the spheroid be that through PP' perpendicular to the plane of the paper, and let the latter plane be that through the axis of the spheroid which intersects the cutting plane in PP' and makes the elliptic section $PAP'A'$.

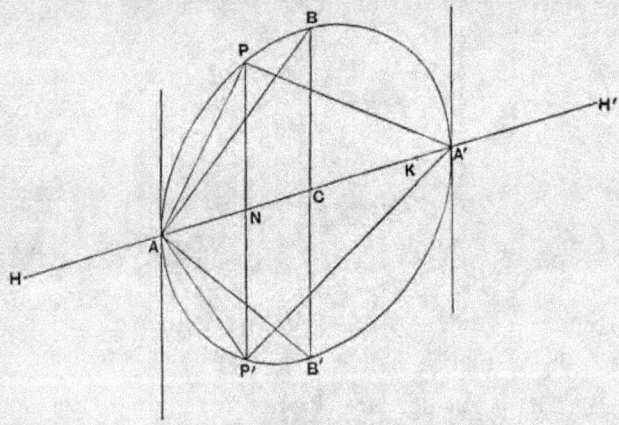

Draw the tangents to the ellipse which are parallel to PP'; let them touch the ellipse at A, A', and through the tangents draw planes parallel to the base of the segments. These planes will touch the spheroid at A, A', the line AA' will pass through the centre C and bisect PP' in N, while AN, $A'N$ will be the axes of the segments.

Then (1) if the cutting plane be perpendicular to the axis of the spheroid, AA' will be that axis, and A, A' will be the vertices of the spheroid as well as of the segments. Also the sections of the spheroid by the cutting plane and all planes parallel to it will be circles.

(2) If the cutting plane be not perpendicular to the axis,

10—2

the base of the segments will be an ellipse of which PP' is an axis, and the sections of the spheroid by all planes parallel to the cutting plane will be similar ellipses.

Draw a plane through C parallel to the base of the segments and meeting the plane of the paper in BB'.

Construct three cones or segments of cones, two having A for their common vertex and the plane sections through PP', BB' for their respective bases, and a third having the plane section through PP' for its base and A' for its vertex.

Produce CA to H and CA' to H' so that

$$AH = A'H' = CA.$$

We have then to prove that

(segment $A'PP'$) : (cone or segment of cone $A'PP'$)

$$= CA + AN : AN$$

$$= NH : AN.$$

Now half the spheroid is double of the cone or segment of a cone ABB' [Props. 27, 28]. Therefore

(the spheroid) $= 4$ (cone or segment of cone ABB').

But

(cone or segmt. of cone ABB') : (cone or segmt. of cone APP')

$$= (CA : AN).(BC^2 : PN^2)$$

$$= (CA : AN).(CA.CA' : AN.A'N)...(\alpha).$$

If we measure AK along AA' so that

$$AK : AC = AC : AN,$$

we have $AK.A'N : AC.A'N = CA : AN,$

and the compound ratio in (α) becomes

$$(AK.A'N : CA.A'N).(CA.CA' : AN.A'N),$$

i.e. $AK.CA' : AN.A'N.$

Thus

(cone or segmt. of cone ABB') : (cone or segmt. of cone APP')

$$= AK.CA' : AN.A'N.$$

But   (cone or segment of cone APP') : (segment APP')

$$= A'N : NH' \qquad [\text{Props. 29, 30}]$$
$$= AN . A'N : AN . NH'.$$

Therefore, *ex aequali,*

(cone or segment of cone ABB') : (segment APP')

$$= AK . CA' : AN . NH',$$

so that   (spheroid) : (segment APP')

$$= HH' . AK : AN . NH',$$

since $HH' = 4CA'.$

Hence   (segment $A'PP'$) : (segment APP')

$$= (HH' . AK - AN . NH') : AN . NH'$$
$$= (AK . NH + NH' . NK) : AN . NH'.$$

Further,

(segment APP') : (cone or segment of cone APP')

$$= NH' : A'N$$
$$= AN . NH' : AN . A'N,$$

and

(cone or segmt. of cone APP') : (cone or segmt. of cone $A'PP'$)

$$= AN : A'N$$
$$= AN . A'N : A'N^2.$$

From the last three proportions we obtain, *ex aequali,*

(segment $A'PP'$) : (cone or segment of cone $A'PP'$)

$$= (AK . NH + NH' . NK) : A'N^2$$
$$= (AK . NH + NH' . NK) : (CA^2 + NH' . CN)$$
$$= (AK . NH + NH' . NK) : (AK . AN + NH' . CN) \dots (\beta).$$

But

$$AK . NH : AK . AN = NH : AN$$
$$= CA + AN : AN$$
$$= AK + CA : CA$$
$$\qquad\qquad (\text{since } AK : AC = AC : AN)$$
$$= HK : CA$$
$$= HK - NH : CA - AN$$
$$= NK : CN$$
$$= NH' . NK : NH' . CN.$$

Hence the ratio in (β) is equal to the ratio

$$AK \cdot NH : AK \cdot AN, \text{ or } NH : AN.$$

Therefore

(segment $A'PP'$) : (cone or segment of cone $A'PP'$)

$$= NH : AN$$

$$= CA + AN : AN.$$

[If (x, y) be the coordinates of P referred to the conjugate diameters AA', BB' as axes of x, y, and if $2a$, $2b$ be the lengths of the diameters respectively, we have, since

(spheroid) $-$ (lesser segment) $=$ (greater segment),

$$4 \cdot ab^2 - \frac{2a + x}{a + x} \cdot y^2 (a - x) = \frac{2a - x}{a - x} \cdot y^2 (a + x);$$

and the above proposition is the geometrical proof of the truth of this equation where x, y are connected by the equation

$$\frac{x^2}{a^2} + \frac{y^2}{b^2} = 1.]$$

ON SPIRALS.

"ARCHIMEDES to Dositheus greeting.

Of most of the theorems which I sent to Conon, and of which you ask me from time to time to send you the proofs, the demonstrations are already before you in the books brought to you by Heracleides; and some more are also contained in that which I now send you. Do not be surprised at my taking a considerable time before publishing these proofs. This has been owing to my desire to communicate them first to persons engaged in mathematical studies and anxious to investigate them. In fact, how many theorems in geometry which have seemed at first impracticable are in time successfully worked out! Now Conon died before he had sufficient time to investigate the theorems referred to; otherwise he would have discovered and made manifest all these things, and would have enriched geometry by many other discoveries besides. For I know well that it was no common ability that he brought to bear on mathematics, and that his industry was extraordinary. But, though many years have elapsed since Conon's death, I do not find that any one of the problems has been stirred by a single person. I wish now to put them in review one by one, particularly as it happens that there are two included among them which are impossible of realisation* [and which may serve as a warning] how those who claim to discover everything but produce no proofs of the same may be confuted as having actually pretended to discover the impossible.

---

* Heiberg reads τέλος δὲ ποθεσόμενα, but F has τέλους, so that the true reading is perhaps τέλους δὲ ποτιδεόμενα. The meaning appears to be simply 'wrong.'

What are the problems I mean, and what are those of which
you have already received the proofs, and those of which the
proofs are contained in this book respectively, I think it proper
to specify. The first of the problems was, Given a sphere, to find
a plane area equal to the surface of the sphere; and this was
first made manifest on the publication of the book concerning the
sphere, for, when it is once proved that the surface of any sphere
is four times the greatest circle in the sphere, it is clear that it
is possible to find a plane area equal to the surface of the sphere.
The second was, Given a cone or a cylinder, to find a sphere
equal to the cone or cylinder; the third, To cut a given sphere
by a plane so that the segments of it have to one another an
assigned ratio; the fourth, To cut a given sphere by a plane so
that the segments of the surface have to one another an assigned
ratio; the fifth, To make a given segment of a sphere similar to
a given segment of a sphere*; the sixth, Given two segments of
either the same or different spheres, to find a segment of a sphere
which shall be similar to one of the segments and have its
surface equal to the surface of the other segment. The seventh
was, From a given sphere to cut off a segment by a plane so
that the segment bears to the cone which has the same base as
the segment and equal height an assigned ratio greater than
that of three to two. Of all the propositions just enumerated
Heracleides brought you the proofs. The proposition stated
next after these was wrong, viz. that, if a sphere be cut by a
plane into unequal parts, the greater segment will have to the
less the duplicate ratio of that which the greater surface has to
the less. That this is wrong is obvious by what I sent you
before; for it included this proposition: If a sphere be cut into
unequal parts by a plane at right angles to any diameter in the
sphere, the greater segment of the surface will have to the less
the same ratio as the greater segment of the diameter has
to the less, while the greater segment of the sphere has to the
less a ratio less than the duplicate ratio of that which the

---

* τὸ δοθὲν τμᾶμα σφαίρας τῷ δοθέντι τμάματι σφαίρας ὁμοιῶσαι, i.e. to make a
segment of a sphere similar to one given segment and equal in content to
another given segment. [Cf. On the Sphere and Cylinder, II. 5.]

greater surface has to the less, but greater than the sesquialterate* of that ratio. The last of the problems was also wrong,
viz. that, if the diameter of any sphere be cut so that the square
on the greater segment is triple of the square on the lesser
segment, and if through the point thus arrived at a plane be
drawn at right angles to the diameter and cutting the sphere,
the figure in such a form as is the greater segment of the sphere
is the greatest of all the segments which have an equal surface.
That this is wrong is also clear from the theorems which I
before sent you. For it was there proved that the hemisphere
is the greatest of all the segments of a sphere bounded by an
equal surface.

After these theorems the following were propounded concerning the cone†. If a section of a right-angled cone [a
parabola], in which the diameter [axis] remains fixed, be made to
revolve so that the diameter [axis] is the axis [of revolution],
let the figure described by the section of the right-angled cone
be called a *conoid*. And if a plane touch the conoidal figure
and another plane drawn parallel to the tangent plane cut off
a segment of the conoid, let the *base* of the segment cut off be
defined as the cutting plane, and the *vertex* as the point in which
the other plane touches the conoid. Now, if the said figure be
cut by a plane at right angles to the axis, it is clear that the
section will be a circle; but it needs to be proved that the
segment cut off will be half as large again as the cone which has
the same base as the segment and equal height. And if two
segments be cut off from the conoid by planes drawn in any
manner, it is clear that the sections will be sections of acute-
angled cones [ellipses] if the cutting planes be not at right
angles to the axis; but it needs to be proved that the
segments will bear to one another the ratio of the squares on
the lines drawn from their vertices parallel to the axis to meet
the cutting planes. The proofs of these propositions are not
yet sent to you.

After these came the following propositions about the *spiral*,

---

* (λόγον) μείζονα ἢ ἡμιόλιον τοῦ, ὃν ἔχει κ.τ.λ., i.e. a ratio greater than (the
ratio of the surfaces)$^{\frac{3}{2}}$. See *On the Sphere and Cylinder*, II. 8.

† This should be presumably 'the *conoid*,' not 'the cone.'

which are as it were another sort of problem having nothing
in common with the foregoing; and I have written out the
proofs of them for you in this book.   They are as follows.   If a
straight line of which one extremity remains fixed be made to
revolve at a uniform rate in a plane until it returns to the
position from which it started, and if, at the same time as the
straight line revolves, a point move at a uniform rate along the
straight line, starting from the fixed extremity, the point will
describe a spiral in the plane.   I say then that the area
bounded by the spiral and the straight line which has returned
to the position from which it started is a third part of the circle
described with the fixed point as centre and with radius the
length traversed by the point along the straight line during the
one revolution.   And, if a straight line touch the spiral at the
extreme end of the spiral, and another straight line be drawn at
right angles to the line which has revolved and resumed its
position from the fixed extremity of it, so as to meet the
tangent, I say that the straight line so drawn to meet it is
equal to the circumference of the circle.   Again, if the revolving
line and the point moving along it make several revolutions
and return to the position from which the straight line started,
I say that the area added by the spiral in the third revolution
will be double of that added in the second, that in the fourth
three times, that in the fifth four times, and generally the areas
added in the later revolutions will be multiples of that added in
the second revolution according to the successive numbers,
while the area bounded by the spiral in the first revolution is a
sixth part of that added in the second revolution.   Also, if on
the spiral described in one revolution two points be taken and
straight lines be drawn joining them to the fixed extremity of
the revolving line, and if two circles be drawn with the fixed
point as centre and radii the lines drawn to the fixed extremity
of the straight line, and the shorter of the two lines be produced,
I say that (1) the area bounded by the circumference of the
greater circle in the direction of (the part of) the spiral included
between the straight lines, the spiral (itself) and the produced
straight line will bear to (2) the area bounded by the circumference of the lesser circle, the same (part of the) spiral and the

straight line joining their extremities the ratio which (3) the radius of the lesser circle together with two thirds of the excess of the radius of the greater circle over the radius of the lesser bears to (4) the radius of the lesser circle together with one third of the said excess.

The proofs then of these theorems and others relating to the spiral are given in the present book. Prefixed to them, after the manner usual in other geometrical works, are the propositions necessary to the proofs of them. And here too, as in the books previously published, I assume the following lemma, that, if there be (two) unequal lines or (two) unequal areas, the excess by which the greater exceeds the less can, by being [continually] added to itself, be made to exceed any given magnitude among those which are comparable with [it and with] one another."

Proposition 1.

If a point move at a uniform rate along any line, and two lengths be taken on it, they will be proportional to the times of describing them.

Two unequal lengths are taken on a straight line, and two lengths on another straight line representing the times; and they are proved to be proportional by taking equimultiples of each length and the corresponding time after the manner of Eucl. V. Def. 5.

Proposition 2.

If each of two points on different lines respectively move along them each at a uniform rate, and if lengths be taken, one on each line, forming pairs, such that each pair are described in equal times, the lengths will be proportionals.

This is proved at once by equating the ratio of the lengths taken on one line to that of the times of description, which must also be equal to the ratio of the lengths taken on the other line.

Proposition 3.

Given any number of circles, it is possible to find a straight line greater than the sum of all their circumferences.

For we have only to describe polygons about each and then take a straight line equal to the sum of the perimeters of the polygons.

Proposition 4.

Given two unequal lines, viz. a straight line and the circumference of a circle, it is possible to find a straight line less than the greater of the two lines and greater than the less.

For, by the Lemma, the excess can, by being added a sufficient number of times to itself, be made to exceed the lesser line.

Thus e.g., if $c > l$ (where c is the circumference of the circle and l the length of the straight line), we can find a number n such that

$$n(c - l) > l.$$

Therefore

$$c - l > \frac{l}{n},$$

and

$$c > l + \frac{l}{n} > l.$$

Hence we have only to divide l into n equal parts and add one of them to l. The resulting line will satisfy the condition.

Proposition 5.

Given a circle with centre O, and the tangent to it at a point A, it is possible to draw from O a straight line OPF, meeting the circle in P and the tangent in F, such that, if c be the circumference of any given circle whatever,

$$FP : OP < (\text{arc } AP) : c.$$

Take a straight line, as D, greater than the circumference c. [Prop. 3]

Through O draw OH parallel to the given tangent, and draw through A a line APH, meeting the circle in P and OH

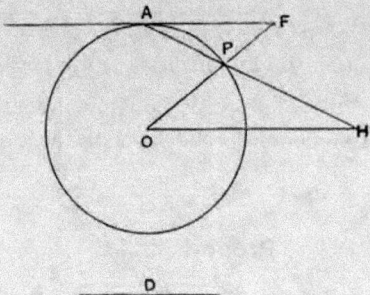

in H, such that the portion PH intercepted between the circle and the line OH may be equal to D*. Join OP and produce it to meet the tangent in F.

Then $\qquad FP : OP = AP : PH$, by parallels,

$$= AP : D$$

$$< (\text{arc } AP) : c.$$

Proposition 6.

Given a circle with centre O, a chord AB less than the diameter, and OM the perpendicular on AB from O, it is possible to draw a straight line OFP, meeting the chord AB in F and the circle in P, such that

$$FP : PB = D : E,$$

where $D : E$ is any given ratio less than $BM : MO$.

Draw OH parallel to AB, and BT perpendicular to BO meeting OH in T.

Then the triangles BMO, OBT are similar, and therefore

$$BM : MO = OB : BT,$$

whence $\qquad D : E < OB : BT.$

---

* This construction, which is assumed without any explanation as to how it is to be effected, is described in the original Greek thus: "let PH be placed ($\kappa\epsilon i\sigma\theta\omega$) equal to D, verging ($\nu\epsilon\acute{u}o\upsilon\sigma\alpha$) towards A." This is the usual phraseology used in the type of problem known by the name of $\nu\epsilon\tilde{u}\sigma\iota\varsigma$.

Suppose that a line PH (greater than BT) is taken such that

$$D : E = OB : PH,$$

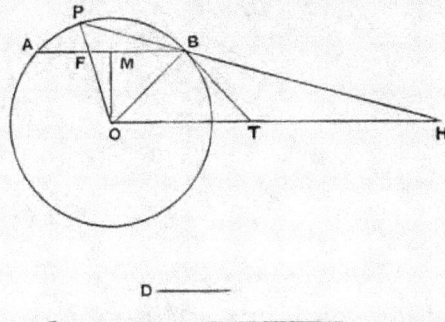

and let PH be so placed that it passes through B and P lies on the circumference of the circle, while H is on the line OH*. (PH will fall outside BT, because $PH > BT$.) Join OP meeting AB in F.

We now have

$$FP : PB = OP : PH$$

$$= OB : PH$$

$$= D : E.$$

Proposition 7.

Given a circle with centre O, a chord AB less than the diameter, and OM the perpendicular on it from O, it is possible to draw from O a straight line OPF, meeting the circle in P and AB produced in F, such that

$$FP : PB = D : E,$$

where $D : E$ is any given ratio greater than $BM : MO$.

Draw OT parallel to AB, and BT perpendicular to BO meeting OT in T.

---

* The Greek phrase is "let PH be placed between the circumference and the straight line (OH) through B." The construction is assumed, like the similar one in the last proposition.

In this case, $D : E > BM : MO$

$> OB : BT$, by similar triangles.

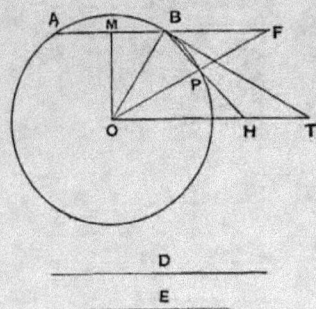

Take a line PH (less than BT) such that

$$D : E = OB : PH,$$

and place PH so that P, H are on the circle and on OT respectively, while HP produced passes through B^*.

Then $FP : PB = OP : PH$

$$= D : E.$$

Proposition 8.

Given a circle with centre O, a chord AB less than the diameter, the tangent at B, and the perpendicular OM from O on AB, it is possible to draw from O a straight line OFP, meeting the chord AB in F, the circle in P and the tangent in G, such that

$$FP : BG = D : E,$$

where D : E is any given ratio less than BM : MO.

If OT be drawn parallel to AB meeting the tangent at B in T,

$$BM : MO = OB : BT,$$

so that $D : E < OB : BT.$

Take a point C on TB produced such that

$$D : E = OB : BC,$$

whence $BC > BT.$

---

* PH is described in the Greek as νεύουσαν ἐπὶ (*verging to*) the point B. As before the construction is assumed.

Through the points O, T, C describe a circle, and let OB be produced to meet this circle in K.

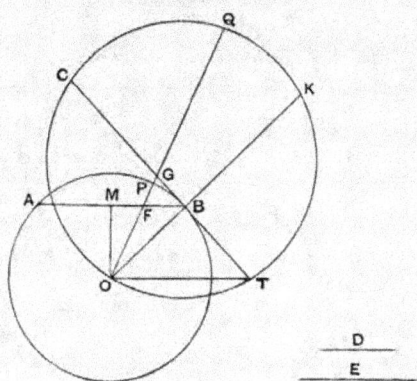

Then, since $BC > BT$, and OB is perpendicular to CT, it is possible to draw from O a straight line OGQ, meeting CT in G and the circle about OTC in Q, such that $GQ = BK^*$.

Let OGQ meet AB in F and the original circle in P.

Now $$CG \cdot GT = OG \cdot GQ;$$

and $$OF : OG = BT : GT,$$

so that $$OF \cdot GT = OG \cdot BT.$$

It follows that

$$CG \cdot GT : OF \cdot GT = OG \cdot GQ : OG \cdot BT,$$

or $$CG : OF = GQ : BT$$

$$= BK : BT, \text{ by construction,}$$

$$= BC : OB$$

$$= BC : OP.$$

Hence $$OP : OF = BC : CG,$$

and therefore $$PF : OP = BG : BC,$$

or $$PF : BG = OP : BC$$

$$= OB : BC$$

$$= D : E.$$

---

* The Greek words used are: "it is possible to place another [straight line] GQ equal to KB verging (νεύουσαν) towards O." This particular νεῦσις is discussed by Pappus (p. 298, ed. Hultsch). See the Introduction, chapter v.

Proposition 9.

Given a circle with centre O, a chord AB less than the diameter, the tangent at B, and the perpendicular OM from O on AB, it is possible to draw from O a straight line $OPGF$, meeting the circle in P, the tangent in G, and AB produced in F, such that

$$FP : BG = D : E,$$

where $D : E$ is any given ratio greater than $BM : MO$.

Let OT be drawn parallel to AB meeting the tangent at B in T.

Then $$D : E > BM : MO$$

$$> OB : BT, \text{ by similar triangles.}$$

Produce TB to C so that

$$D : E = OB : BC,$$

whence $$BC < BT.$$

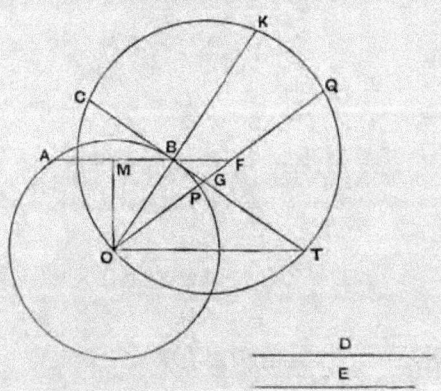

Describe a circle through the points O, T, C, and produce OB to meet this circle in K.

Then, since $TB > BC$, and OB is perpendicular to CT, it is possible to draw from O a line OGQ, meeting CT in G, and the circle about OTC in Q, such that $GQ = BK^*$. Let OQ meet the original circle in P and AB produced in F.

We now prove, exactly as in the last proposition, that

$$CG : OF = BK : BT$$
$$= BC : OP.$$

Thus, as before,

$$OP : OF = BC : CG,$$

and

$$OP : PF = BC : BG,$$

whence

$$PF : BG = OP : BC$$
$$= OB : BC$$
$$= D : E.$$

Proposition 10.

If $A_1, A_2, A_3, \ldots A_n$ be n lines forming an ascending arithmetical progression in which the common difference is equal to A_1, the least term, then

$$(n+1) A_n^2 + A_1 (A_1 + A_2 + \ldots + A_n) = 3 (A_1^2 + A_2^2 + \ldots + A_n^2).$$

[Archimedes' proof of this proposition is given above, p. 107–9, and it is there pointed out that the result is equivalent to

$$1^2 + 2^2 + 3^2 + \ldots + n^2 = \frac{n(n+1)(2n+1)}{6}.]$$

COR. 1. *It follows from this proposition that*

$$n \cdot A_n^2 < 3 (A_1^2 + A_2^2 + \ldots + A_n^2),$$

and also that

$$n \cdot A_n^2 > 3 (A_1^2 + A_2^2 + \ldots + A_{n-1}^2).$$

[For the proof of the latter inequality see p. 109 above.]

COR. 2. *All the results will equally hold if similar figures are substituted for squares.*

---

* See the note on the last proposition.

Proposition 11.

If A_1, A_2,...A_n be n lines forming an ascending arith-metical progression [in which the common difference is equal to the least term A_1]*, then

$$(n-1) A_n^2 : (A_n^2 + A_{n-1}^2 + \ldots + A_2^2)$$
$$< A_n^2 : \{A_n . A_1 + \tfrac{1}{3}(A_n - A_1)^2\} ;$$

but

$$(n-1) A_n^2 : (A_{n-1}^2 + A_{n-2}^2 + \ldots + A_1^2)$$
$$> A_n^2 : \{A_n . A_1 + \tfrac{1}{3}(A_n - A_1)^2\}.$$

[Archimedes sets out the terms side by side in the manner shown in the figure, where $BC = A_n$, $DE = A_{n-1}$,... $RS = A_1$, and produces DE, FG,...RS until they are respectively equal to BC or A_n, so that EH, GI,...SU in the figure are re-spectively equal to A_1, A_2...A_{n-1}. He further measures lengths BK, DL, FM,...PV along BC, DE, FG,...PQ re-spectively each equal to RS.

The figure makes the relations between the terms easier to see with the eye, but the use of so large a number of letters makes the proof somewhat difficult to follow, and it may be more clearly represented as follows.]

It is evident that $(A_n - A_1) = A_{n-1}$.

The following proportion is therefore obviously true, viz.

$$(n-1) A_n^2 : (n-1)(A_n . A_1 + \tfrac{1}{3} A_{n-1}^2)$$
$$= A_n^2 : \{A_n . A_1 + \tfrac{1}{3}(A_n - A_1)^2\}.$$

---

* The proposition is true even when the common difference is not equal to A_1, and is assumed in the more general form in Props. 25 and 26. But, as Archimedes' proof assumes the equality of A_1 and the common difference, the words are here inserted to prevent misapprehension.

In order therefore to prove the desired result, we have only to show that

$$(n-1) A_n . A_1 + \tfrac{1}{3}(n-1) A_{n-1}{}^2 < (A_n{}^2 + A_{n-1}{}^2 + \ldots + A_2{}^2)$$

but $$> (A_{n-1}{}^2 + A_{n-2}{}^2 + \ldots + A_1{}^2).$$

I.   To prove the first inequality, we have

$$(n-1) A_n . A_1 + \tfrac{1}{3}(n-1) A_{n-1}{}^2$$
$$= (n-1) A_1{}^2 + (n-1) A_1 . A_{n-1} + \tfrac{1}{3}(n-1) A_{n-1}{}^2 \ldots (1).$$

And

$$A_n{}^2 + A_{n-1}{}^2 + \ldots + A_2{}^2$$
$$= (A_{n-1} + A_1)^2 + (A_{n-2} + A_1)^2 + \ldots + (A_1 + A_1)^2$$
$$= (A_{n-1}{}^2 + A_{n-2}{}^2 + \ldots + A_1{}^2)$$
$$\quad + (n-1) A_1{}^2$$
$$\quad + 2A_1 (A_{n-1} + A_{n-2} + \ldots + A_1)$$
$$= (A_{n-1}{}^2 + A_{n-2}{}^2 + \ldots + A_1{}^2)$$
$$\quad + (n-1) A_1{}^2$$
$$\quad + A_1 \{ A_{n-1} + A_{n-2} + A_{n-3} + \ldots + A_1$$
$$\quad\quad\quad + A_1 \ + A_2 \ + \ldots + A_{n-2} + A_{n-1} \}$$
$$= (A_{n-1}{}^2 + A_{n-2}{}^2 + \ldots + A_1{}^2)$$
$$\quad + (n-1) A_1{}^2$$
$$\quad + n A_1 . A_{n-1} \ldots\ldots\ldots\ldots\ldots\ldots\ldots\ldots\ldots (2).$$

Comparing the right-hand sides of (1) and (2), we see that $(n-1) A_1{}^2$ is common to both sides, and

$$(n-1) A_1 . A_{n-1} < n A_1 . A_{n-1},$$

while, by Prop. 10, Cor. 1,

$$\tfrac{1}{3}(n-1) A_{n-1}{}^2 < A_{n-1}{}^2 + A_{n-2}{}^2 + \ldots + A_1{}^2.$$

It follows therefore that

$$(n-1) A_n . A_1 + \tfrac{1}{3}(n-1) A_{n-1}{}^2 < (A_n{}^2 + A_{n-1}{}^2 + \ldots + A_2{}^2);$$

and hence the first part of the proposition is proved.

II.   We have now, in order to prove the second result, to show that

$$(n-1) A_n . A_1 + \tfrac{1}{3}(n-1) A_{n-1}{}^2 > (A_{n-1}{}^2 + A_{n-2}{}^2 + \ldots + A_1{}^2).$$

The right-hand side is equal to

$$(A_{n-2} + A_1)^2 + (A_{n-3} + A_1)^2 + \ldots + (A_1 + A_1)^2 + A_1^2$$
$$= A_{n-2}^2 + A_{n-3}^2 + \ldots + A_1^2$$
$$+ (n-1) A_1^2$$
$$+ 2A_1 (A_{n-2} + A_{n-3} + \ldots + A_1)$$
$$= (A_{n-2}^2 + A_{n-3}^2 + \ldots + A_1^2)$$
$$+ (n-1) A_1^2$$
$$+ A_1 \left\{ \begin{matrix} A_{n-2} + A_{n-3} + \ldots + A_1 \\ + A_1 \quad + A_2 \quad + \ldots + A_{n-2} \end{matrix} \right\}$$
$$= (A_{n-2}^2 + A_{n-3}^2 + \ldots + A_1^2)$$
$$+ (n-1) A_1^2$$
$$+ (n-2) A_1 . A_{n-1} \ldots\ldots\ldots\ldots\ldots\ldots\ldots(3).$$

Comparing this expression with the right-hand side of (1) above, we see that $(n-1) A_1^2$ is common to both sides, and

$$(n-1) A_1 . A_{n-1} > (n-2) A_1 . A_{n-1},$$

while, by Prop. 10, Cor. 1,

$$\tfrac{1}{3} (n-1) A_{n-1}^2 > (A_{n-2}^2 + A_{n-3}^2 + \ldots + A_1^2).$$

Hence

$$(n-1) A_n . A_1 + \tfrac{1}{3} (n-1) A_{n-1}^2 > (A_{n-1}^2 + A_{n-2}^2 + \ldots + A_1^2);$$

and the second required result follows.

COR. *The results in the above proposition are equally true if similar figures be substituted for squares on the several lines.*

DEFINITIONS.

1. If a straight line drawn in a plane revolve at a uniform rate about one extremity which remains fixed and return to the position from which it started, and if, at the same time as the line revolves, a point move at a uniform rate along the straight line beginning from the extremity which remains fixed, the point will describe a **spiral** (ἕλιξ) in the plane.

2. Let the extremity of the straight line which remains

fixed while the straight line revolves be called the **origin***
(ἀρχά) of the spiral.

3. And let the position of the line from which the straight
line began to revolve be called the **initial line*** in the
revolution (ἀρχὰ τᾶς περιφορᾶς).

4. Let the length which the point that moves along the
straight line describes in one revolution be called the **first
distance**, that which the same point describes in the second
revolution the **second distance**, and similarly let the distances
described in further revolutions be called after the number of
the particular revolution.

5. Let the area bounded by the spiral described in the
first revolution and the *first distance* be called the **first area**,
that bounded by the spiral described in the second revolution
and the *second distance* the **second area**, and similarly for the
rest in order.

6. If from the origin of the spiral any straight line be
drawn, let that side of it which is in the same direction as that
of the revolution be called **forward** (προαγούμενα), and that
which is in the other direction **backward** (ἑπόμενα).

7. Let the circle drawn with the *origin* as centre and the
first distance as radius be called the **first circle**, that drawn
with the same centre and twice the radius the **second circle**,
and similarly for the succeeding circles.

Proposition 12.

*If any number of straight lines drawn from the origin to
meet the spiral make equal angles with one another, the lines will
be in arithmetical progression.*

[The proof is obvious.]

---

* The literal translation would of course be the "beginning of the spiral"
and "the beginning of the revolution" respectively. But the modern names
will be more suitable for use later on, and are therefore employed here.

Proposition 13.

If a straight line touch the spiral, it will touch it in one point only.

Let O be the origin of the spiral, and BC a tangent to it.

If possible, let BC touch the spiral in two points P, Q. Join OP, OQ, and bisect the angle POQ by the straight line OR meeting the spiral in R.

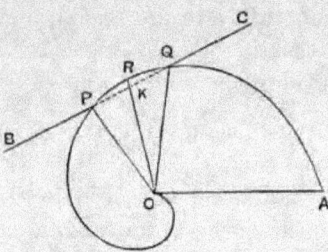

Then [Prop. 12] OR is an arithmetic mean between OP and OQ, or

$$OP + OQ = 2OR.$$

But in any triangle POQ, if the bisector of the angle POQ meets PQ in K,

$$OP + OQ > 2OK *.$$

Therefore $OK < OR$, and it follows that some point on BC between P and Q lies within the spiral. Hence BC cuts the spiral; which is contrary to the hypothesis.

Proposition 14.

If O be the origin, and P, Q two points on the first turn of the spiral, and if OP, OQ produced meet the 'first circle' $AKP'Q'$ in P', Q' respectively, OA being the initial line, then

$$OP : OQ = (\text{arc } AKP') : (\text{arc } AKQ').$$

For, while the revolving line OA moves about O, the point A on it moves uniformly along the circumference of the circle

---

* This is assumed as a known proposition ; but it is easily proved.

$AKP'Q'$, and at the same time the point describing the spiral moves uniformly along OA.

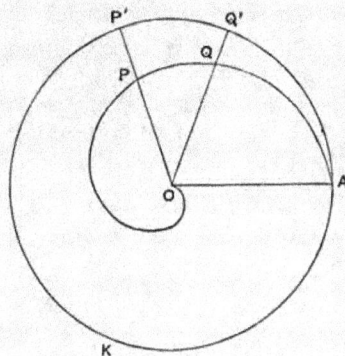

Thus, while A describes the arc AKP', the moving point on OA describes the length OP, and, while A describes the arc AKQ', the moving point on OA describes the distance OQ.

Hence $OP : OQ = $ (arc AKP') : (arc AKQ').   [Prop. 2]

Proposition 15.

If P, Q be points on the second turn of the spiral, and OP, OQ meet the 'first circle' $AKP'Q'$ in P', Q', as in the last proposition, and if c be the circumference of the first circle, then

$$OP : OQ = c + (\text{arc } AKP') : c + (\text{arc } AKQ').$$

For, while the moving point on OA describes the distance OP, the point A describes the whole of the circumference of the 'first circle' together with the arc AKP'; and, while the moving point on OA describes the distance OQ, the point A describes the whole circumference of the 'first circle' together with the arc AKQ'.

Cor. Similarly, if P, Q are on the nth turn of the spiral,

$$OP : OQ = (n-1)c + (\text{arc } AKP') : (n-1)c + (\text{arc } AKQ').$$

Propositions 16, 17.

If BC be the tangent at P, any point on the spiral, PC being the 'forward' part of BC, and if OP be joined, the angle OPC is obtuse while the angle OPB is acute.

I.   Suppose P to be on the first turn of the spiral.

Let OA be the initial line, AKP' the 'first circle.' Draw the circle DLP with centre O and radius OP, meeting OA in D.   This circle must then, in the 'forward' direction from P,

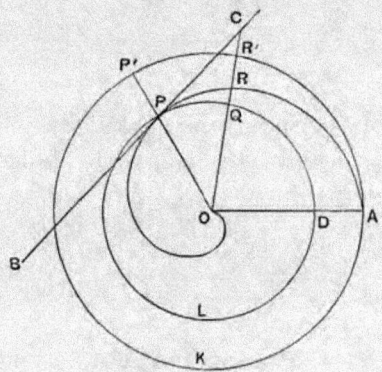

fall within the spiral, and in the 'backward' direction outside it, since the radii vectores of the spiral are on the 'forward' side greater, and on the 'backward' side less, than OP.   Hence the angle OPC cannot be acute, since it cannot be less than the angle between OP and the tangent to the circle at P, which is a right angle.

It only remains therefore to prove that OPC is not a right angle.

If possible, let it be a right angle.   BC will then touch the circle at P.

Therefore [Prop. 5] it is possible to draw a line OQC meeting the circle through P in Q and BC in C, such that

$$CQ : OQ < (\text{arc } PQ) : (\text{arc } DLP) \ldots\ldots\ldots\ldots(1).$$

Suppose that OC meets the spiral in R and the 'first circle' in R'; and produce OP to meet the 'first circle' in P'.

From (1) it follows, *componendo*, that

$$CO : OQ < (\text{arc } DLQ) : (\text{arc } DLP)$$

$$< (\text{arc } AKR') : (\text{arc } AKP')$$

$$< OR : OP. \qquad\qquad \text{[Prop. 14]}$$

But this is impossible, because $OQ = OP$, and $OR < OC$.

Hence the angle OPC is not a right angle. It was also proved not to be acute.

Therefore the angle OPC is obtuse, and the angle OPB consequently acute.

II. If P is on the second, or the nth turn, the proof is the same, except that in the proportion (1) above we have to substitute for the arc DLP an arc equal to $(p + \text{arc } DLP)$ or $(n-1 \cdot p + \text{arc } DLP)$, where p is the perimeter of the circle

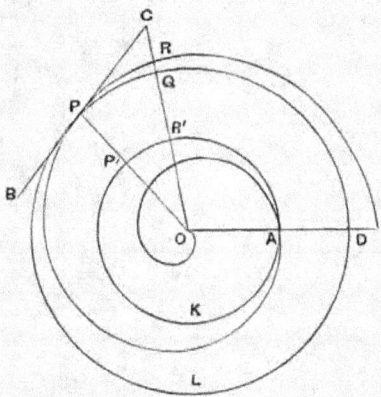

DLP through P. Similarly, in the later steps, p or $(n-1) p$ will be added to each of the arcs DLQ and DLP, and c or $(n-1) c$ to each of the arcs AKR', AKP', where c is the circumference of the 'first circle' AKP'.

Propositions 18, 19.

I. *If OA be the initial line, A the end of the first turn of the spiral, and if the tangent to the spiral at A be drawn, the straight line OB drawn from O perpendicular to OA will meet the said tangent in some point B, and OB will be equal to the circumference of the 'first circle.'*

II. *If A' be the end of the second turn, the perpendicular OB will meet the tangent at A' in some point B', and OB' will be equal to 2 (circumference of 'second circle').*

III. *Generally, if A_n be the end of the nth turn, and OB meet the tangent at A_n in B_n, then*

$$OB_n = nc_n,$$

where c_n is the circumference of the 'nth circle.'

I. Let *AKC* be the 'first circle.' Then, since the 'backward' angle between *OA* and the tangent at *A* is acute [Prop. 16], the tangent will meet the 'first circle' in a second point *C*. And the angles *CAO, BOA* are together less than two right angles; therefore *OB* will meet *AC* produced in some point *B*.

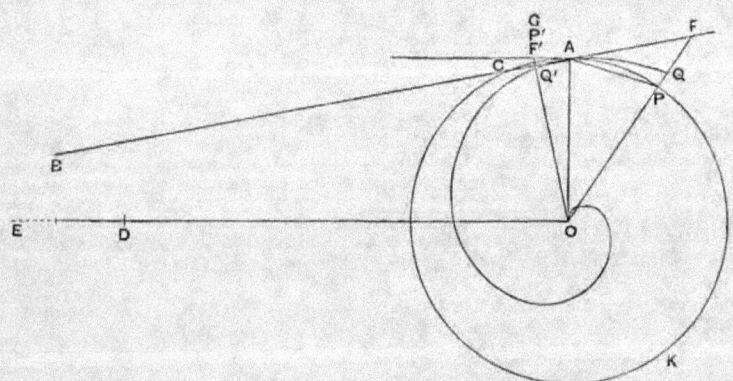

Then, if *c* be the circumference of the first circle, we have to prove that

$$OB = c.$$

If not, *OB* must be either greater or less than *c*.

(1) If possible, suppose $OB > c$.

Measure along OB a length OD less than OB but greater than c.

We have then a circle AKC, a chord AC in it less than the diameter, and a ratio $AO:OD$ which is greater than the ratio $AO:OB$ or (what is, by similar triangles, equal to it) the ratio of $\frac{1}{2}AC$ to the perpendicular from O on AC. Therefore [Prop. 7] we can draw a straight line OPF, meeting the circle in P and CA produced in F, such that

$$FP : PA = AO : OD.$$

Thus, alternately, since $AO = PO$,

$$FP : PO = PA : OD$$
$$< (\text{arc } PA) : c,$$

since $(\text{arc } PA) > PA$, and $OD > c$.

Componendo,

$$FO : PO < (c + \text{arc } PA) : c$$
$$< OQ : OA,$$

where OF meets the spiral in Q.   [Prop. 15]

Therefore, since $OA = OP$, $FO < OQ$; which is impossible.

Hence $OB \ngtr c$.

(2) If possible, suppose $OB < c$.

Measure OE along OB so that OE is greater than OB but less than c.

In this case, since the ratio $AO : OE$ is less than the ratio $AO : OB$ (or the ratio of $\frac{1}{2}AC$ to the perpendicular from O on AC), we can [Prop. 8] draw a line $OF''P'G$, meeting AC in F', the circle in P', and the tangent at A to the circle in G, such that

$$F'P' : AG = AO : OE.$$

Let $OP'G$ cut the spiral in Q'.

Then we have, alternately,

$$F'P' : P'O = AG : OE$$
$$> (\text{arc } AP') : c,$$

because $AG > (\text{arc } AP')$, and $OE < c$.

Therefore

$$F'O : P'O < (\text{arc } AKP') : c$$

$$< OQ' : OA. \qquad \text{[Prop. 14]}$$

But this is impossible, since $OA = OP'$, and $OQ' < OF'$.

Hence $\qquad\qquad\qquad OB \not< c.$

Since therefore OB is neither greater nor less than c,

$$OB = c.$$

II. Let $A'K'C'$ be the 'second circle,' $A'C'$ being the tangent to the spiral at A' (which will cut the second circle, since the 'backward' angle $OA'C'$ is acute). Thus, as before, the perpendicular OB' to OA' will meet $A'C'$ produced in some point B'.

If then c' is the circumference of the 'second circle,' we have to prove that $OB' = 2c'$.

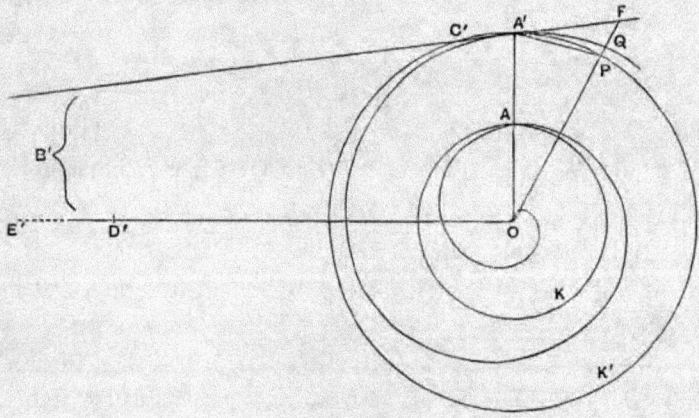

For, if not, OB' must be either greater or less than $2c'$.

(1) If possible, suppose $OB' > 2c'$.

Measure OD' along OB' so that OD' is less than OB' but greater than $2c'$.

Then, as in the case of the 'first circle' above, we can draw a straight line OPF meeting the 'second circle' in P and $C'A'$ produced in F, such that

$$FP : PA' = A'O : OD'.$$

Let OF meet the spiral in Q.

We now have, since $A'O = PO$,

$$FP : PO = PA' : OD'$$
$$< (\text{arc } A'P) : 2c',$$

because $(\text{arc } A'P) > A'P$ and $OD' > 2c'$.

Therefore $FO : PO < (2c' + \text{arc } A'P) : 2c'$
$$< OQ : OA'.\qquad\qquad \text{[Prop. 15, Cor.]}$$

Hence $FO < OQ$; which is impossible.

Thus $OB' \not> 2c'$.

Similarly, as in the case of the 'first circle', we can prove that

$$OB' \not< 2c'.$$

Therefore $OB' = 2c'$.

III.   Proceeding, in like manner, to the 'third' and succeeding circles, we shall prove that

$$OB_n = nc_n.$$

Proposition 20.

I.   *If P be any point on the first turn of the spiral and OT be drawn perpendicular to OP, OT will meet the tangent at P to the spiral in some point T; and, if the circle drawn with centre O and radius OP meet the initial line in K, then OT is equal to the arc of this circle between K and P measured in the 'forward' direction of the spiral.*

II.   *Generally, if P be a point on the nth turn, and the notation be as before, while p represents the circumference of the circle with radius OP,*

$$OT = (n - 1)\,p + \text{arc } KP\ (measured\ 'forward').$$

I.   Let P be a point on the first turn of the spiral, OA the initial line, PR the tangent at P taken in the 'backward' direction.

Then [Prop. 16] the angle OPR is acute.   Therefore PR

meets the circle through P in some point R; and also OT will meet PR produced in some point T.

If now OT is not equal to the arc KRP, it must be either greater or less.

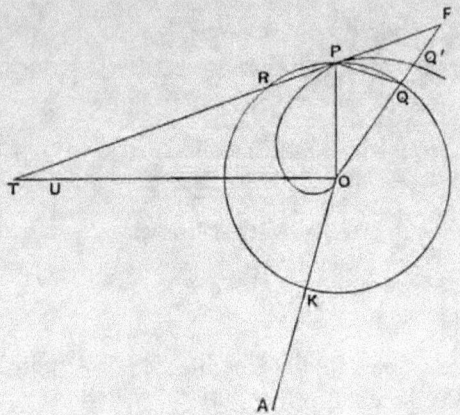

(1) If possible, let OT be greater than the arc KRP.

Measure OU along OT less than OT but greater than the arc KRP.

Then, since the ratio $PO : OU$ is greater than the ratio $PO : OT$, or (what is, by similar triangles, equal to it) the ratio of $\frac{1}{2}PR$ to the perpendicular from O on PR, we can draw a line OQF, meeting the circle in Q and RP produced in F, such that

$$FQ : PQ = PO : OU. \qquad \text{[Prop. 7]}$$

Let OF meet the spiral in Q'.

We have then

$$FQ : QO = PQ : OU$$
$$< (\text{arc } PQ) : (\text{arc } KRP), \text{ by hypothesis.}$$

Componendo,

$$FO : QO < (\text{arc } KRQ) : (\text{arc } KRP)$$
$$< OQ' : OP. \qquad \text{[Prop. 14]}$$

But $QO = OP$.

Therefore $FO < OQ'$; which is impossible.

Hence $\qquad\qquad OT \not> (\text{arc } KRP)$.

(2) The proof that $OT \not< (\text{arc } KRP)$ follows the method of Prop. 18, I. (2), exactly as the above follows that of Prop. 18, I. (1).

Since then OT is neither greater nor less than the arc KRP, it is equal to it.

II. If P be on the second turn, the same method shows that

$$OT = p + (\text{arc } KRP);$$

and, similarly, we have, for a point P on the nth turn,

$$OT = (n-1)\,p + (\text{arc } KRP).$$

Propositions 21, 22, 23.

Given an area bounded by any arc of a spiral and the lines joining the extremities of the arc to the origin, it is possible to circumscribe about the area one figure, and to inscribe in it another figure, each consisting of similar sectors of circles, and such that the circumscribed figure exceeds the inscribed by less than any assigned area.

For let BC be any arc of the spiral, O the origin. Draw the circle with centre O and radius OC, where C is the 'forward' end of the arc.

Then, by bisecting the angle BOC, bisecting the resulting angles, and so on continually, we shall ultimately arrive at an angle COr cutting off a sector of the circle less than any assigned area. Let COr be this sector.

Let the other lines dividing the angle BOC into equal parts meet the spiral in P, Q, and let Or meet it in R. With O as centre and radii OB, OP, OQ, OR respectively describe arcs of circles Bp', bBq', pQr', qRc', each meeting the adjacent radii as shown in the figure. In each case the arc in the 'forward' direction from each point will fall within, and the arc in the 'backward' direction outside, the spiral.

We have now a circumscribed figure and an inscribed figure each consisting of similar sectors of circles. To compare their areas, we take the successive sectors of each, beginning from OC, and compare them.

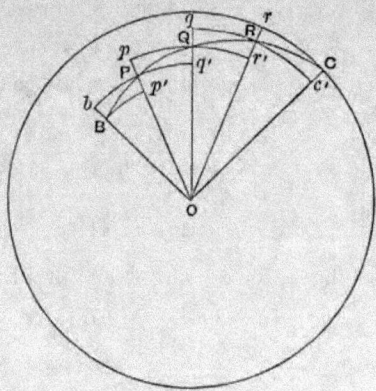

The sector OCr in the circumscribed figure stands alone.

And
$$(\text{sector } ORq) = (\text{sector } ORc'),$$
$$(\text{sector } OQp) = (\text{sector } OQr'),$$
$$(\text{sector } OPb) = (\text{sector } OPq'),$$

while the sector OBp' in the inscribed figure stands alone.

Hence, if the equal sectors be taken away, the difference between the circumscribed and inscribed figures is equal to the difference between the sectors OCr and OBp'; and this difference is less than the sector OCr, which is itself less than any assigned area.

The proof is exactly the same whatever be the number of angles into which the angle BOC is divided, the only difference being that, when the arc begins from the origin, the smallest sectors OPb, OPq' in each figure are equal, and there is therefore no inscribed sector standing by itself, so that the difference between the circumscribed and inscribed figures is equal to the sector OCr itself.

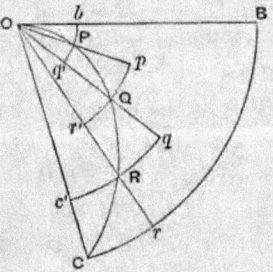

Thus the proposition is universally true.

COR. Since the area bounded by the spiral is intermediate in magnitude between the circumscribed and inscribed figures, it follows that

(1) *a figure can be circumscribed to the area such that it exceeds the area by less than any assigned space*,

(2) *a figure can be inscribed such that the area exceeds it by less than any assigned space*.

Proposition 24.

The area bounded by the first turn of the spiral and the initial line is equal to one-third of the 'first circle' $[= \frac{1}{3}\pi(2\pi a)^2$, where the spiral is $r = a\theta]$.

[The same proof shows equally that, *if OP be any radius vector in the first turn of the spiral, the area of the portion of the spiral bounded thereby is equal to one-third of that sector of the circle drawn with radius OP which is bounded by the initial line and OP, measured in the 'forward' direction from the initial line.*]

Let O be the origin, OA the initial line, A the extremity of the first turn.

Draw the 'first circle,' i.e. the circle with O as centre and OA as radius.

Then, if C_1 be the area of the first circle, R_1 that of the first turn of the spiral bounded by OA, we have to prove that

$$R_1 = \frac{1}{3}C_1.$$

For, if not, R_1 must be either greater or less than C_1.

I. If possible, suppose $R_1 < \frac{1}{3}C_1$.

We can then circumscribe a figure about R_1 made up of similar sectors of circles such that, if F be the area of this figure,

$$F - R_1 < \frac{1}{3}C_1 - R_1,$$

whence $F < \frac{1}{3}C_1$.

Let OP, OQ, ... be the radii of the circular sectors, beginning from the smallest. The radius of the largest is of course OA.

The radii then form an ascending arithmetical progression in which the common difference is equal to the least term OP. If n be the number of the sectors, we have [by Prop. 10, Cor. 1]

$$n \cdot OA^2 < 3 (OP^2 + OQ^2 + \ldots + OA^2);$$

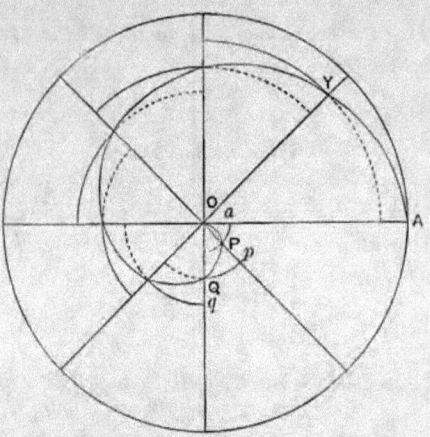

and, since the similar sectors are proportional to the squares on their radii, it follows that

$$C_1 < 3F,$$

or

$$F > \tfrac{1}{3} C_1.$$

But this is impossible, since F was less than $\tfrac{1}{3} C_1$.

Therefore $R_1 \not< \tfrac{1}{3} C_1$.

II. If possible, suppose $R_1 > \tfrac{1}{3} C_1$.

We can then *inscribe* a figure made up of similar sectors of circles such that, if f be its area,

$$R_1 - f < R_1 - \tfrac{1}{3} C_1,$$

whence $f > \tfrac{1}{3} C_1$.

If there are $(n - 1)$ sectors, their radii, as OP, OQ, ..., form an ascending arithmetical progression in which the least term is equal to the common difference, and the greatest term, as OY, is equal to $(n - 1) OP$.

12—2

Thus [Prop. 10, Cor. 1]
$$n \cdot OA^2 > 3\,(OP^2 + OQ^2 + \ldots + OY^2),$$
whence $\qquad\qquad C_1 > 3f,$

or $\qquad\qquad\qquad f < \tfrac{1}{3}C_1;$

which is impossible, since $f > \tfrac{1}{3}C_1.$

Therefore $\qquad\qquad R_1 \not> \tfrac{1}{3}C_1.$

Since then R_1 is neither greater nor less than $\tfrac{1}{3}C_1,$
$$R_1 = \tfrac{1}{3}C_1.$$

[Archimedes does not actually find the area of the spiral cut off by the radius vector OP, where P is any point on the first turn; but, in order to do this, we have only to substitute

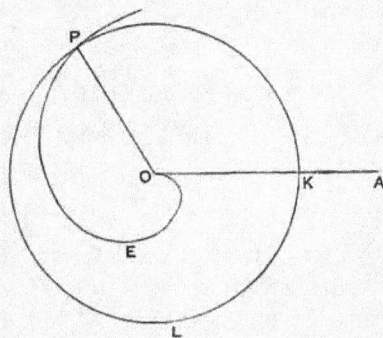

in the above proof the area of the sector KLP of the circle drawn with O as centre and OP as radius for the area C_1 of the 'first circle', while the two figures made up of similar sectors have to be circumscribed about and inscribed in the portion OEP of the spiral. The same method of proof then applies exactly, and the area of OEP is seen to be $\tfrac{1}{3}$ (sector KLP).

We can prove also, by the same method, that, if P be a point on the second, or any later turn, as the nth, the *complete area described by the radius vector* from the beginning up to the time when it reaches the position OP is, if C denote the area of the complete circle with O as centre and OP as radius, $\tfrac{1}{3}\,(C + \text{sector } KLP)$ or $\tfrac{1}{3}\,(\overline{n-1} \cdot C + \text{sector } KLP)$ respectively.

The area so described by the radius vector is of course not the same thing as the area bounded by the last complete turn of the spiral ending at P and the intercepted portion of the radius vector OP. Thus, suppose R_1 to be the area bounded by the first turn of the spiral and OA_1 (the first turn ending at A_1 on the initial line), R_2 the area *added* to this by the second complete turn ending at A_2 on the initial line, and so on. R_1 has then been described *twice* by the radius vector when it arrives at the position OA_2; when the radius vector arrives at the position OA_3, it has described R_1 three times, the ring R_2 twice, and the ring R_3 once; and so on.

Thus, generally, if C_n denote the area of the 'nth circle,' we shall have

$$\tfrac{1}{3}nC_n = R_n + 2R_{n-1} + 3R_{n-2} + \ldots + nR_1,$$

while the actual area bounded by the outside, or the complete nth, turn and the intercepted portion of OA_n will be equal to

$$R_n + R_{n-1} + R_{n-2} + \ldots + R_1.$$

It can now be seen that the results of the later Props. 25 and 26 may be obtained from the extension of Prop. 24 just given.

To obtain the general result of Prop. 26, suppose BC to be an arc on any turn whatever of the spiral, being itself less than a complete turn, and suppose B to be beyond A_n the extremity of the nth complete turn, while C is 'forward' from B.

Let $\dfrac{p}{q}$ be the fraction of a turn between the end of the nth turn and the point B.

Then the *area described by the radius vector* up to the position OB (starting from the beginning of the spiral) is equal to

$$\tfrac{1}{3}\left(n + \frac{p}{q}\right)(\text{circle with rad. } OB).$$

Also the *area described by the radius vector* from the beginning up to the position OC is

$$\tfrac{1}{3}\left\{\left(n + \frac{p}{q}\right)(\text{circle with rad. } OC) + (\text{sector } B'MC)\right\}.$$

The area bounded by OB, OC and the portion BEC of the spiral is equal to the difference between these two expressions; and, since the circles are to one another as OB^2 to OC^2, the difference may be expressed as

$$\tfrac{1}{3}\left\{\left(n+\frac{p}{q}\right)\left(1-\frac{OB^2}{OC^2}\right)(\text{circle with rad. } OC)+(\text{sector } B'MC)\right\}.$$

But, by Prop. 15, Cor.,

$$\left(n+\frac{p}{q}\right)(\text{circle } B'MC):\left\{\left(n+\frac{p}{q}\right)(\text{circle } B'MC)+(\text{sector } B'MC)\right\}$$
$$=OB:OC,$$

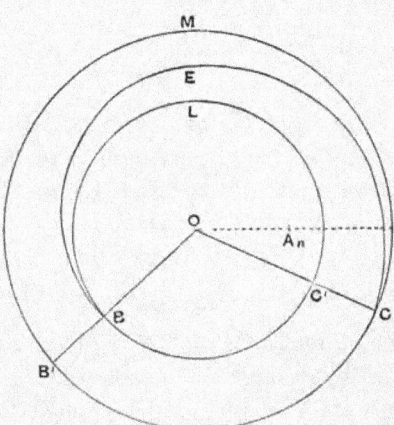

so that

$$\left(n+\frac{p}{q}\right)(\text{circle } B'MC):(\text{sector } B'MC)=OB:(OC-OB).$$

Thus $\quad\dfrac{\text{area } BEC}{\text{sector } B'MC}=\tfrac{1}{3}\left\{\left(\dfrac{OB}{OC-OB}\right)\left(1-\dfrac{OB^2}{OC^2}\right)+1\right\}$

$$=\tfrac{1}{3}\cdot\dfrac{OB(OC+OB)+OC^2}{OC^2}$$

$$=\dfrac{OC.OB+\tfrac{1}{3}(OC-OB)^2}{OC^2}.$$

The result of Prop. 25 is a particular case of this, and the result of Prop. 27 follows immediately, as shown under that proposition.]

Propositions 25, 26, 27.

[Prop. 25.] *If A_2 be the end of the second turn of the spiral, the area bounded by the second turn and OA_2 is to the area of the 'second circle' in the ratio of 7 to 12, being the ratio of $\{r_2r_1 + \frac{1}{3}(r_2 - r_1)^2\}$ to r_2^2, where r_1, r_2 are the radii of the 'first' and 'second' circles respectively.*

[Prop. 26.] *If BC be any arc measured in the 'forward' direction on any turn of a spiral, not being greater than the complete turn, and if a circle be drawn with O as centre and OC as radius meeting OB in B', then*

(area of spiral between OB, OC) : (sector $OB'C$)

$$= \{OC \cdot OB + \tfrac{1}{3}(OC - OB)^2\} : OC^2.$$

[Prop. 27.] *If R_1 be the area of the first turn of the spiral bounded by the initial line, R_2 the area of the ring added by the second complete turn, R_3 that of the ring added by the third turn, and so on, then*

$$R_3 = 2R_2, \ R_4 = 3R_2, \ R_5 = 4R_2, \dots, R_n = (n-1)R_2.$$

Also $$R_2 = 6R_1.$$

[Archimedes' proof of Prop. 25 is, *mutatis mutandis*, the same as his proof of the more general Prop. 26. The latter will accordingly be given here, and applied to Prop. 25 as a particular case.]

Let BC be an arc measured in the 'forward' direction on any turn of the spiral, CKB' the circle drawn with O as centre and OC as radius.

Take a circle such that the square of its radius is equal to $OC \cdot OB + \frac{1}{3}(OC - OB)^2$, and let σ be a sector in it whose central angle is equal to the angle BOC.

Thus σ : (sector $OB'C$) = $\{OC \cdot OB + \frac{1}{3}(OC - OB)^2\} : OC^2$,

and we have therefore to prove that

(area of spiral OBC) $= \sigma$.

For, if not, the area of the spiral OBC (which we will call S) must be either greater or less than σ.

I. Suppose, if possible, $S < \sigma$.

Circumscribe to the area S a figure made up of similar sectors of circles, such that, if F be the area of the figure,

$$F - S < \sigma - S,$$

whence $F < \sigma.$

Let the radii of the successive sectors, starting from OB, be OP, $OQ, \ldots OC$.   Produce OP, $OQ, \ldots$ to meet the circle $CKB', \ldots$

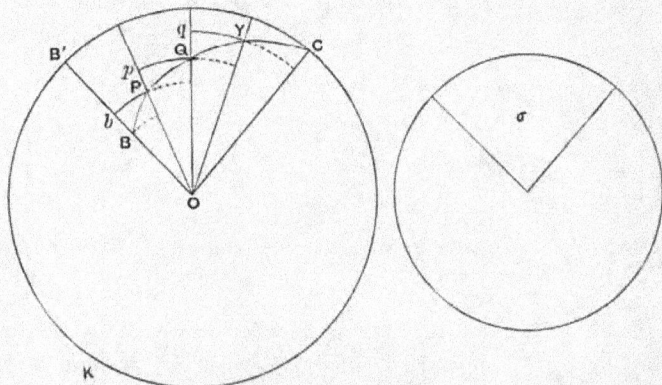

If then the lines OB, OP, $OQ, \ldots OC$ be n in number, the number of sectors in the circumscribed figure will be $(n-1)$, and the sector $OB'C$ will also be divided into $(n-1)$ equal sectors.   Also OB, OP, $OQ, \ldots OC$ will form an ascending arithmetical progression of n terms.

Therefore [see Prop. 11 and Cor.]

$$(n-1)\,OC^2 : (OP^2 + OQ^2 + \ldots + OC^2)$$
$$< OC^2 : \{OC . OB + \tfrac{1}{3}(OC - OB)^2\}$$
$$< (\text{sector } OB'C) : \sigma, \text{ by hypothesis.}$$

Hence, since similar sectors are as the squares of their radii,

$$(\text{sector } OB'C) : F < (\text{sector } OB'C) : \sigma,$$

so that $F > \sigma.$

But this is impossible, because $F < \sigma$.

Therefore $S \not< \sigma.$

II. Suppose, if possible, $S > \sigma$.

Inscribe in the area S a figure made up of similar sectors of circles such that, if f be its area,

$$S - f < S - \sigma,$$

whence $\qquad\qquad\qquad\qquad f > \sigma.$

Suppose $OB, OP, \ldots OY$ to be the radii of the successive sectors making up the figure f, being $(n - 1)$ in number.

We shall have in this case [see Prop. 11 and Cor.]

$$(n - 1)\, OC^2 : (OB^2 + OP^2 + \ldots + OY^2)$$
$$> OC^2 : \{OC.OB + \tfrac{1}{3}(OC - OB)^2\},$$

whence $\qquad$ (sector $OB'C$) $: f >$ (sector $OB'C$) $: \sigma$,

so that $\qquad\qquad\qquad\qquad f < \sigma.$

But this is impossible, because $f > \sigma$.

Therefore $\qquad\qquad\qquad\qquad S \not> \sigma.$

Since then S is neither greater nor less than σ, it follows that

$$S = \sigma.$$

In the particular case where B coincides with A_1, the end of the first turn of the spiral, and C with A_2, the end of the second turn, the sector $OB'C$ becomes the complete 'second circle,' that, namely, with OA_2 (or r_2) as radius.

Thus

$$(\text{area of spiral bounded by } OA_2) : (\text{'second circle'})$$
$$= \{r_2 r_1 + \tfrac{1}{3}(r_2 - r_1)^2\} : r_2^2$$
$$= (2 + \tfrac{1}{3}) : 4 \quad (\text{since } r_2 = 2r_1)$$
$$= 7 : 12.$$

Again, the area of the spiral bounded by OA_2 is equal to $R_1 + R_2$ (i.e. the area bounded by the first turn and OA_1, together with the ring added by the second turn). Also the 'second circle' is four times the 'first circle,' and therefore equal to $12\,R_1$.

Hence $\qquad\qquad (R_1 + R_2) : 12R_1 = 7 : 12,$

or $\qquad\qquad\qquad\qquad R_1 + R_2 = 7R_1.$

Thus $\qquad\qquad\qquad\qquad R_2 = 6R_1 \ldots\ldots\ldots\ldots\ldots\ldots(1).$

Next, for the third turn, we have

$$(R_1 + R_2 + R_3) : (\text{'third circle'}) = \{r_3 r_2 + \tfrac{1}{3}(r_3 - r_2)^2\} : r_3^2$$
$$= (3 \cdot 2 + \tfrac{1}{3}) : 3^2$$
$$= 19 : 27,$$

and $(\text{'third circle'}) = 9\,(\text{'first circle'})$
$$= 27 R_1;$$

therefore $R_1 + R_2 + R_3 = 19 R_1,$

and, by (1) above, it follows that

$$R_3 = 12 R_1$$
$$= 2 R_2 \dots\dots\dots\dots\dots\dots\dots\dots(2),$$

and so on.

Generally, we have

$$(R_1 + R_2 + \dots + R_n) : (n\text{th circle}) = \{r_n r_{n-1} + \tfrac{1}{3}(r_n - r_{n-1})^2\} : r_n^2,$$

$$(R_1 + R_2 + \dots + R_{n-1}) : (n - 1\text{th circle})$$
$$= \{r_{n-1} r_{n-2} + \tfrac{1}{3}(r_{n-1} - r_{n-2})^2\} : r_{n-1}^2,$$

and $(n\text{th circle}) : (\overline{n - 1}\text{th circle}) = r_n^2 : r_{n-1}^2.$

Therefore

$$(R_1 + R_2 + \dots + R_n) : (R_1 + R_2 + \dots + R_{n-1})$$
$$= \{n(n - 1) + \tfrac{1}{3}\} : \{(n - 1)(n - 2) + \tfrac{1}{3}\}$$
$$= \{3n(n - 1) + 1\} : \{3(n - 1)(n - 2) + 1\}.$$

Dirimendo,

$$R_n : (R_1 + R_2 + \dots + R_{n-1})$$
$$= 6(n - 1) : \{3(n - 1)(n - 2) + 1\} \dots\dots\dots(\alpha).$$

Similarly

$$R_{n-1} : (R_1 + R_2 + \dots + R_{n-2}) = 6(n - 2) : \{3(n - 2)(n - 3) + 1\},$$

from which we derive

$$R_{n-1} : (R_1 + R_2 + \dots + R_{n-1})$$
$$= 6(n - 2) : \{6(n - 2) + 3(n - 2)(n - 3) + 1\}$$
$$= 6(n - 2) : \{3(n - 1)(n - 2) + 1\}\dots\dots\dots\dots(\beta).$$

Combining (α) and (β), we obtain

$$R_n : R_{n-1} = (n-1) : (n-2).$$

Thus

$R_2, R_3, R_4, \ldots R_n$ are in the ratio of the successive numbers 1, 2, 3 ... $(n-1)$.

Proposition 28.

If O be the origin and BC any arc measured in the 'forward' direction on any turn of the spiral, let two circles be drawn (1) with centre O, and radius OB, meeting OC in C', and (2) with centre O and radius OC, meeting OB produced in B'. Then, if E denote the area bounded by the larger circular arc B'C, the line B'B, and the spiral BC, while F denotes the area bounded by the smaller arc BC'', the line CC' and the spiral BC,

$$E : F = \{OB + \tfrac{2}{3}(OC - OB)\} : \{OB + \tfrac{1}{3}(OC - OB)\}.$$

Let σ denote the area of the lesser sector OBC''; then the larger sector $OB'C$ is equal to $\sigma + F + E$.

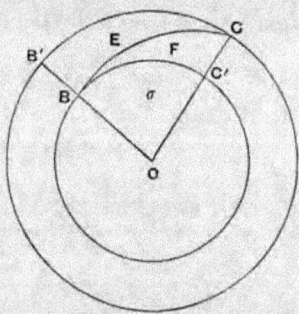

Thus [Prop. 26]

$$(\sigma + F) : (\sigma + F + E) = \{OC . OB + \tfrac{1}{3}(OC - OB)^2\} : OC^2 \ldots (1),$$

whence

$$E : (\sigma + F) = \{OC (OC - OB) - \tfrac{1}{3}(OC - OB)^2\}$$
$$: \{OC . OB + \tfrac{1}{3}(OC - OB)^2\}$$
$$= \{OB (OC - OB) + \tfrac{2}{3}(OC - OB)^2\}$$
$$: \{OC . OB + \tfrac{1}{3}(OC - OB)^2\} \ldots \ldots \ldots (2).$$

Again
$$(\sigma + F + E) : \sigma = OC^2 : OB^2.$$

Therefore, by the first proportion above, *ex aequali*,
$$(\sigma + F) : \sigma = \{OC \cdot OB + \tfrac{1}{3}(OC - OB)^2\} : OB^2,$$
whence

$$(\sigma + F) : F = \{OC \cdot OB + \tfrac{1}{3}(OC - OB)^2\}$$
$$: \{OB(OC - OB) + \tfrac{1}{3}(OC - OB)^2\}.$$

Combining this with (2) above, we obtain

$$E : F = \{OB(OC - OB) + \tfrac{2}{3}(OC - OB)^2\}$$
$$: \{OB(OC - OB) + \tfrac{1}{3}(OC - OB)^2\}$$
$$= \{OB + \tfrac{2}{3}(OC - OB)\} : \{OB + \tfrac{1}{3}(OC - OB)\}.$$

ON THE EQUILIBRIUM OF PLANES

OR

THE CENTRES OF GRAVITY OF PLANES.

BOOK I.

"I POSTULATE the following:

1. Equal weights at equal distances are in equilibrium, and equal weights at unequal distances are not in equilibrium but incline towards the weight which is at the greater distance.

2. If, when weights at certain distances are in equilibrium, something be added to one of the weights, they are not in equilibrium but incline towards that weight to which the addition was made.

3. Similarly, if anything be taken away from one of the weights, they are not in equilibrium but incline towards the weight from which nothing was taken.

4. When equal and similar plane figures coincide if applied to one another, their centres of gravity similarly coincide.

5. In figures which are unequal but similar the centres of gravity will be similarly situated. By points similarly situated in relation to similar figures I mean points such that, if straight lines be drawn from them to the equal angles, they make equal angles with the corresponding sides.

6. If magnitudes at certain distances be in equilibrium, (other) magnitudes equal to them will also be in equilibrium at the same distances.

7. In any figure whose perimeter is concave in (one and) the same direction the centre of gravity must be within the figure."

Proposition 1.

Weights which balance at equal distances are equal.

For, if they are unequal, take away from the greater the difference between the two. The remainders will then not balance [*Post.* 3]; which is absurd.

Therefore the weights cannot be unequal.

Proposition 2.

Unequal weights at equal distances will not balance but will incline towards the greater weight.

For take away from the greater the difference between the two. The equal remainders will therefore balance [*Post.* 1]. Hence, if we add the difference again, the weights will not balance but incline towards the greater [*Post.* 2].

Proposition 3.

Unequal weights will balance at unequal distances, the greater weight being at the lesser distance.

Let A, B be two unequal weights (of which A is the greater) balancing about C at distances AC, BC respectively.

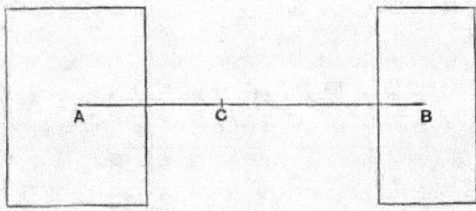

Then shall AC be less than BC. For, if not, take away from A the weight $(A - B.)$ The remainders will then incline towards B [*Post.* 3]. But this is impossible, for (1) if $AC = CB$, the equal remainders will balance, or (2) if $AC > CB$, they will incline towards A at the greater distance [*Post.* 1].

Hence $AC < CB$.

Conversely, if the weights balance, and $AC < CB$, then $A > B$.

Proposition 4.

If two equal weights have not the same centre of gravity, the centre of gravity of both taken together is at the middle point of the line joining their centres of gravity.

[Proved from Prop. 3 by *reductio ad absurdum*. Archimedes assumes that the centre of gravity of both together is on the straight line joining the centres of gravity of each, saying that this had been proved before (προδέδεικται). The allusion is no doubt to the lost treatise *On levers* (περὶ ζυγῶν).]

Proposition 5.

If three equal magnitudes have their centres of gravity on a straight line at equal distances, the centre of gravity of the system will coincide with that of the middle magnitude.

[This follows immediately from Prop. 4.]

Cor 1. *The same is true of any odd number of magnitudes if those which are at equal distances from the middle one are equal, while the distances between their centres of gravity are equal.*

Cor. 2. *If there be an even number of magnitudes with their centres of gravity situated at equal distances on one straight line, and if the two middle ones be equal, while those which are equidistant from them (on each side) are equal respectively, the centre of gravity of the system is the middle point of the line joining the centres of gravity of the two middle ones.*

Propositions 6, 7.

Two magnitudes, whether commensurable [Prop. 6] *or incommensurable* [Prop. 7], *balance at distances reciprocally proportional to the magnitudes.*

I. Suppose the magnitudes A, B to be commensurable, and the points A, B to be their centres of gravity. Let DE be a straight line so divided at C that

$$A : B = DC : CE.$$

We have then to prove that, if A be placed at E and B at D, C is the centre of gravity of the two taken together.

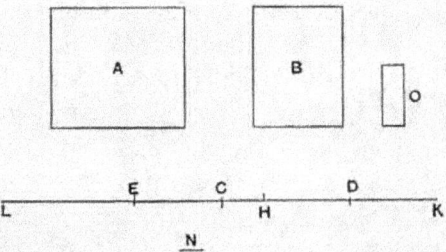

Since A, B are commensurable, so are DC, CE. Let N be a common measure of DC, CE. Make DH, DK each equal to CE, and EL (on CE produced) equal to CD. Then $EH = CD$, since $DH = CE$. Therefore LH is bisected at E, as HK is bisected at D.

Thus LH, HK must each contain N an even number of times.

Take a magnitude O such that O is contained as many times in A as N is contained in LH, whence

$$A : O = LH : N.$$

But $\qquad\qquad B : A = CE : DC$

$$= HK : LH.$$

Hence, *ex aequali*, $B : O = HK : N$, or O is contained in B as many times as N is contained in HK.

Thus O is a common measure of A, B.

Divide LH, HK into parts each equal to N, and A, B into parts each equal to O. The parts of A will therefore be equal in number to those of LH, and the parts of B equal in number to those of HK. Place one of the parts of A at the middle point of each of the parts N of LH, and one of the parts of B at the middle point of each of the parts N of HK.

Then the centre of gravity of the parts of A placed at equal distances on LH will be at E, the middle point of LH [Prop. 5, Cor. 2], and the centre of gravity of the parts of B placed at equal distances along HK will be at D, the middle point of HK.

Thus we may suppose A itself applied at E, and B itself applied at D.

But the system formed by the parts O of A and B together is a system of equal magnitudes even in number and placed at equal distances along LK. And, since $LE = CD$, and $EC = DK$, $LC = CK$, so that C is the middle point of LK. Therefore C is the centre of gravity of the system ranged along LK.

Therefore A acting at E and B acting at D balance about the point C.

II. Suppose the magnitudes to be incommensurable, and let them be $(A + a)$ and B respectively. Let DE be a line divided at C so that

$$(A + a) : B = DC : CE.$$

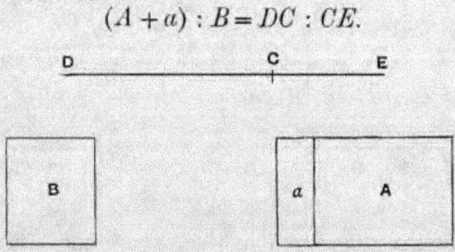

Then, if $(A + a)$ placed at E and B placed at D do not balance about C, $(A + a)$ is either too great to balance B, or not great enough.

Suppose, if possible, that $(A + a)$ is too great to balance B. Take from $(A + a)$ a magnitude a smaller than the deduction which would make the remainder balance B, but such that the remainder A and the magnitude B are commensurable.

Then, since A, B are commensurable, and

$$A : B < DC : CE,$$

A and B will not balance [Prop. 6], but D will be depressed.

But this is impossible, since the deduction a was an insufficient deduction from $(A + a)$ to produce equilibrium, so that E was still depressed.

Therefore $(A + a)$ is not too great to balance B; and similarly it may be proved that B is not too great to balance $(A + a)$.

Hence $(A + a)$, B taken together have their centre of gravity at C.

Proposition 8.

If AB be a magnitude whose centre of gravity is C, and AD a part of it whose centre of gravity is F, then the centre of gravity of the remaining part will be a point G on FC produced such that

$$GC : CF = (AD) : (DE).$$

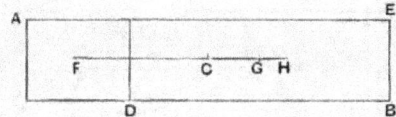

For, if the centre of gravity of the remainder (DE) be not G, let it be a point H. Then an absurdity follows at once from Props. 6, 7.

Proposition 9.

The centre of gravity of any parallelogram lies on the straight line joining the middle points of opposite sides.

Let $ABCD$ be a parallelogram, and let EF join the middle points of the opposite sides AD, BC.

If the centre of gravity does not lie on EF, suppose it to be H, and draw HK parallel to AD or BC meeting EF in K.

Then it is possible, by bisecting *ED*, then bisecting the halves, and so on continually, to arrive at a length *EL* less

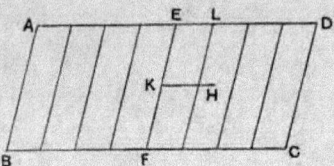

than *KH*. Divide both *AE* and *ED* into parts each equal to *EL*, and through the points of division draw parallels to *AB* or *CD*.

We have then a number of equal and similar parallelograms, and, if any one be applied to any other, their centres of gravity coincide [*Post.* 4]. Thus we have an even number of equal magnitudes whose centres of gravity lie at equal distances along a straight line. Hence the centre of gravity of the whole parallelogram will lie on the line joining the centres of gravity of the two middle parallelograms [Prop. 5, Cor. 2].

But this is impossible, for *H* is outside the middle parallelograms.

Therefore the centre of gravity cannot but lie on *EF*.

Proposition 10.

The centre of gravity of a parallelogram is the point of intersection of its diagonals.

For, by the last proposition, the centre of gravity lies on each of the lines which bisect opposite sides. Therefore it is at the point of their intersection; and this is also the point of intersection of the diagonals.

Alternative proof.

Let *ABCD* be the given parallelogram, and *BD* a diagonal. Then the triangles *ABD*, *CDB* are equal and similar, so that [*Post.* 4], if one be applied to the other, their centres of gravity will fall one upon the other.

13—2

Suppose F to be the centre of gravity of the triangle ABD.
Let G be the middle point of BD.
Join FG and produce it to H, so
that $FG = GH$.

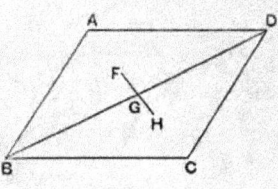

If we then apply the triangle
ABD to the triangle CDB so that
AD falls on CB and AB on CD, the
point F will fall on H.

But [by *Post.* 4] F will fall on the centre of gravity of
CDB. Therefore H is the centre of gravity of CDB.

Hence, since F, H are the centres of gravity of the two
equal triangles, the centre of gravity of the whole parallelogram
is at the middle point of FH, i.e. at the middle point of BD,
which is the intersection of the two diagonals.

Proposition 11.

*If abc, ABC be two similar triangles, and g, G two points in
them similarly situated with respect to them respectively, then, if
g be the centre of gravity of the triangle abc, G must be the centre
of gravity of the triangle ABC.*

Suppose $ab : bc : ca = AB : BC : CA$.

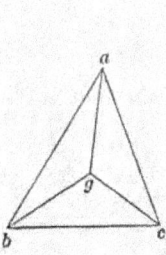
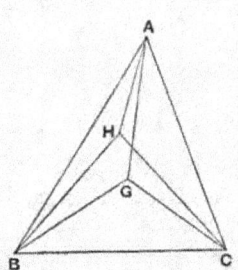

The proposition is proved by an obvious *reductio ad
absurdum*. For, if G be not the centre of gravity of the
triangle ABC, suppose H to be its centre of gravity.

Post. 5 requires that g, H shall be similarly situated with
respect to the triangles respectively; and this leads at once
to the absurdity that the angles HAB, GAB are equal.

Proposition 12.

Given two similar triangles abc, ABC, and d, D the middle points of bc, BC respectively, then, if the centre of gravity of abc lie on ad, that of ABC will lie on AD.

Let g be the point on ad which is the centre of gravity of abc.

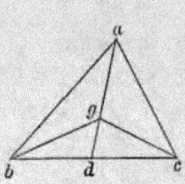 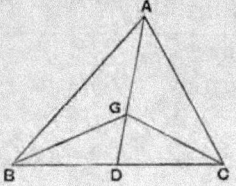

Take G on AD such that

$$ad : ag = AD : AG,$$

and join gb, gc, GB, GC.

Then, since the triangles are similar, and bd, BD are the halves of bc, BC respectively,

$$ab : bd = AB : BD,$$

and the angles abd, ABD are equal.

Therefore the triangles abd, ABD are similar, and

$$\angle bad = \angle BAD.$$

Also $\qquad ba : ad = BA : AD,$

while, from above, $\qquad ad : ag = AD : AG.$

Therefore $ba : ag = BA : AG$, while the angles bag, BAG are equal.

Hence the triangles bag, BAG are similar, and

$$\angle abg = \angle ABG.$$

And, since the angles abd, ABD are equal, it follows that

$$\angle gbd = \angle GBD.$$

In exactly the same manner we prove that

$$\angle gac = \angle GAC,$$
$$\angle acg = \angle ACG,$$
$$\angle gcd = \angle GCD.$$

Therefore g, G are similarly situated with respect to the triangles respectively; whence [Prop. 11] G is the centre of gravity of ABC.

Proposition 13.

In any triangle the centre of gravity lies on the straight line joining any angle to the middle point of the opposite side.

Let ABC be a triangle and D the middle point of BC. Join AD. Then shall the centre of gravity lie on AD.

For, if possible, let this not be the case, and let H be the centre of gravity. Draw HI parallel to CB meeting AD in I.

Then, if we bisect DC, then bisect the halves, and so on, we shall at length arrive at a length, as DE, less than HI.

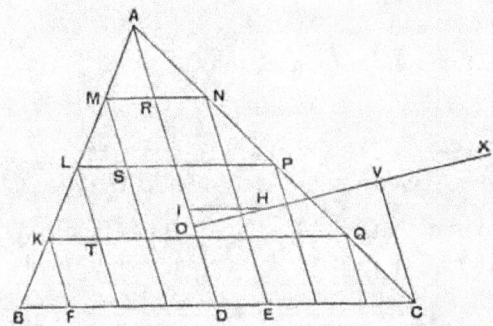

Divide both BD and DC into lengths each equal to DE, and through the points of division draw lines each parallel to DA meeting BA and AC in points as K, L, M and N, P, Q respectively.

Join MN, LP, KQ, which lines will then be each parallel to BC.

We have now a series of parallelograms as FQ, TP, SN, and AD bisects opposite sides in each. Thus the centre of gravity of each parallelogram lies on AD [Prop. 9], and therefore the centre of gravity of the figure made up of them all lies on AD.

Let the centre of gravity of all the parallelograms taken together be O. Join OH and produce it; also draw CV parallel to DA meeting OH produced in V.

Now, if n be the number of parts into which AC is divided,

$\triangle ADC$: (sum of triangles on $AN, NP, \ldots$)

$$= AC^2 : (AN^2 + NP^2 + \ldots)$$
$$= n^2 : n$$
$$= n : 1$$
$$= AC : AN.$$

Similarly

$\triangle ABD$: (sum of triangles on $AM, ML, \ldots$) $= AB : AM.$

And $\qquad\qquad AC : AN = AB : AM.$

It follows that

$\triangle ABC$: (sum of all the small $\triangle$s) $= CA : AN$

$\qquad\qquad\qquad\qquad > VO : OH$, by parallels.

Suppose OV produced to X so that

$$\triangle ABC : \text{(sum of small } \triangle\text{s)} = XO : OH,$$

whence, *dividendo*,

(sum of parallelograms) : (sum of small $\triangle$s) $= XH : HO.$

Since then the centre of gravity of the triangle ABC is at H, and the centre of gravity of the part of it made up of the parallelograms is at O, it follows from Prop. 8 that the centre of gravity of the remaining portion consisting of all the small triangles taken together is at X.

But this is impossible, since all the triangles are on one side of the line through X parallel to AD.

Therefore the centre of gravity of the triangle cannot but lie on AD.

Alternative proof.

Suppose, if possible, that H, not lying on AD, is the centre of gravity of the triangle ABC. Join AH, BH, CH. Let E, F be the middle points of CA, AB respectively, and join DE, EF, FD. Let EF meet AD in M.

Draw *FK*, *EL* parallel to *AH* meeting *BH*, *CH* in *K*, *L* respectively. Join *KD*, *HD*, *LD*, *KL*. Let *KL* meet *DH* in *N*, and join *MN*.

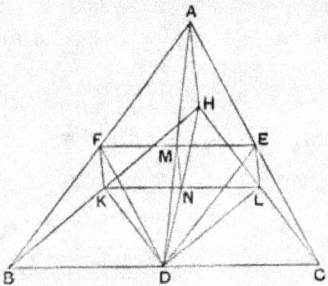

Since *DE* is parallel to *AB*, the triangles *ABC*, *EDC* are similar.

And, since *CE = EA*, and *EL* is parallel to *AH*, it follows that *CL = LH*. And *CD = DB*. Therefore *BH* is parallel to *DL*.

Thus in the similar and similarly situated triangles *ABC*, *EDC* the straight lines *AH*, *BH* are respectively parallel to *EL*, *DL*; and it follows that *H*, *L* are similarly situated with respect to the triangles respectively.

But *H* is, by hypothesis, the centre of gravity of *ABC*. Therefore *L* is the centre of gravity of *EDC*.   [Prop. 11]

Similarly the point *K* is the centre of gravity of the triangle *FBD*.

And the triangles *FBD*, *EDC* are equal, so that the centre of gravity of both together is at the middle point of *KL*, i.e. at the point *N*.

The remainder of the triangle *ABC*, after the triangles *FBD*, *EDC* are deducted, is the parallelogram *AFDE*, and the centre of gravity of this parallelogram is at *M*, the intersection of its diagonals.

It follows that the centre of gravity of the whole triangle *ABC* must lie on *MN*; that is, *MN* must pass through *H*, which is impossible (since *MN* is parallel to *AH*).

Therefore the centre of gravity of the triangle *ABC* cannot but lie on *AD*.

Proposition 14.

It follows at once from the last proposition that *the centre of gravity of any triangle is at the intersection of the lines drawn from any two angles to the middle points of the opposite sides respectively.*

Proposition 15.

If AD, BC be the two parallel sides of a trapezium ABCD, AD being the smaller, and if AD, BC be bisected at E, F respectively, then the centre of gravity of the trapezium is at a point G on EF such that

$$GE : GF = (2BC + AD) : (2AD + BC).$$

Produce *BA, CD* to meet at *O*. Then *FE* produced will also pass through *O*, since *AE = ED*, and *BF = FC*.

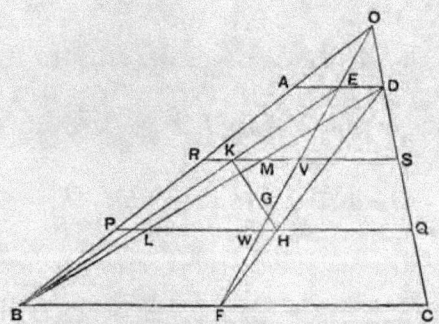

Now the centre of gravity of the triangle *OAD* will lie on *OE*, and that of the triangle *OBC* will lie on *OF*.   [Prop. 13]

It follows that the centre of gravity of the remainder, the trapezium *ABCD*, will also lie on *OF*.   [Prop. 8]

Join *BD*, and divide it at *L, M* into three equal parts. Through *L, M* draw *PQ, RS* parallel to *BC* meeting *BA* in *P, R, FE* in *W, V*, and *CD* in *Q, S* respectively.

Join *DF, BE* meeting *PQ* in *H* and *RS* in *K* respectively.

Now, since $BL = \frac{1}{3} BD,$

$$FH = \frac{1}{3} FD.$$

Therefore H is the centre of gravity of the triangle $DBC*$.

Similarly, since $EK = \frac{1}{3} BE$, it follows that K is the centre of gravity of the triangle ADB.

Therefore the centre of gravity of the triangles DBC, ADB together, i.e. of the trapezium, lies on the line HK.

But it also lies on OF.

Therefore, if OF, HK meet in G, G is the centre of gravity of the trapezium.

Hence [Props. 6, 7]
$$\triangle DBC : \triangle ABD = KG : GH$$
$$= VG : GW.$$

But $\qquad \triangle DBC : \triangle ABD = BC : AD.$

Therefore $\qquad\qquad BC : AD = VG : GW.$

It follows that
$$(2BC + AD) : (2AD + BC) = (2VG + GW) : (2GW + VG)$$
$$= EG : GF.$$

Q. E. D.

* This easy deduction from Prop. 14 is assumed by Archimedes without proof.

ON THE EQUILIBRIUM OF PLANES.

BOOK II.

Proposition 1.

If P, P′ be two parabolic segments and D, E their centres of gravity respectively, the centre of gravity of the two segments taken together will be at a point C on DE determined by the relation

$$P : P' = CE : CD^*.$$

In the same straight line with *DE* measure *EH, EL* each equal to *DC*, and *DK* equal to *DH*; whence it follows at once that *DK = CE*, and also that *KC = CL*.

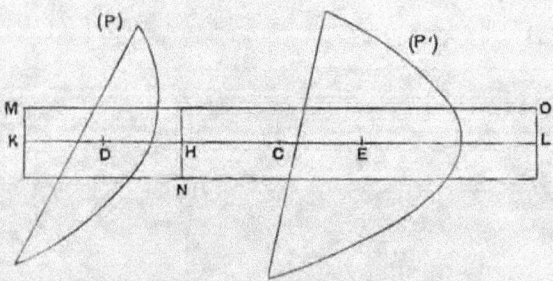

---

* This proposition is really a particular case of Props. 6, 7 of Book I. and is therefore hardly necessary. As, however, Book II. relates exclusively to parabolic segments, Archimedes' object was perhaps to emphasize the fact that the magnitudes in I. 6, 7 might be parabolic segments as well as rectilinear figures. His procedure is to substitute for the segments rectangles of equal area, a substitution which is rendered possible by the results obtained in his separate treatise on the *Quadrature of the Parabola*.

Apply a rectangle MN equal in area to the parabolic segment P to a base equal to KH, and place the rectangle so that KH bisects it, and is parallel to its base.

Then D is the centre of gravity of MN, since $KD = DH$.

Produce the sides of the rectangle which are parallel to KH, and complete the rectangle NO whose base is equal to HL. Then E is the centre of gravity of the rectangle NO.

Now
$$(MN) : (NO) = KH : HL$$
$$= DH : EH$$
$$= CE : CD$$
$$= P : P'.$$

But
$$(MN) = P.$$

Therefore
$$(NO) = P'.$$

Also, since C is the middle point of KL, C is the centre of gravity of the whole parallelogram made up of the two parallelograms (MN), (NO), which are equal to, and have the same centres of gravity as, P, P' respectively.

Hence C is the centre of gravity of P, P' taken together.

Definition and lemmas preliminary to Proposition 2.

"If in a segment bounded by a straight line and a section of a right-angled cone [a parabola] a triangle be inscribed having the same base as the segment and equal height, if again triangles be inscribed in the remaining segments having the same bases as the segments and equal height, and if in the remaining segments triangles be inscribed in the same manner, let the resulting figure be said to be **inscribed in the recognised manner** (γνωρίμως ἐγγράφεσθαι) in the segment.

And it is plain

(1) that *the lines joining the two angles of the figure so inscribed which are nearest to the vertex of the segment, and the next*

pairs of angles in order, will be parallel to the base of the segment,

(2) that *the said lines will be bisected by the diameter of the segment,* and

(3) that *they will cut the diameter in the proportions of the successive odd numbers, the number one having reference to [the length adjacent to] the vertex of the segment.*

And these properties will have to be proved in their proper places (ἐν ταῖς τάξεσιν)."

[The last words indicate an intention to give these propositions in their proper connexion with systematic proofs; but the intention does not appear to have been carried out, or at least we know of no lost work of Archimedes in which they could have appeared. The results can however be easily derived from propositions given in the *Quadrature of the Parabola* as follows.

(1) Let *BRQPApqrb* be a figure inscribed 'in the recognised manner' in the parabolic segment *BAb* of which *Bb* is the base, *A* the vertex and *AO* the diameter.

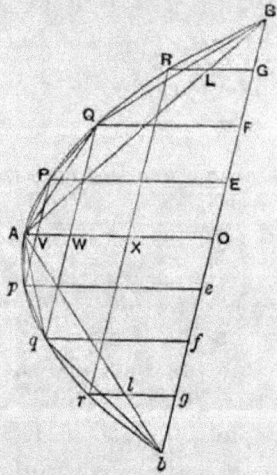

Bisect each of the lines *BQ, BA, QA, Aq, Ab, qb,* and through the middle points draw lines parallel to *AO* meeting *Bb* in *G, F, E, e, f, g* respectively.

These lines will then pass through the vertices R, Q, P, p, q, r of the respective parabolic segments [*Quadrature of the Parabola*, Prop. 18], i.e. through the angular points of the inscribed figure (since the triangles and segments are of equal height).

Also $BG = GF = FE = EO$, and $Oe = ef = fg = gb$. But $BO = Ob$, and therefore all the parts into which Bb is divided are equal.

If now AB, RG meet in L, and Ab, rg in l, we have

$$BG : GL = BO : OA, \text{ by parallels,}$$

$$= bO : OA$$

$$= bg : gl,$$

whence $GL = gl$.

Again [*ibid.*, Prop. 4]

$$GL : LR = BO : OG$$

$$= bO : Og$$

$$= gl : lr;$$

and, since $GL = gl$, $LR = lr$.

Therefore GR, gr are equal as well as parallel.

Hence $GRrg$ is a parallelogram, and Rr is parallel to Bb.

Similarly it may be shown that Pp, Qq are each parallel to Bb.

(2) Since $RGgr$ is a parallelogram, and RG, rg are parallel to AO, while $GO = Og$, it follows that Rr is bisected by AO.

And similarly for Pp, Qq.

(3) Lastly, if V, W, X be the points of bisection of Pp, Qq, Rr,

$$AV : AW : AX : AO = PV^2 : QW^2 : RX^2 : BO^2$$

$$= 1 : 4 : 9 : 16,$$

whence $AV : VW : WX : XO = 1 : 3 : 5 : 7.$]

Proposition 2.

If a figure be 'inscribed in the recognised manner' in a parabolic segment, the centre of gravity of the figure so inscribed will lie on the diameter of the segment.

For, in the figure of the foregoing lemmas, the centre of gravity of the trapezium $BRrb$ must lie on XO, that of the trapezium $RQqr$ on WX, and so on, while the centre of gravity of the triangle PAp lies on AV.

Hence the centre of gravity of the whole figure lies on AO.

Proposition 3.

If BAB', bab' be two similar parabolic segments whose diameters are AO, ao respectively, and if a figure be inscribed in each segment ' in the recognised manner,' the number of sides in each figure being equal, the centres of gravity of the inscribed figures will divide AO, ao in the same ratio.

[Archimedes enunciates this proposition as true of *similar* segments, but it is equally true of segments which are not similar, as the course of the proof will show.]

Suppose $BRQPAP'Q'R'B'$, $brqpap'q'r'b'$ to be the two figures inscribed 'in the recognised manner.' Join PP', QQ', RR' meeting AO in L, M, N, and pp', qq', rr' meeting ao in l, m, n.

Then [Lemma (3)]
$$AL : LM : MN : NO$$
$$= 1 : 3 : 5 : 7$$
$$= al : lm : mn : no,$$
so that AO, ao are divided in the same proportion.

Also, by reversing the proof of Lemma (3), we see that
$$PP' : pp' = QQ' : qq' = RR' : rr' = BB' : bb'.$$

Since then $RR' : BB' = rr' : bb'$, and these ratios respectively determine the proportion in which NO, no are divided

by the centres of gravity of the trapezia $BRR'B'$, $brr'b'$ [I. 15],
it follows that the centres of gravity of the trapezia divide NO,
no in the same ratio.

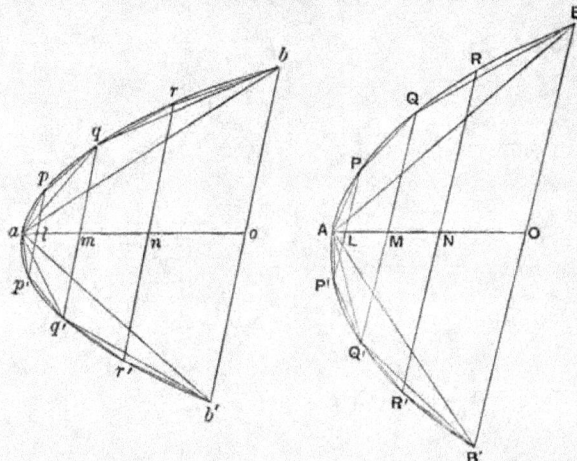

Similarly the centres of gravity of the trapezia $RQQ'R'$,
$rqq'r'$ divide MN, mn in the same ratio respectively, and so on.

Lastly, the centres of gravity of the triangles PAP', pap'
divide AL, al respectively in the same ratio.

Moreover the corresponding trapezia and triangles are, each
to each, in the same proportion (since their sides and heights
are respectively proportional), while AO, ao are divided in
the same proportion.

Therefore the centres of gravity of the complete inscribed
figures divide AO, ao in the same proportion.

Proposition 4.

*The centre of gravity of any parabolic segment cut off by a
straight line lies on the diameter of the segment.*

Let BAB' be a parabolic segment, A its vertex and AO its
diameter.

Then, if the centre of gravity of the segment does not lie on
AO, suppose it to be, if possible, the point F. Draw FE
parallel to AO meeting BB' in E.

Inscribe in the segment the triangle ABB' having the same vertex and height as the segment, and take an area S such that

$$\triangle ABB' : S = BE : EO.$$

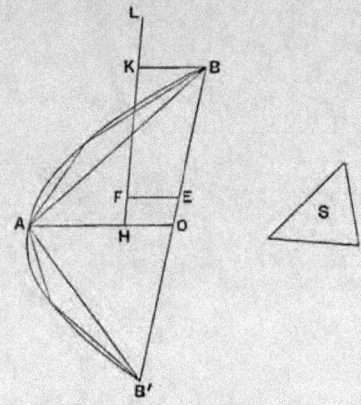

We can then inscribe in the segment 'in the recognised manner' a figure such that the segments of the parabola left over are together less than S. [For Prop. 20 of the *Quadrature of the Parabola* proves that, if in any segment the triangle with the same base and height be inscribed, the triangle is greater than half the segment; whence it appears that, each time that we increase the number of the sides of the figure inscribed 'in the recognised manner,' we take away more than half of the remaining segments.]

Let the inscribed figure be drawn accordingly; its centre of gravity then lies on AO [Prop. 2]. Let it be the point H.

Join HF and produce it to meet in K the line through B parallel to AO.

Then we have

(inscribed figure) : (remainder of segmt.) $> \triangle ABB' : S$

$$> BE : EO$$

$$> KF : FH.$$

Suppose L taken on HK produced so that the former ratio is equal to the ratio $LF : FH$.

Then, since H is the centre of gravity of the inscribed figure, and F that of the segment, L must be the centre of gravity of all the segments taken together which form the remainder of the original segment.   [I. 8]

But this is impossible, since all these segments lie on one side of the line drawn through L parallel to AO [Cf. *Post.* 7].

Hence the centre of gravity of the segment cannot but lie on AO.

Proposition 5.

If in a parabolic segment a figure be inscribed 'in the recognised manner,' the centre of gravity of the segment is nearer to the vertex of the segment than the centre of gravity of the inscribed figure is.

Let BAB' be the given segment, and AO its diameter. *First,* let ABB' be the *triangle* inscribed 'in the recognised manner.'

Divide AO in F so that $AF = 2FO$; F is then the centre of gravity of the triangle ABB'.

Bisect AB, AB' in D, D' respectively, and join DD' meeting AO in E. Draw DQ, $D'Q'$ parallel to OA to meet the curve. QD, $Q'D'$ will then be the diameters of the segments whose bases are AB, AB', and the centres of gravity of those segments will lie respectively on QD, $Q'D'$ [Prop. 4]. Let them be H, H', and join HH' meeting AO in K.

Now QD, $Q'D'$ are equal*, and therefore the segments of which they are the diameters are equal [*On Conoids and Spheroids*, Prop. 3].

---

* This may either be inferred from Lemma (1) above (since QQ', DD' are both parallel to BB'), or from Prop. 19 of the *Quadrature of the Parabola*, which applies equally to Q or Q'.

Also, since $QD, Q'D'$ are parallel*, and $DE = ED'$, K is the middle point of HH'.

Hence the centre of gravity of the equal segments AQB, $AQ'B'$ taken together is K, where K lies between E and A. And the centre of gravity of the triangle ABB' is F.

It follows that the centre of gravity of the whole segment BAB' lies between K and F, and is therefore nearer to the vertex A than F is.

Secondly, take the *five-sided* figure $BQAQ'B'$ inscribed 'in the recognised manner,' $QD, Q'D'$ being, as before, the diameters of the segments $AQB, AQ'B'$.

Then, by the first part of this proposition, the centre of gravity of the segment AQB (lying of course on QD) is nearer to Q than the centre of gravity of the triangle AQB is. Let the centre of gravity of the segment be H, and that of the triangle I.

Similarly let H' be the centre of gravity of the segment $AQ'B'$, and I' that of the triangle $AQ'B'$.

It follows that the centre of gravity of the two segments $AQB, AQ'B'$ taken together is K, the middle point of HH', and that of the two triangles $AQB, AQ'B'$ is L, the middle point of II'.

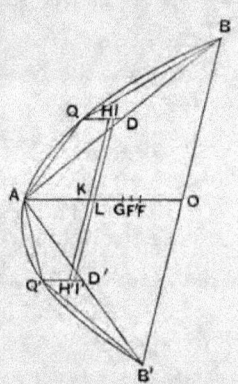

If now the centre of gravity of the triangle ABB' be F, the centre of gravity of the whole segment BAB' (i.e. that of the triangle ABB' and the two segments $AQB, AQ'B'$ taken together) is a point G on KF determined by the proportion

(sum of segments $AQB, AQ'B'$) : $\triangle ABB' = FG : GK$.   [I. 6, 7]

---

* There is clearly some interpolation in the text here, which has the words καὶ ἐπεὶ παραλληλόγραμμόν ἐστι τὸ ΘΖΗΙ. It is not yet proved that $H'D'DH$ is a *parallelogram*; this can only be inferred from the fact that H, H' divide QD, $Q'D'$ respectively in the same ratio. But this latter property does not appear till Prop. 7, and is then only enunciated of *similar* segments. The interpolation must have been made before Eutocius' time, because he has a note on the phrase, and explains it by gravely assuming that H, H' divide $QD, Q'D'$ respectively in the same ratio.

And the centre of gravity of the inscribed figure $BQAQ'B'$ is a point F' on LF determined by the proportion

$$(\triangle AQB + \triangle AQ'B') : \triangle ABB' = FF' : F'L. \quad \text{[I. 6, 7]}$$

[Hence $FG : GK > FF' : F'L,$

or $GK : FG < F'L : FF',$

and, *componendo*, $FK : FG < FL : FF'$, while $FK > FL.$]

Therefore $FG > FF'$, or G lies nearer than F' to the vertex A.

Using this last result, and proceeding in the same way, we can prove the proposition for *any* figure inscribed 'in the recognised manner.'

Proposition 6.

Given a segment of a parabola cut off by a straight line, it is possible to inscribe in it ' in the recognised manner' a figure such that the distance between the centres of gravity of the segment and of the inscribed figure is less than any assigned length.

Let BAB' be the segment, AO its diameter, G its centre of gravity, and ABB' the triangle inscribed 'in the recognised manner.'

Let D be the assigned length and S an area such that

$$AG : D = \triangle ABB' : S.$$

In the segment inscribe 'in the recognised manner' a figure such that the sum of the segments left over is less than S. Let F be the centre of gravity of the inscribed figure.

We shall prove that $FG < D$.

For, if not, FG must be either equal to, or greater than, D.

And clearly

(inscribed fig.) : (sum of remaining segmts.)

$$> \triangle ABB' : S$$

$$> AG : D$$

$$> AG : FG, \text{ by hypothesis (since } FG \not< D).$$

Let the first ratio be equal to the ratio $KG : FG$ (where K lies on GA produced); and it follows that K is the centre of gravity of the small segments taken together.   [I. 8]

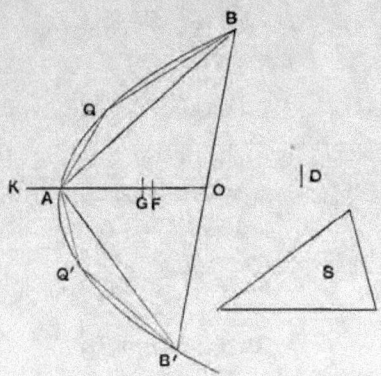

But this is impossible, since the segments are all on the same side of a line drawn through K parallel to BB'.

Hence FG cannot but be less than D.

Proposition 7.

If there be two similar parabolic segments, their centres of gravity divide their diameters in the same ratio.

[This proposition, though enunciated of *similar* segments only, like Prop. 3 on which it depends, is equally true of *any* segments. This fact did not escape Archimedes, who uses the proposition in its more general form for the proof of Prop. 8 immediately following.]

Let BAB', bab' be the two similar segments, AO, ao their diameters, and G, g their centres of gravity respectively.

Then, if G, g do not divide AO, ao respectively in the same ratio, suppose H to be such a point on AO that

$$AH : HO = ag : go ;$$

and inscribe in the segment BAB' 'in the recognised manner' a figure such that, if F be its centre of gravity,

$$GF < GH. \qquad\qquad \text{[Prop. 6]}$$

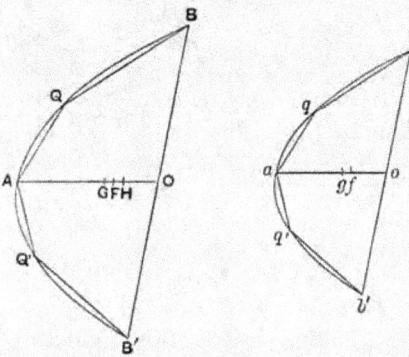

Inscribe in the segment bab' 'in the recognised manner' a similar figure; then, if f be the centre of gravity of this figure,

$$ag < af. \qquad\qquad \text{[Prop. 5]}$$

And, by Prop. 3, $af : fo = AF : FO.$

But $AF : FO < AH : HO$

$$< ag : go, \text{ by hypothesis.}$$

Therefore $af : fo < ag : go$; which is impossible.

It follows that G, g cannot but divide AO, ao in the same ratio.

Proposition 8.

If AO be the diameter of a parabolic segment, and G its centre of gravity, then

$$AG = \tfrac{3}{2} GO.$$

Let the segment be BAB'. Inscribe the triangle ABB' 'in the recognised manner,' and let F be its centre of gravity.

Bisect AB, AB' in D, D', and draw $DQ, D'Q'$ parallel to OA to meet the curve, so that $QD, Q'D'$ are the diameters of the segments $AQB, AQ'B'$ respectively.

Let H, H' be the centres of gravity of the segments AQB, $AQ'B'$ respectively. Join QQ', HH' meeting AO in V, K respectively.

K is then the centre of gravity of the two segments AQB, $AQ'B'$ taken together.

Now $\qquad AG : GO = QH : HD$,

$\qquad\qquad\qquad\qquad$ [Prop. 7]

whence $\qquad AO : OG = QD : HD$.

But $AO = 4QD$ [as is easily proved by means of Lemma (3), p. 206].

Therefore $\qquad\qquad OG = 4HD$;

and, by subtraction, $\qquad AG = 4QH$.

Also, by Lemma (2), QQ' is parallel to BB' and therefore to DD'. It follows from Prop. 7 that HH' is also parallel to QQ' or DD',

and hence $\qquad\qquad QH = VK$.

Therefore $\qquad\qquad AG = 4VK$,

and $\qquad\qquad AV + KG = 3VK$.

Measuring VL along VK so that $VL = \frac{1}{3}AV$, we have

$$KG = 3LK \dots\dots\dots\dots\dots\dots\dots(1).$$

Again $\qquad\qquad AO = 4AV \qquad\qquad$ [Lemma (3)]

$\qquad\qquad\qquad = 3AL$, since $AV = 3VL$,

whence $\qquad\qquad AL = \frac{1}{3}AO = OF \dots\dots\dots\dots\dots (2).$

Now, by I. 6, 7,

$\qquad \triangle ABB' : (\text{sum of segmts. } AQB, AQ'B') = KG : GF$,

and $\qquad \triangle ABB' = 3 (\text{sum of segments } AQB, AQ'B')$

[since the segment ABB' is equal to $\frac{4}{3} \triangle ABB'$ (*Quadrature of the Parabola*, Props. 17, 24)].

Hence $\qquad\qquad KG = 3GF$.

But $\qquad\qquad KG = 3LK$, from (1) above.

Therefore $\qquad\qquad LF = LK + KG + GF$

$\qquad\qquad\qquad\quad = 5GF.$

And, from (2),
$$LF = (AO - AL - OF) = \tfrac{1}{3}AO = OF.$$
Therefore $OF = 5GF,$
and $OG = 6GF.$
But $AO = 3OF = 15GF.$
Therefore, by subtraction,
$$AG = 9GF$$
$$= \tfrac{3}{2}GO.$$

Proposition 9 (Lemma).

If a, b, c, d be four lines in continued proportion and in descending order of magnitude, and if
$$d : (a - d) = x : \tfrac{3}{5}(a - c),$$
and $(2a + 4b + 6c + 3d) : (5a + 10b + 10c + 5d) = y : (a - c),$
it is required to prove that
$$x + y = \tfrac{2}{5}a.$$

[The following is the proof given by Archimedes, with the only difference that it is set out in algebraical instead of geometrical notation. This is done in the particular case simply in order to make the proof easier to follow. Archimedes exhibits his lines in the figure reproduced in the margin, but, now that it is possible to use algebraical notation, there is no advantage in using the figure and the more cumbrous notation which only obscures the course of the proof. The relation between Archimedes' figure and the letters used below is as follows;

$AB = a,\ \Gamma B = b,\ \Delta B = c,\ EB = d,\ ZH = x,\ H\Theta = y,\ \Delta O = z.$]

We have $\dfrac{a}{b} = \dfrac{b}{c} = \dfrac{c}{d}$ ....(1),

whence $\dfrac{a - b}{b} = \dfrac{b - c}{c} = \dfrac{c - d}{d},$

and therefore $\dfrac{a - b}{b - c} = \dfrac{b - c}{c - d} = \dfrac{a}{b} = \dfrac{b}{c} = \dfrac{c}{d}$ .... (2).

Now $\dfrac{2(a + b)}{2c} = \dfrac{a + b}{c} = \dfrac{a + b}{b} \cdot \dfrac{b}{c} = \dfrac{a - c}{b - c} \cdot \dfrac{b - c}{c - d} = \dfrac{a - c}{c - d}.$

And, in like manner,

$$\frac{b+c}{d} = \frac{b+c}{c} \cdot \frac{c}{d} = \frac{a-c}{c-d}.$$

It follows from the last two relations that

$$\frac{a-c}{c-d} = \frac{2a+3b+c}{2c+d} \quad \dots \dots \dots \dots (3).$$

Suppose z to be so taken that

$$\frac{2a+4b+4c+2d}{2c+d} = \frac{a-c}{z} \quad \dots \dots \dots \dots (4),$$

so that $z < (c-d)$.

Therefore $\quad \dfrac{a-c+z}{a-c} = \dfrac{2a+4b+6c+3d}{2(a+d)+4(b+c)}.$

And, by hypothesis,

$$\frac{a-c}{y} = \frac{5(a+d)+10(b+c)}{2a+4b+6c+3d},$$

so that $\quad \dfrac{a-c+z}{y} = \dfrac{5(a+d)+10(b+c)}{2(a+d)+4(b+c)} = \dfrac{5}{2} \quad \dots \dots \dots (5).$

Again, dividing (3) by (4) crosswise, we obtain

$$\frac{z}{c-d} = \frac{2a+3b+c}{2(a+d)+4(b+c)},$$

whence $\quad \dfrac{c-d-z}{c-d} = \dfrac{b+3c+2d}{2(a+d)+4(b+c)} \quad \dots \dots \dots \dots (6).$

But, by (2),

$$\frac{c-d}{d} = \frac{a-b}{b} = \frac{3(b-c)}{3c} = \frac{2(c-d)}{2d},$$

so that $\quad \dfrac{c-d}{d} = \dfrac{(a-b)+3(b-c)+2(c-d)}{b+3c+2d} \quad \dots \dots \dots (7).$

Combining (6) and (7), we have

$$\frac{c-d-z}{d} = \frac{(a-b)+3(b-c)+2(c-d)}{2(a+d)+4(b+c)},$$

whence $\quad \dfrac{c-z}{d} = \dfrac{3a+6b+3c}{2(a+d)+4(b+c)} \quad \dots \dots \dots \dots (8).$

And, since [by (1)]

$$\frac{c-d}{c+d} = \frac{b-c}{b+c} = \frac{a-b}{a+b},$$

we have
$$\frac{c-d}{a-c} = \frac{c+d}{b+c+a+b},$$

whence
$$\frac{a-d}{a-c} = \frac{a+2b+2c+d}{a+2b+c} = \frac{2(a+d)+4(b+c)}{2(a+c)+4b} \quad\ldots\ldots(9).$$

Thus
$$\frac{a-d}{\frac{3}{5}(a-c)} = \frac{2(a+d)+4(b+c)}{\frac{3}{5}\{2(a+c)+4b\}},$$

and therefore, by hypothesis,
$$\frac{d}{x} = \frac{2(a+d)+4(b+c)}{\frac{3}{5}\{2(a+c)+4b\}}.$$

But, by (8),
$$\frac{c-z}{d} = \frac{3a+6b+3c}{2(a+d)+4(b+c)};$$

and it follows, *ex aequali*, that
$$\frac{c-z}{x} = \frac{3(a+c)+6b}{\frac{3}{5}\{2(a+c)+4b\}} = \frac{5}{3}\cdot\frac{3}{2} = \frac{5}{2}.$$

And, by (5),
$$\frac{a-c+z}{y} = \frac{5}{2}.$$

Therefore
$$\frac{5}{2} = \frac{a}{x+y},$$

or
$$x+y = \tfrac{2}{5}a.$$

Proposition 10.

If $PP'B'B$ be the portion of a parabola intercepted between two parallel chords PP', BB' bisected respectively in N, O by the diameter ANO (N being nearer than O to A, the vertex of the segments), and if NO be divided into five equal parts of which LM is the middle one (L being nearer than M to N), then, if G be a point on LM such that

$$LG : GM = BO^2.(2PN+BO) : PN^2.(2BO+PN),$$

G will be the centre of gravity of the area $PP'B'B$.

Take a line ao equal to AO, and an on it equal to AN. Let p, q be points on the line ao such that

$$ao : aq = aq : an \quad\ldots\ldots\ldots\ldots\ldots\ldots (1),$$
$$ao : an = aq : ap \quad\ldots\ldots\ldots\ldots\ldots (2),$$

[whence $ao : aq = aq : an = an : ap$, or ao, aq, an, ap are lines in continued proportion and in descending order of magnitude].

Measure along GA a length GF such that

$$op : ap = OL : GF \ldots\ldots\ldots\ldots\ldots(3).$$

Then, since PN, BO are ordinates to ANO,

$$BO^2 : PN^2 = AO : AN$$
$$= ao : an$$
$$= ao^2 : aq^2, \text{ by (1)},$$

so that $\qquad BO : PN = ao : aq$ .... (4),

and $\qquad BO^3 : PN^3 = ao^3 : aq^3$

$$= (ao : aq).(aq : an).(an : ap)$$
$$= ao : ap \dots\dots\dots\dots\dots\dots\dots (5).$$

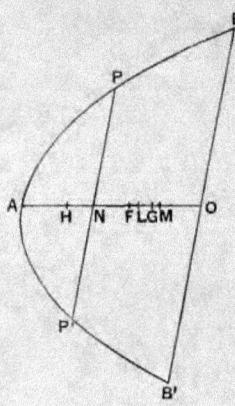 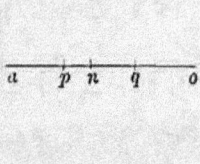

Thus (segment BAB') : (segment PAP')

$$= \triangle BAB' : \triangle PAP'$$
$$= BO^3 : PN^3$$
$$= ao : ap,$$

whence

$\qquad$ (area $PP'B'B$) : (segment PAP') $= op : ap$

$$= OL : GF, \text{ by (3)},$$
$$= \tfrac{2}{3}ON : GF \dots\dots\dots (6).$$

Now $\qquad BO^2.(2PN+BO) : BO^3 = (2PN+BO) : BO$

$$= (2aq+ao) : ao, \text{ by (4)},$$
$$BO^3 : PN^3 = ao : ap, \text{ by (5)},$$

and $\qquad PN^3 : PN^2.(2BO+PN) = PN : (2BO+PN)$

$$= aq : (2ao+aq), \text{ by (4)},$$
$$= ap : (2an+ap), \text{ by (2)}.$$

Hence, *ex aequali,*

$$BO^2 . (2PN + BO) : PN^2 . (2BO + PN) = (2aq + ao) : (2an + ap),$$

so that, by hypothesis,

$$LG : GM = (2aq + ao) : (2an + ap).$$

Componendo, and multiplying the antecedents by 5,

$$ON : GM = \{5 (ao + ap) + 10 (aq + an)\} : (2an + ap).$$

But $ON : OM = 5 : 2$

$$= \{5 (ao + ap) + 10 (aq + an)\} : \{2 (ao + ap) + 4 (aq + an)\}.$$

It follows that

$$ON : OG = \{5 (ao + ap) + 10 (aq + an)\} : (2ao + 4aq + 6an + 3ap).$$

Therefore

$$(2ao + 4aq + 6an + 3ap) : \{5 (ao + ap) + 10 (aq + an)\} = OG : ON$$
$$= OG : on.$$

And $\qquad ap : (ao - ap) = ap : op$

$$= GF : OL, \text{ by hypothesis,}$$

$$= GF : \tfrac{3}{5} on,$$

while ao, aq, an, ap are in continued proportion.

Therefore, by Prop. 9,

$$GF + OG = OF = \tfrac{3}{5} ao = \tfrac{3}{5} OA.$$

Thus F is the centre of gravity of the segment BAB'. [Prop. 8]

Let H be the centre of gravity of the segment PAP', so that $AH = \tfrac{3}{5} AN$.

And, since $\qquad\qquad AF = \tfrac{3}{5} AO,$

we have, by subtraction, $HF = \tfrac{3}{5} ON.$

But, by (6) above,

$$(\text{area } PP'B'B) : (\text{segment } PAP') = \tfrac{3}{5} ON : GF$$
$$= HF : FG.$$

Thus, since F, H are the centres of gravity of the segments BAB', PAP' respectively, it follows [by I. 6, 7] that G is the centre of gravity of the area $PP'B'B$.

THE SAND-RECKONER.

"THERE are some, king Gelon, who think that the number of the sand is infinite in multitude; and I mean by the sand not only that which exists about Syracuse and the rest of Sicily but also that which is found in every region whether inhabited or uninhabited. Again there are some who, without regarding it as infinite, yet think that no number has been named which is great enough to exceed its multitude. And it is clear that they who hold this view, if they imagined a mass made up of sand in other respects as large as the mass of the earth, including in it all the seas and the hollows of the earth filled up to a height equal to that of the highest of the mountains, would be many times further still from recognising that any number could be expressed which exceeded the multitude of the sand so taken. But I will try to show you by means of geometrical proofs, which you will be able to follow, that, of the numbers named by me and given in the work which I sent to Zeuxippus, some exceed not only the number of the mass of sand equal in magnitude to the earth filled up in the way described, but also that of a mass equal in magnitude to the universe. Now you are aware that 'universe' is the name given by most astronomers to the sphere whose centre is the centre of the earth and whose radius is equal to the straight line between the centre of the sun and the centre of the earth. This is the common account ($\tau\grave{a}\ \gamma\rho\alpha\phi\acute{o}\mu\epsilon\nu\alpha$), as you have heard from astronomers. But Aristarchus of Samos brought out a

book consisting of some hypotheses, in which the premisses lead
to the result that the universe is many times greater than that
now so called.   His hypotheses are that the fixed stars and the
sun remain unmoved, that the earth revolves about the sun in
the circumference of a circle, the sun lying in the middle of the
orbit, and that the sphere of the fixed stars, situated about. the
same centre as the sun, is so great that the circle in which he
supposes the earth to revolve bears such a proportion to the
distance of the fixed stars as the centre of the sphere bears to
its surface.   Now it is easy to see that this is impossible; for,
since the centre of the sphere has no magnitude, we cannot
conceive it to bear any ratio whatever to the surface of the
sphere.   We must however take Aristarchus to mean this:
since we conceive the earth to be, as it were, the centre of
the universe, the ratio which the earth bears to what we
describe as the 'universe' is the same as the ratio which the
sphere containing the circle in which he supposes the earth to
revolve bears to the sphere of the fixed stars.   For he adapts
the proofs of his results to a hypothesis of this kind, and in
particular he appears to suppose the magnitude of the sphere
in which he represents the earth as moving to be equal to what
we call the 'universe.'

I say then that, even if a sphere were made up of the sand,
as great as Aristarchus supposes the sphere of the fixed stars
to be, I shall still prove that, of the numbers named in the
Principles*, some exceed in multitude the number of the
sand which is equal in magnitude to the sphere referred to,
provided that the following assumptions be made.

1.   *The perimeter of the earth is about* 3,000,000 *stadia and
not greater.*

It is true that some have tried, as you are of course aware,
to prove that the said perimeter is about 300,000 stadia.   But
I go further and, putting the magnitude of the earth at ten
times the size that my predecessors thought it, I suppose its
perimeter to be about 3,000,000 stadia and not greater.

---

* Ἀρχαί was apparently the title of the work sent to Zeuxippus.   Cf. the
note attached to the enumeration of lost works of Archimedes in the Introduction,
Chapter II., *ad fin.*

2. *The diameter of the earth is greater than the diameter of the moon, and the diameter of the sun is greater than the diameter of the earth.*

In this assumption I follow most of the earlier astronomers.

3. *The diameter of the sun is about 30 times the diameter of the moon and not greater.*

It is true that, of the earlier astronomers, Eudoxus declared it to be about nine times as great, and Pheidias my father* twelve times, while Aristarchus tried to prove that the diameter of the sun is greater than 18 times but less than 20 times the diameter of the moon. But I go even further than Aristarchus, in order that the truth of my proposition may be established beyond dispute, and I suppose the diameter of the sun to be about 30 times that of the moon and not greater.

4. *The diameter of the sun is greater than the side of the chiliagon inscribed in the greatest circle in the (sphere of the) universe.*

I make this assumption† because Aristarchus discovered that the sun appeared to be about $\frac{1}{720}$th part of the circle of the zodiac, and I myself tried, by a method which I will now describe, to find experimentally (ὀργανικῶς) the angle subtended by the sun and having its vertex at the eye (τὰν γωνίαν, εἰς ἂν ὁ ἅλιος ἐναρμόζει τὰν κορυφὰν ἔχουσαν ποτὶ τᾷ ὄψει)."

[Up to this point the treatise has been literally translated because of the historical interest attaching to the *ipsissima verba* of Archimedes on such a subject. The rest of the work can now be more freely reproduced, and, before proceeding to the mathematical contents of it, it is only necessary to remark that Archimedes next describes how he arrived at a higher and a lower limit for the angle subtended by the sun. This he did

---

* τοῦ ἁμοῦ πατρὸς is the correction of Blass for τοῦ Ἀκούπατρος (*Jahrb. f. Philol.* cxxvii. 1883).

† This is not, strictly speaking, an assumption; it is a proposition proved later (pp. 224—6) by means of the result of an experiment about to be described.

by taking a long rod or ruler (κανών), fastening on the end of it a small cylinder or disc, pointing the rod in the direction of the sun just after its rising (so that it was possible to look directly at it), then putting the cylinder at such a distance that it just concealed, and just failed to conceal, the sun, and lastly measuring the angles subtended by the cylinder. He explains also the correction which he thought it necessary to make because " the eye does not see from one point but from a certain area " (ἐπεὶ αἱ ὄψιες οὐκ ἀφ' ἑνὸς σαμείου βλέποντι, ἀλλὰ ἀπό τινος μεγέθεος).]

The result of the experiment was to show that the angle subtended by the diameter of the sun was less than $\frac{1}{164}$th part, and greater than $\frac{1}{200}$th part, of a right angle.

To prove that (on this assumption) the diameter of the sun is greater than the side of a chiliagon, or figure with 1000 equal sides, inscribed in a great circle of the ' universe.'

Suppose the plane of the paper to be the plane passing through the centre of the sun, the centre of the earth and the eye, at the time when the sun has just risen above the horizon. Let the plane cut the earth in the circle *EHL* and the sun in the circle *FKG*, the centres of the earth and sun being *C*, *O* respectively, and *E* being the position of the eye.

Further, let the plane cut the sphere of the ' universe ' (i.e. the sphere whose centre is *C* and radius *CO*) in the great circle *AOB*.

Draw from *E* two tangents to the circle *FKG* touching it at *P*, *Q*, and from *C* draw two other tangents to the same circle touching it in *F*, *G* respectively.

Let *CO* meet the sections of the earth and sun in *H*, *K* respectively; and let *CF*, *CG* produced meet the great circle *AOB* in *A*, *B*.

Join *EO*, *OF*, *OG*, *OP*, *OQ*, *AB*, and let *AB* meet *CO* in *M*.

Now *CO* > *EO*, since the sun is just above the horizon. Therefore $\angle PEQ > \angle FCG.$

And $\angle PEQ > \frac{1}{200}R$ where R represents a right angle.
but $< \frac{1}{164}R$

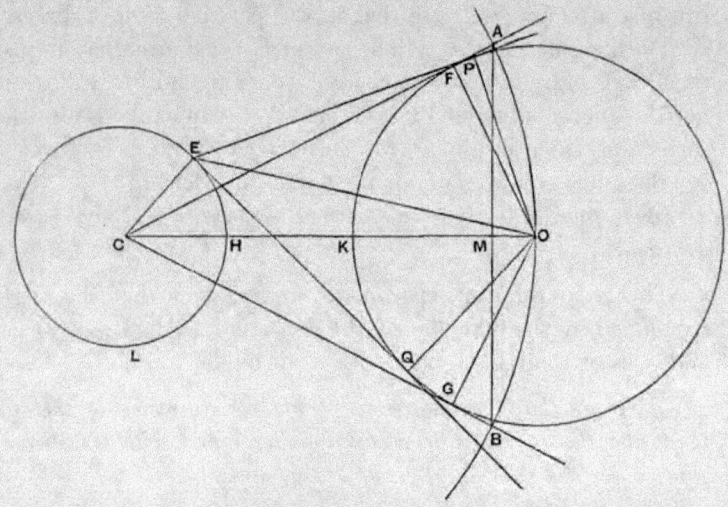

Thus $\angle FCG < \frac{1}{164}R$, *a fortiori*,

and the chord AB subtends an arc of the great circle which is less than $\frac{1}{656}$th of the circumference of that circle, i.e.

$AB <$ (side of 656-sided polygon inscribed in the circle).

Now the perimeter of any polygon inscribed in the great circle is less than $\frac{44}{7}CO$. [Cf. *Measurement of a circle*, Prop. 3.]

Therefore $AB : CO < 11 : 1148$,

and, *a fortiori*, $AB < \frac{1}{100}CO$.........................(α).

Again, since $CA = CO$, and AM is perpendicular to CO, while OF is perpendicular to CA,

$$AM = OF.$$

Therefore $AB = 2AM =$ (diameter of sun).

Thus (diameter of sun) $< \frac{1}{100}CO$, by (α),

and, *a fortiori*,

(diameter of earth) $< \frac{1}{100}CO$. [Assumption 2]

H. A. 15

Hence $\qquad CH + OK < \frac{1}{100}CO,$

so that $\qquad HK > \frac{99}{100}CO,$

or $\qquad CO : HK < 100 : 99.$

And $\qquad CO > CF,$

while $\qquad HK < EQ.$

Therefore $\qquad CF : EQ < 100 : 99 \dots\dots\dots\dots\dots(\beta).$

Now in the right-angled triangles CFO, EQO, of the sides about the right angles,

$$OF = OQ, \text{ but } EQ < CF \text{ (since } EO < CO).$$

Therefore $\qquad \angle OEQ : \angle OCF > CO : EO,$

but $\qquad\qquad\qquad\qquad < CF : EQ^*.$

Doubling the angles,

$$\angle PEQ : \angle ACB < CF : EQ$$
$$< 100 : 99, \text{ by } (\beta) \text{ above.}$$

But $\qquad \angle PEQ > \frac{1}{200}R, \text{ by hypothesis.}$

Therefore $\qquad \angle ACB > \frac{99}{20000}R$
$$> \frac{1}{203}R.$$

It follows that the arc AB is greater than $\frac{1}{812}$th of the circumference of the great circle AOB.

Hence, *a fortiori*,

$$AB > (\text{side of chiliagon inscribed in great circle}),$$

and AB is equal to the diameter of the sun, as proved above.

The following results can now be proved :

(diameter of 'universe') < 10,000 (diameter of earth),

and (diameter of 'universe') < 10,000,000,000 stadia.

---

* The proposition here assumed is of course equivalent to the trigonometrical formula which states that, if a, β are the circular measures of two angles, each less than a right angle, of which a is the greater, then

$$\frac{\tan a}{\tan \beta} > \frac{a}{\beta} > \frac{\sin a}{\sin \beta}.$$

(1) Suppose, for brevity, that d_u represents the diameter of the 'universe,' d_s that of the sun, d_e that of the earth, and d_m that of the moon.

By hypothesis, $d_s \not> 30d_m,$ [Assumption 3]

and $d_e > d_m;$ [Assumption 2]

therefore $d_s < 30d_e.$

Now, by the last proposition,

$d_s >$ (side of chiliagon inscribed in great circle),

so that   (perimeter of chiliagon) $< 1000d_s$

$< 30,000d_e.$

But the perimeter of any regular polygon with more sides than 6 inscribed in a circle is greater than that of the inscribed regular hexagon, and therefore greater than three times the diameter. Hence

(perimeter of chiliagon) $> 3d_u.$

It follows that $d_u < 10,000d_e.$

(2) (Perimeter of earth) $\not> 3,000,000$ stadia.

[Assumption 1]

and   (perimeter of earth) $> 3d_e.$

Therefore $d_e < 1,000,000$ stadia,

whence $d_u < 10,000,000,000$ stadia.

Assumption 5.

Suppose a quantity of sand taken not greater than a poppy-seed, and suppose that it contains not more than 10,000 grains.

Next suppose the diameter of the poppy-seed to be not less than $\frac{1}{40}$th of a finger-breadth.

Orders and periods of numbers.

I. We have traditional names for numbers up to a myriad (10,000); we can therefore express numbers up to a myriad myriads (100,000,000). Let these numbers be called numbers of the *first order.*

Suppose the 100,000,000 to be the unit of the *second order*, and let the *second order* consist of the numbers from that unit up to $(100,000,000)^2.$

Let this again be the unit of the *third order* of numbers ending with $(100,000,000)^3$; and so on, until we reach the $100,000,000th$ *order* of numbers ending with $(100,000,000)^{100,000,000}$, which we will call P.

II. Suppose the numbers from 1 to P just described to form the *first period*.

Let P be the unit of the *first order of the second period*, and let this consist of the numbers from P up to $100,000,000 P$.

Let the last number be the unit of the *second order of the second period*, and let this end with $(100,000,000)^2 P$.

We can go on in this way till we reach the $100,000,000th$ *order of the second period* ending with $(100,000,000)^{100,000,000} P$, or P^2.

III. Taking P^2 as the unit of the *first order of the third period*, we proceed in the same way till we reach the $100,000,000th$ *order of the third period* ending with P^3.

IV. Taking P^3 as the unit of the *first order of the fourth period*, we continue the same process until we arrive at the $100,000,000th$ *order of the* $100,000,000th$ *period* ending with $P^{100,000,000}$. This last number is expressed by Archimedes as "a myriad-myriad units of the myriad-myriad-th order of the myriad-myriad-th period (αἱ μυριακισμυριοστᾶς περιόδου μυρια-κισμυριοστῶν ἀριθμῶν μυρίαι μυριάδες)," which is easily seen to be 100,000,000 times the product of $(100,000,000)^{99,999,999}$ and $P^{99,999,999}$, i.e. $P^{100,000,000}$.

[The scheme of numbers thus described can be exhibited more clearly by means of *indices* as follows.

FIRST PERIOD.

    First order.   Numbers from 1 to 10^8.

    Second order.   ,,   ,, 10^8 to 10^{16}.

    Third order.   ,,   ,, 10^{16} to 10^{24}.

              $\vdots$

    $(10^8)th$ *order.*   ,,   ,, $10^{8.(10^8-1)}$ to $10^{8.10^8}$ (P, say).

SECOND PERIOD.

First order.   Numbers from $P.1$ to $P.10^8$.

Second order.   ,,   ,, $P.10^8$ to $P.10^{16}$.

$\vdots$

$(10^8)th$ *order.*   ,,   ,, $P.10^{8.(10^8-1)}$ to
$$P.10^{9.10^8} \text{ (or } P^2).$$

$\vdots$

(10^8)TH PERIOD.

First order.   ,,   ,, $P^{10^8-1}.1$ to $P^{10^8-1}.10^8$.

Second order.   ,,   ,, $P^{10^8-1}.10^8$ to $P^{10^8-1}.10^{16}$.

$\vdots$

$(10^8)th$ *order.*   ,,   ,, $P^{10^8-1}.10^{8.(10^8-1)}$ to
$$P^{10^8-1}.10^{8.10^8} \text{ (i.e. } P^{10^8}).$$

The prodigious extent of this scheme will be appreciated when it is considered that the last number in the *first period* would be represented now by 1 followed by 800,000,000 ciphers, while the last number of the $(10^8)th$ *period* would require 100,000,000 times as many ciphers, i.e. 80,000 million millions of ciphers.]

Octads.

Consider the series of terms in continued proportion of which the first is 1 and the second 10 [i.e. the geometrical progression $1, 10^1, 10^2, 10^3, \ldots$]. The *first octad* of these terms [*i.e.* $1, 10^1, 10^2, \ldots 10^7$] fall accordingly under the *first order of the first period* above described, the *second octad* [i.e. $10^8, 10^9, \ldots 10^{15}$] under the *second order of the first period*, the first term of the octad being the unit of the corresponding order in each case. Similarly for the *third octad*, and so on. We can, in the same way, place any number of octads.

Theorem.

If there be any number of terms of a series in continued proportion, say $A_1, A_2, A_3, \ldots A_m, \ldots A_n, \ldots A_{m+n-1}, \ldots$ *of which* $A_1 = 1, A_2 = 10$ [*so that the series forms the geometrical progression* $1, 10^1, 10^2, \ldots 10^{m-1}, \ldots 10^{n-1}, \ldots 10^{m+n-2}, \ldots$], *and if any two terms as* A_m, A_n *be taken and multiplied, the product*

$A_m . A_n$ will be a term in the same series and will be as many terms distant from A_n as A_m is distant from A_1; also it will be distant from A_1 by a number of terms less by one than the sum of the numbers of terms by which A_m and A_n respectively are distant from A_1.

Take the term which is distant from A_n by the same number of terms as A_m is distant from A_1. This number of terms is m (the first and last being both counted). Thus the term to be taken is m terms distant from A_n, and is therefore the term A_{m+n-1}.

We have therefore to prove that

$$A_m . A_n = A_{m+n-1}.$$

Now terms equally distant from other terms in the continued proportion are proportional.

Thus $\dfrac{A_m}{A_1} = \dfrac{A_{m+n-1}}{A_n}.$

But $A_m = A_m . A_1$, since $A_1 = 1$.

Therefore $A_{m+n-1} = A_m . A_n$ ....(1).

The second result is now obvious, since A_m is m terms distant from A_1, A_n is n terms distant from A_1, and A_{m+n-1} is $(m + n - 1)$ terms distant from A_1.

Application to the number of the sand.

By Assumption 5 [p. 227],

(diam. of poppy-seed) $\not< \frac{1}{40}$ (finger-breadth);

and, since spheres are to one another in the triplicate ratio of their diameters, it follows that

(sphere of diam. 1 finger-breadth) $\not> 64,000$ poppy-seeds

$\not> 64,000 \times 10,000$

$\not> 640,000,000$

$\not> 6$ units of *second order* $+ 40,000,000$ units of *first order*

$(a\ fortiori) < 10$ units of *second order* of numbers.

⎫ grains ⎬ of ⎭ sand.

We now gradually increase the diameter of the supposed sphere, multiplying it by 100 each time. Thus, remembering that the sphere is thereby multiplied by 100^3 or 1,000,000, the number of grains of sand which would be contained in a sphere with each successive diameter may be arrived at as follows.

Diameter of sphere.	Corresponding number of grains of sand.
(1) 100 finger-breadths	$<$1,000,000 × 10 units of *second order* $<$(7th term of series) × (10th term of series) $<$16th term of series [i.e. 10^{15}] $<$[10^7 or] 10,000,000 units of the *second order*.
(2) 10,000 finger-breadths	$<$1,000,000 × (last number) $<$(7th term of series) × (16th term) $<$22nd term of series [i.e. 10^{21}] $<$[10^5 or] 100,000 units of *third order*.
(3) 1 stadium ($<$ 10,000 finger-breadths)	$<$100,000 units of *third order*.
(4) 100 stadia	$<$1,000,000 × (last number) $<$(7th term of series) × (22nd term) $<$28th term of series [10^{27}] $<$[10^3 or] 1,000 units of *fourth order*.
(5) 10,000 stadia	$<$1,000,000 × (last number) $<$(7th term of series) × (28th term) $<$34th term of series [10^{33}] $<$10 units of *fifth order*.
(6) 1,000,000 stadia	$<$(7th term of series) × (34th term) $<$40th term [10^{39}] $<$[10^7 or] 10,000,000 units of *fifth order*.
(7) 100,000,000 stadia	$<$(7th term of series) × (40th term) $<$46th term [10^{45}] $<$[10^5 or] 100,000 units of *sixth order*.
(8) 10,000,000,000 stadia	$<$(7th term of series) × (46th term) $<$52nd term of series [10^{51}] $<$[10^3 or] 1,000 units of *seventh order*.

But, by the proposition above [p. 227],

(diameter of 'universe') < 10,000,000,000 stadia.

Hence *the number of grains of sand which could be contained in a sphere of the size of our 'universe' is less than* 1,000 *units of the seventh order of numbers* [or 10^{51}].

From this we can prove further that *a sphere of the size attributed by Aristarchus to the sphere of the fixed stars would contain a number of grains of sand less than 10,000,000 units of the eighth order of numbers* [or $10^{56+7} = 10^{63}$].

For, by hypothesis,

(earth) : ('universe') = ('universe') : (sphere of fixed stars).

And [p. 227]

(diameter of 'universe') < 10,000 (diam. of earth);

whence

(diam. of sphere of fixed stars) < 10,000 (diam. of 'universe').

Therefore

(sphere of fixed stars) < $(10,000)^3$. ('universe').

It follows that the number of grains of sand which would be contained in a sphere equal to the sphere of the fixed stars

$< (10,000)^3 \times 1,000$ units of *seventh order*

$< $ (13th term of series) $\times$ (52nd term of series)

$< $ 64th term of series   [i.e. 10^{63}]

$< [10^7$ or] 10,000,000 units of *eighth order* of numbers.

Conclusion.

"I conceive that these things, king Gelon, will appear incredible to the great majority of people who have not studied mathematics, but that to those who are conversant therewith and have given thought to the question of the distances and sizes of the earth the sun and moon and the whole universe the proof will carry conviction. And it was for this reason that I thought the subject would be not inappropriate for your consideration."

QUADRATURE OF THE PARABOLA.

"ARCHIMEDES to Dositheus greeting.

" When I heard that Conon, who was my friend in his life-time, was dead, but that you were acquainted with Conon and withal versed in geometry, while I grieved for the loss not only of a friend but of an admirable mathematician, I set myself the task of communicating to you, as I had intended to send to Conon, a certain geometrical theorem which had not been investigated before but has now been investigated by me, and which I first discovered by means of mechanics and then exhibited by means of geometry. Now some of the earlier geometers tried to prove it possible to find a rectilineal area equal to a given circle and a given segment of a circle; and after that they endeavoured to square the area bounded by the section of the whole cone* and a straight line, assuming lemmas not easily conceded, so that it was recognised by most people that the problem was not solved. But I am not aware that any one of my predecessors has attempted to square the segment bounded by a straight line and a section of a right-angled cone [a parabola], of which problem I have now dis-covered the solution. For it is here shown that every segment bounded by a straight line and a section of a right-angled cone [a parabola] is four-thirds of the triangle which has the same base and equal height with the segment, and for the demonstration

---

* There appears to be some corruption here : the expression in the text is τᾶς ὅλου τοῦ κώνου τομᾶς, and it is not easy to give a natural and intelligible meaning to it. The section of 'the whole cone' might perhaps mean a section cutting right through it, i.e. an ellipse, and the 'straight line' might be an axis or a diameter. But Heiberg objects to the suggestion to read τᾶς ὀξυγωνίου κώνου τομᾶς, in view of the addition of καὶ εὐθείας, on the ground that the former expression always signifies the whole of an ellipse, never a segment of it (*Quaestiones Archimedeae*, p. 149).

of this property the following lemma is assumed: that the excess by which the greater of (two) unequal areas exceeds the less can, by being added to itself, be made to exceed any given finite area. The earlier geometers have also used this lemma; for it is by the use of this same lemma that they have shown that circles are to one another in the duplicate ratio of their diameters, and that spheres are to one another in the triplicate ratio of their diameters, and further that every pyramid is one third part of the prism which has the same base with the pyramid and equal height; also, that every cone is one third part of the cylinder having the same base as the cone and equal height they proved by assuming a certain lemma similar to that aforesaid. And, in the result, each of the aforesaid theorems has been accepted* no less than those proved without the lemma. As therefore my work now published has satisfied the same test as the propositions referred to, I have written out the proof and send it to you, first as investigated by means of mechanics, and afterwards too as demonstrated by geometry. Prefixed are, also, the elementary propositions in conics which are of service in the proof (στοιχεῖα κωνικὰ χρείαν ἔχοντα ἐς τὰν ἀπόδειξιν). Farewell."

Proposition 1.

If from a point on a parabola a straight line be drawn which is either itself the axis or parallel to the axis, as PV, and if QQ' be a chord parallel to the tangent to the parabola at P and meeting PV in V, then

$$QV = VQ'.$$

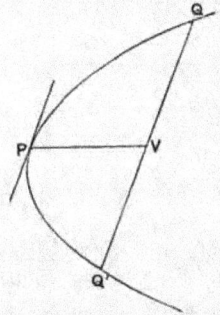

Conversely, if QV = VQ', the chord QQ' will be parallel to the tangent at P.

---

* The Greek of this passage is: συμβαίνει δὲ τῶν προειρημένων θεωρημάτων ἕκαστον μηδὲν ἧσσον τῶν ἄνευ τούτου τοῦ λήμματος ἀποδεδειγμένων πεπιστευκέναι. Here it would seem that πεπιστευκέναι must be wrong and that the passive should have been used.

Proposition 2.

If in a parabola QQ' be a chord parallel to the tangent at P, and if a straight line be drawn through P which is either itself the axis or parallel to the axis, and which meets QQ' in V and the tangent at Q to the parabola in T, then

$$PV = PT.$$

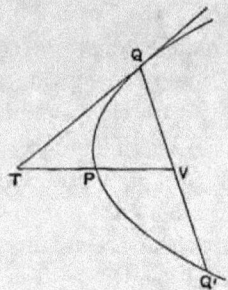

Proposition 3.

If from a point on a parabola a straight line be drawn which is either itself the axis or parallel to the axis, as PV, and if from two other points Q, Q' on the parabola straight lines be drawn parallel to the tangent at P and meeting PV in V, V' respectively, then

$$PV : PV' = QV^2 : Q'V'^2.$$

"*And these propositions are proved in the elements of conics.*[*]"

Proposition 4.

If Qq be the base of any segment of a parabola, and P the vertex of the segment, and if the diameter through any other point R meet Qq in O and QP (produced if necessary) in F, then

$$QV : VO = OF : FR.$$

Draw the ordinate RW to PV, meeting QP in K.

---

[*] i.e. in the treatises on conics by Euclid and Aristaeus.

Then $PV : PW = QV^2 : RW^2;$

whence, by parallels,

$$PQ : PK = PQ^2 : PF^2.$$

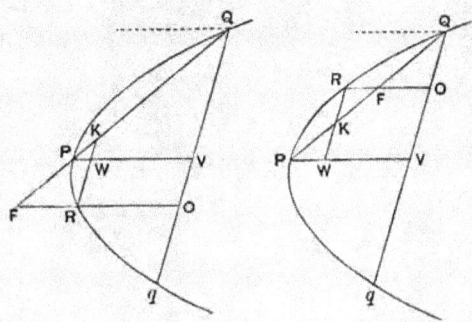

In other words, PQ, PF, PK are in continued proportion; therefore

$$PQ : PF = PF : PK$$
$$= PQ \pm PF : PF \pm PK$$
$$= QF : KF.$$

Hence, by parallels,

$$QV : VO = OF : FR.$$

[It is easily seen that this equation is equivalent to a change of axes of coordinates from the tangent and diameter to new axes consisting of the chord Qq (as axis of x, say) and the diameter through Q (as axis of y).

For, if $QV = a$, $PV = \dfrac{a^2}{p}$, where p is the parameter of the ordinates to PV.

Thus, if $QO = x$, and $RO = y$, the above result gives

$$\frac{a}{x - a} = \frac{OF}{OF - y},$$

whence $$\frac{a}{2a - x} = \frac{OF}{y} = \frac{x \cdot \dfrac{a}{p}}{y},$$

or $$py = x (2a - x).]$$

Proposition 5.

If Qq be the base of any segment of a parabola, P the vertex of the segment, and PV its diameter, and if the diameter of the parabola through any other point R meet Qq in O and the tangent at Q in E, then

$$QO : Oq = ER : RO.$$

Let the diameter through R meet QP in F.

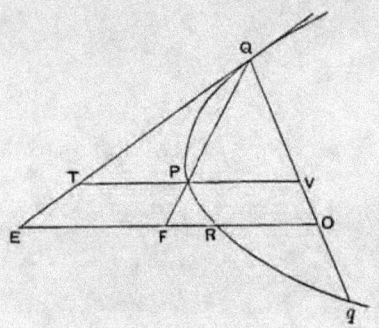

Then, by Prop. 4,

$$QV : VO = OF : FR.$$

Since $QV = Vq$, it follows that

$$QV : qO = OF : OR \dots\dots\dots\dots(1).$$

Also, if VP meet the tangent in T,

$$PT = PV, \text{ and therefore } EF = OF.$$

Accordingly, doubling the antecedents in (1), we have

$$Qq : qO = OE : OR,$$

whence $\qquad QO : Oq = ER : RO.$

Propositions 6, 7*.

Suppose a lever AOB placed horizontally and supported at its middle point O. Let a triangle BCD in which the angle C is right or obtuse be suspended from B and O, so that C is attached to O and CD is in the same vertical line with O. Then, if P be such an area as, when suspended from A, will keep the system in equilibrium,

$$P = \tfrac{1}{3} \triangle BCD.$$

Take a point E on OB such that $BE = 2OE$, and draw EFH parallel to OCD meeting BC, BD in F, H respectively. Let G be the middle point of FH.

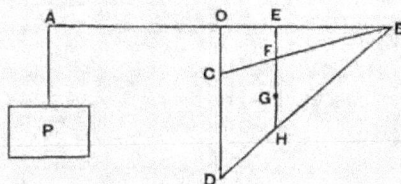

Then G is the centre of gravity of the triangle BCD.

Hence, if the angular points B, C be set free and the triangle be suspended by attaching F to E, the triangle will hang in the same position as before, because EFG is a vertical straight line. "For this is proved†."

Therefore, as before, there will be equilibrium.

Thus $P : \triangle BCD = OE : AO$

$$= 1 : 3,$$

or $P = \tfrac{1}{3} \triangle BCD.$

---

* In Prop. 6 Archimedes takes the separate case in which the angle BCD of the triangle is a right angle so that C coincides with O in the figure and F with E. He then proves, in Prop. 7, the same property for the triangle in which BCD is an obtuse angle, by treating the triangle as the difference between two right-angled triangles BOD, BOC and using the result of Prop. 6. I have combined the two propositions in one proof, for the sake of brevity. The same remark applies to the propositions following Props. 6, 7.

† Doubtless in the lost book περὶ ζυγῶν. Cf. the Introduction, Chapter II., *ad fin.*

Propositions 8, 9.

Suppose a lever AOB placed horizontally and supported at its middle point O. Let a triangle BCD, right-angled or obtuse-angled at C, be suspended from the points B, E on OB, the angular point C being so attached to E that the side CD is in the same vertical line with E. Let Q be an area such that

$$AO : OE = \triangle BCD : Q.$$

Then, if an area P suspended from A keep the system in equilibrium,

$$P < \triangle BCD \text{ but } > Q.$$

Take G the centre of gravity of the triangle BCD, and draw GH parallel to DC, i.e. vertically, meeting BO in H.

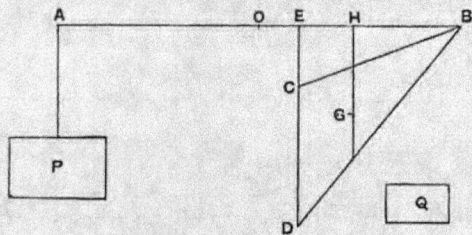

We may now suppose the triangle BCD suspended from H, and, since there is equilibrium,

$$\triangle BCD : P = AO : OH \dots\dots\dots\dots(1),$$

whence $P < \triangle BCD.$

Also $\triangle BCD : Q = AO : OE.$

Therefore, by (1), $\triangle BCD : Q > \triangle BCD : P,$

and $P > Q.$

Propositions 10, 11.

Suppose a lever AOB placed horizontally and supported at O, its middle point. Let CDEF be a trapezium which can be so placed that its parallel sides CD, FE are vertical, while C is vertically below O, and the other sides CF, DE meet in B. Let EF meet BO in H, and let the trapezium be suspended by attaching F to H and C to O. Further, suppose Q to be an area such that

$$AO : OH = (\text{trapezium } CDEF) : Q.$$

Then, if P be the area which, when suspended from A, keeps the system in equilibrium,

$$P < Q.$$

The same is true in the particular case where the angles at C, F are right, and consequently C, F coincide with O, H respectively.

Divide OH in K so that

$$(2CD + FE) : (2FE + CD) = HK : KO.$$

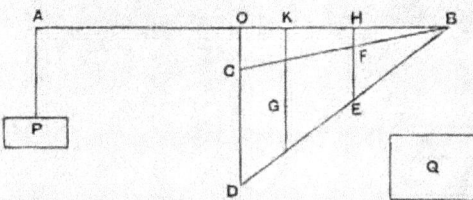

Draw KG parallel to OD, and let G be the middle point of the portion of KG intercepted within the trapezium. Then G is the centre of gravity of the trapezium [*On the equilibrium of planes,* I. 15].

Thus we may suppose the trapezium suspended from K, and the equilibrium will remain undisturbed.

Therefore

$$AO : OK = (\text{trapezium } CDEF) : P,$$

and, by hypothesis,

$$AO : OH = (\text{trapezium } CDEF) : Q.$$

Since $OK < OH$, it follows that

$$P < Q.$$

Propositions 12, 13.

If the trapezium CDEF be placed as in the last propositions, except that CD is vertically below a point L on OB instead of being below O, and the trapezium is suspended from L, H, suppose that Q, R are areas such that

$$AO : OH = (\text{trapezium } CDEF) : Q,$$

and

$$AO : OL = (\text{trapezium } CDEF) : R.$$

If then an area P suspended from A keep the system in equilibrium,

$$P > R \text{ but } < Q.$$

Take the centre of gravity G of the trapezium, as in the last propositions, and let the line through G parallel to DC meet OB in K.

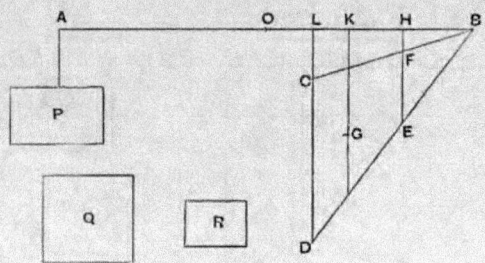

Then we may suppose the trapezium suspended from K, and there will still be equilibrium.

Therefore   (trapezium $CDEF$) : $P = AO : OK$.

Hence

(trapezium $CDEF$) : $P >$ (trapezium $CDEF$) : Q,

but $<$ (trapezium $CDEF$) : R.

It follows that $P < Q$ but $> R$.

Propositions 14, 15.

Let Qq be the base of any segment of a parabola. Then, if two lines be drawn from Q, q, each parallel to the axis of the parabola and on the same side of Qq as the segment is, either (1) the angles so formed at Q, q are both right angles, or (2) one is acute and the other obtuse. In the latter case let the angle at q be the obtuse angle.

Divide Qq into any number of equal parts at the points $O_1, O_2, \ldots O_n$. Draw through q, $O_1, O_2, \ldots O_n$ diameters of the parabola meeting the tangent at Q in E, $E_1, E_2, \ldots E_n$ and the parabola itself in q, $R_1, R_2, \ldots R_n$. Join QR_1, QR_2, $\ldots QR_n$ meeting qE, O_1E_1, O_2E_2, $\ldots O_{n-1}E_{n-1}$ in F, F_1, F_2, $\ldots F_{n-1}$.

Let the diameters Eq, E_1O_1, ... E_nO_n meet a straight line QOA drawn through Q perpendicular to the diameters in the points O, H_1, H_2, ... H_n respectively. (In the particular case where Qq is itself perpendicular to the diameters q will coincide with O, O_1 with H_1, and so on.)

It is required to prove that

(1) $\triangle EqQ < 3$(*sum of trapezia* FO_1, F_1O_2, ... $F_{n-1}O_n$ *and* $\triangle E_nO_nQ$),

(2) $\triangle EqQ > 3$(*sum of trapezia* R_1O_2, R_2O_3, ... $R_{n-1}O_n$ *and* $\triangle R_nO_nQ$).

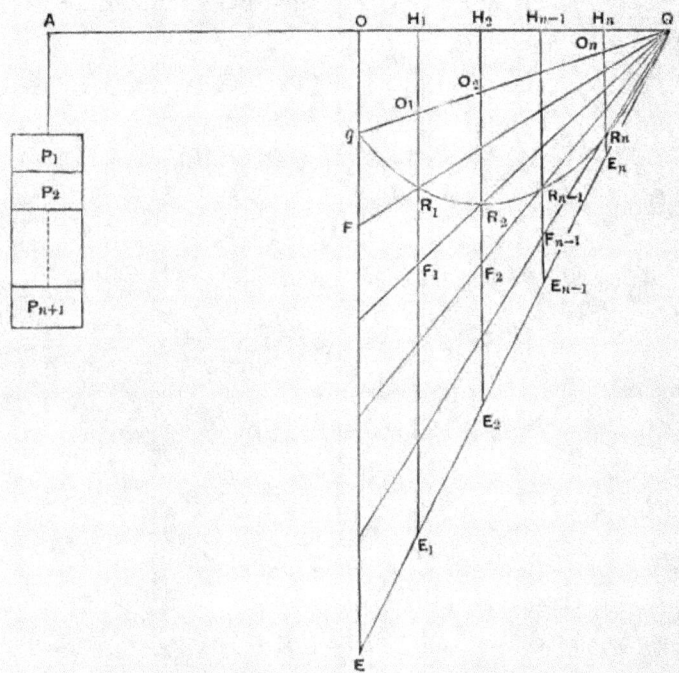

Suppose AO made equal to OQ, and conceive QOA as a lever placed horizontally and supported at O. Suppose the triangle EqQ suspended from OQ in the position drawn, and suppose that the trapezium EO_1 in the position drawn is balanced by an area P_1 suspended from A, the trapezium E_1O_2 in the position drawn is balanced by the area P_2 suspended from A, and so on, the triangle $E_n O_n Q$ being in like manner balanced by P_{n+1}.

Then $P_1 + P_2 + \ldots + P_{n+1}$ will balance the whole triangle EqQ as drawn, and therefore

$$P_1 + P_2 + \ldots + P_{n+1} = \tfrac{1}{3} \triangle EqQ. \qquad \text{[Props. 6, 7]}$$

Again $\qquad AO : OH_1 = QO : OH_1$

$$= Qq : qO_1$$

$$= E_1 O_1 : O_1 R_1 \text{ [by means of Prop. 5]}$$

$$= (\text{trapezium } EO_1) : (\text{trapezium } FO_1):$$

whence [Props. 10, 11]

$$(FO_1) > P_1.$$

Next $\qquad AO : OH_1 = E_1 O_1 : O_1 R_1$

$$= (E_1 O_2) : (R_1 O_2) \ldots\ldots\ldots\ldots(\alpha),$$

while $\qquad AO : OH_2 = E_2 O_2 : O_2 R_2$

$$= (E_1 O_2) : (F_1 O_2) \ldots\ldots\ldots\ldots(\beta);$$

and, since (α) and (β) are simultaneously true, we have, by Props. 12, 13,

$$(F_1 O_2) > P_2 > (R_1 O_2).$$

Similarly it may be proved that

$$(F_2 O_3) > P_3 > (R_2 O_3),$$

and so on.

Lastly [Props. 8, 9]

$$\triangle E_n O_n Q > P_{n+1} > \triangle R_n O_n Q.$$

By addition, we obtain

(1) $(FO_1) + (F_1 O_2) + \ldots + (F_{n-1} O_n) + \triangle E_n O_n Q > P_1 + P_2 + \ldots + P_{n+1}$

$$> \tfrac{1}{3} \triangle EqQ,$$

or $\qquad \triangle EqQ < 3 (FO_1 + F_1 O_2 + \ldots + F_{n-1} O_n + \triangle E_n O_n Q).$

(2) $(R_1 O_2) + (R_2 O_3) + \ldots + (R_{n-1} O_n) + \triangle R_n O_n Q < P_2 + P_3 + \ldots + P_{n+1}$

$$< P_1 + P_2 + \ldots + P_{n+1}, \text{ a fortiori,}$$

$$< \tfrac{1}{3} \triangle EqQ,$$

or $\qquad \triangle EqQ > 3 (R_1 O_2 + R_2 O_3 + \ldots + R_{n-1} O_n + \triangle R_n O_n Q).$

16—2

Proposition 16.

Suppose Qq to be the base of a parabolic segment, q being not more distant than Q from the vertex of the parabola. Draw through q the straight line qE parallel to the axis of the parabola to meet the tangent at Q in E. It is required to prove that

$$(area\ of\ segment) = \tfrac{1}{3}\,\triangle EqQ.$$

For, if not, the area of the segment must be either greater or less than $\frac{1}{3}\triangle EqQ$.

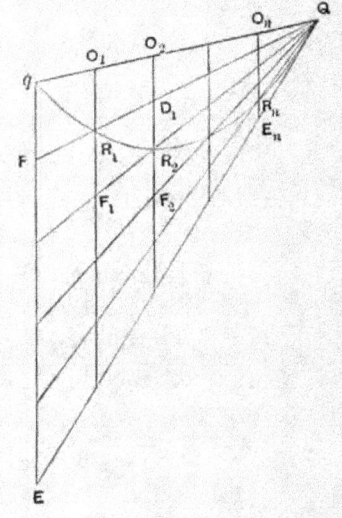

I. Suppose the area of the segment greater than $\frac{1}{3}\triangle EqQ$. Then the excess can, if continually added to itself, be made to exceed $\triangle EqQ$. And it is possible to find a submultiple of the triangle EqQ less than the said excess of the segment over $\frac{1}{3}\triangle EqQ$.

Let the triangle FqQ be such a submultiple of the triangle EqQ. Divide Eq into equal parts each equal to qF, and let all the points of division including F be joined to Q meeting the parabola in R_1, R_2, ... R_n respectively. Through R_1, R_2, ... R_n draw diameters of the parabola meeting qQ in O_1, O_2, ... O_n respectively.

Let O_1R_1 meet QR_2 in F_1.

Let O_2R_2 meet QR_1 in D_1 and QR_3 in F_2.

Let O_3R_3 meet QR_2 in D_2 and QR_4 in F_3, and so on.

We have, by hypothesis,

$$\triangle FqQ < (area\ of\ segment) - \tfrac{1}{3}\triangle EqQ,$$

or $(area\ of\ segment) - \triangle FqQ > \tfrac{1}{3}\triangle EqQ$ .... (α).

Now, since all the parts of qE, as qF and the rest, are equal, $O_1R_1 = R_1F_1$, $O_2D_1 = D_1R_2 = R_2F_2$, and so on; therefore

$$\triangle FqQ = (FO_1 + R_1O_2 + D_1O_3 + \ldots)$$

$$= (FO_1 + F_1D_1 + F_2D_2 + \ldots + F_{n-1}D_{n-1} + \triangle E_nR_nQ)\ldots(\beta).$$

But

(area of segment) $< (FO_1 + F_1O_2 + \ldots + F_{n-1}O_n + \triangle E_nO_nQ)$.

Subtracting, we have

(area of segment) $- \triangle FqQ < (R_1O_2 + R_2O_3 + \ldots$

$$+ R_{n-1}O_n + \triangle R_nO_nQ),$$

whence, *a fortiori*, by (α),

$$\tfrac{1}{3}\triangle EqQ < (R_1O_2 + R_2O_3 + \ldots + R_{n-1}O_n + \triangle R_nO_nQ).$$

But this is impossible, since [Props. 14, 15]

$$\tfrac{1}{3}\triangle EqQ > (R_1O_2 + R_2O_3 + \ldots + R_{n-1}O_n + \triangle R_nO_nQ).$$

Therefore

$$\text{(area of segment)} \not< \tfrac{1}{3}\triangle EqQ.$$

II. If possible, suppose the area of the segment less than $\tfrac{1}{3}\triangle EqQ$.

Take a submultiple of the triangle EqQ, as the triangle FqQ, less than the excess of $\tfrac{1}{3}\triangle EqQ$ over the area of the segment, and make the same construction as before.

Since $\quad \triangle FqQ < \tfrac{1}{3}\triangle EqQ - \text{(area of segment)}$,

it follows that

$$\triangle FqQ + \text{(area of segment)} < \tfrac{1}{3}\triangle EqQ$$

$$< (FO_1 + F_1O_2 + \ldots + F_{n-1}O_n + \triangle E_nO_nQ).$$
$$\text{[Props. 14, 15]}$$

Subtracting from each side the area of the segment, we have

$$\triangle FqQ < \text{(sum of spaces } qFR_1, R_1F_1R_2, \ldots E_nR_nQ)$$

$$< (FO_1 + F_1D_1 + \ldots + F_{n-1}D_{n-1} + \triangle E_nR_nQ), \textit{a fortiori};$$

which is impossible, because, by (β) above,

$$\triangle FqQ = FO_1 + F_1D_1 + \ldots + F_{n-1}D_{n-1} + \triangle E_nR_nQ.$$

Hence $\quad$ (area of segment) $\not< \tfrac{1}{3}\triangle EqQ$.

Since then the area of the segment is neither less nor greater than $\tfrac{1}{3}\triangle EqQ$, it is equal to it.

Proposition 17.

It is now manifest that *the area of any segment of a parabola is four-thirds of the triangle which has the same base as the segment and equal height.*

Let Qq be the base of the segment, P its vertex. Then PQq is the inscribed triangle with the same base as the segment and equal height.

Since P is the vertex* of the segment, the diameter through P bisects Qq. Let V be the point of bisection.

Let VP, and qE drawn parallel to it, meet the tangent at Q in T, E respectively.

Then, by parallels,

$$qE = 2VT,$$

and $\qquad\qquad PV = PT, \qquad$ [Prop. 2]

so that $\qquad\qquad VT = 2PV.$

Hence $\triangle EqQ = 4\triangle PQq.$

But, by Prop. 16, the area of the segment is equal to $\frac{1}{3}\triangle EqQ.$

Therefore $\qquad$ (area of segment) $= \frac{4}{3}\triangle PQq.$

DEF. "In segments bounded by a straight line and any curve I call the straight line the **base**, and the **height** the greatest perpendicular drawn from the curve to the base of the segment, and the **vertex** the point from which the greatest perpendicular is drawn."

* It is curious that Archimedes uses the terms *base* and *vertex* of a segment here, but gives the definition of them later (at the end of the proposition). Moreover he assumes the converse of the property proved in Prop. 18.

Proposition 18.

If Qq be the base of a segment of a parabola, and V the middle point of Qq, and if the diameter through V meet the curve in P, then P is the vertex of the segment.

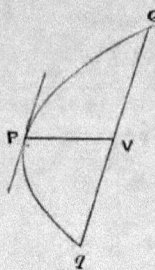

For Qq is parallel to the tangent at P [Prop. 1]. Therefore, of all the perpendiculars which can be drawn from points on the segment to the base Qq, that from P is the greatest. Hence, by the definition, P is the vertex of the segment.

Proposition 19.

If Qq be a chord of a parabola bisected in V by the diameter PV, and if RM be a diameter bisecting QV in M, and RW be the ordinate from R to PV, then

$$PV = \tfrac{4}{3}RM.$$

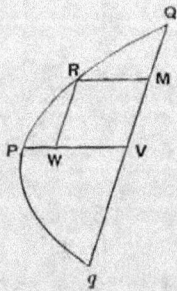

For, by the property of the parabola,

$$PV : PW = QV^2 : RW^2$$
$$= 4RW^2 : RW^2,$$

so that $\qquad\qquad PV = 4PW,$

whence $\qquad\qquad PV = \tfrac{4}{3}RM.$

Proposition 20.

If Qq be the base, and P the vertex, of a parabolic segment, then the triangle PQq is greater than half the segment PQq.

For the chord Qq is parallel to the tangent at P, and the triangle PQq is half the parallelogram formed by Qq, the tangent at P, and the diameters through Q, q.

Therefore the triangle PQq is greater than half the segment.

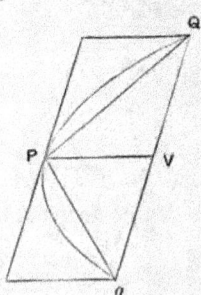

Cor. It follows that *it is possible to inscribe in the segment a polygon such that the segments left over are together less than any assigned area.*

Proposition 21.

If Qq be the base, and P the vertex, of any parabolic segment, and if R be the vertex of the segment cut off by PQ, then

$$\triangle PQq = 8 \triangle PRQ.$$

The diameter through R will bisect the chord PQ, and therefore also QV, where PV is the diameter bisecting Qq. Let the diameter through R bisect PQ in Y and QV in M. Join PM.

By Prop. 19,
$$PV = \tfrac{4}{3}RM.$$

Also $\qquad\qquad PV = 2YM.$

Therefore $\qquad\quad YM = 2RY,$

and $\qquad\quad \triangle PQM = 2 \triangle PRQ.$

Hence $\qquad\quad \triangle PQV = 4 \triangle PRQ,$

and $\qquad\qquad \triangle PQq = 8 \triangle PRQ.$

Also, if RW, the ordinate from R to PV, be produced to meet the curve again in r,

$$RW = rW,$$

and the same proof shows that

$$\triangle PQq = 8\triangle Prq.$$

Proposition 22.

If there be a series of areas A, B, C, D, ... each of which is four times the next in order, and if the largest, A, be equal to the triangle PQq inscribed in a parabolic segment PQq and having the same base with it and equal height, then

$$(A + B + C + D + ...) < (area\ of\ segment\ PQq).$$

For, since $\triangle PQq = 8\triangle PRQ = 8\triangle Pqr$, where R, r are the vertices of the segments cut off by PQ, Pq, as in the last proposition,

$$\triangle PQq = 4\,(\triangle PQR + \triangle Pqr).$$

Therefore, since $\triangle PQq = A$,

$$\triangle PQR + \triangle Pqr = B.$$

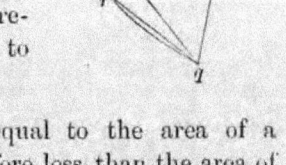

In like manner we prove that the triangles similarly inscribed in the remaining segments are together equal to the area C, and so on.

Therefore $A + B + C + D + ...$ is equal to the area of a certain inscribed polygon, and is therefore less than the area of the segment.

Proposition 23.

Given a series of areas A, B, C, D, ... Z, of which A is the greatest, and each is equal to four times the next in order, then

$$A + B + C + ... + Z + \tfrac{1}{3}Z = \tfrac{4}{3}A.$$

Take areas $b, c, d, \ldots$ such that

$$b = \tfrac{1}{3}B,$$

$$c = \tfrac{1}{3}C,$$

$$d = \tfrac{1}{3}D, \text{ and so on.}$$

Then, since $\qquad b = \tfrac{1}{3}B,$

and $\qquad\qquad\quad B = \tfrac{1}{4}A,$

$$B + b = \tfrac{1}{3}A.$$

Similarly $\qquad\quad C + c = \tfrac{1}{3}B.$

$$\cdots\cdots\cdots\cdots$$

Therefore

$$B + C + D + \ldots + Z + b + c + d + \ldots + z = \tfrac{1}{3}(A + B + C + \ldots + Y).$$

But $\qquad b + c + d + \ldots + y = \tfrac{1}{3}(B + C + D + \ldots + Y).$

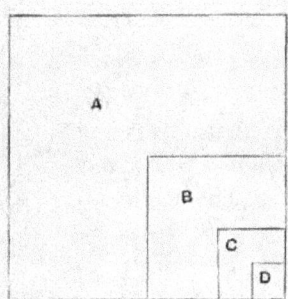

Therefore, by subtraction,

$$B + C + D + \ldots + Z + z = \tfrac{1}{3}A$$

or $\qquad\quad A + B + C + \ldots + Z + \tfrac{1}{3}Z = \tfrac{4}{3}A.$

[The algebraical equivalent of this result is of course

$$1 + \tfrac{1}{4} + (\tfrac{1}{4})^2 + \ldots + (\tfrac{1}{4})^{n-1} = \tfrac{4}{3} - \tfrac{1}{3}(\tfrac{1}{4})^{n-1}$$

$$= \frac{1 - (\tfrac{1}{4})^n}{1 - \tfrac{1}{4}} . \Big]$$

Proposition 24.

Every segment bounded by a parabola and a chord Qq is equal to four-thirds of the triangle which has the same base as the segment and equal height.

Suppose $\qquad\qquad K = \tfrac{4}{3}\triangle PQq,$

where P is the vertex of the segment; and we have then to prove that the area of the segment is equal to K.

For, if the segment be not equal to K, it must either be greater or less.

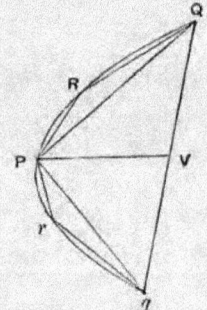

I. Suppose the area of the segment greater than K.

If then we inscribe in the segments cut off by PQ, Pq triangles which have the same base and equal height, i.e. triangles with the same vertices R, r as those of the segments, and if in the remaining segments we inscribe triangles in the same manner, and so on, we shall finally have segments remaining whose sum is less than the area by which the segment PQq exceeds K.

Therefore the polygon so formed must be greater than the area K; which is impossible, since [Prop. 23]

$$A + B + C + \ldots + Z < \tfrac{4}{3}A,$$

where $\qquad\qquad A = \triangle PQq.$

Thus the area of the segment cannot be greater than K.

II. Suppose, if possible, that the area of the segment is less than K.

If then $\triangle PQq = A$, $B = \frac{1}{4}A$, $C = \frac{1}{4}B$, and so on, until we arrive at an area X such that X is less than the difference between K and the segment, we have

$$A + B + C + \ldots + X + \tfrac{1}{3}X = \tfrac{4}{3}A \qquad \text{[Prop. 23]}$$

$$= K.$$

Now, since K exceeds $A + B + C + \ldots + X$ by an area less than X, and the area of the segment by an area greater than X, it follows that

$$A + B + C + \ldots + X > \text{(the segment)};$$

which is impossible, by Prop. 22 above.

Hence the segment is not less than K.

Thus, since the segment is neither greater nor less than K,

$$\text{(area of segment } PQq) = K = \tfrac{4}{3}\triangle PQq.$$

ON FLOATING BODIES.

BOOK I.

Postulate 1.

"Let it be supposed that a fluid is of such a character that, its parts lying evenly and being continuous, that part which is thrust the less is driven along by that which is thrust the more; and that each of its parts is thrust by the fluid which is above it in a perpendicular direction if the fluid be sunk in anything and compressed by anything else."

Proposition 1.

If a surface be cut by a plane always passing through a certain point, and if the section be always a circumference [of a circle] whose centre is the aforesaid point, the surface is that of a sphere.

For, if not, there will be some two lines drawn from the point to the surface which are not equal.

Suppose O to be the fixed point, and A, B to be two points on the surface such that OA, OB are unequal. Let the surface be cut by a plane passing through OA, OB. Then the section is, by hypothesis, a circle whose centre is O.

Thus $OA = OB$; which is contrary to the assumption. Therefore the surface cannot but be a sphere.

Proposition 2.

The surface of any fluid at rest is the surface of a sphere whose centre is the same as that of the earth.

Suppose the surface of the fluid cut by a plane through O, the centre of the earth, in the curve $ABCD$.

$ABCD$ shall be the circumference of a circle.

For, if not, some of the lines drawn from O to the curve will be unequal. Take one of them, OB, such that OB is greater than some of the lines from O to the curve and less than others. Draw a circle with OB as radius. Let it be EBF, which will therefore fall partly within and partly without the surface of the fluid.

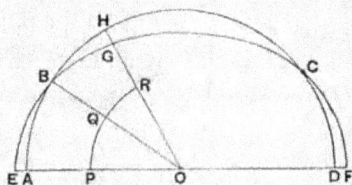

Draw OGH making with OB an angle equal to the angle EOB, and meeting the surface in H and the circle in G. Draw also in the plane an arc of a circle PQR with centre O and within the fluid.

Then the parts of the fluid along PQR are uniform and continuous, and the part PQ is compressed by the part between it and AB, while the part QR is compressed by the part between QR and BH. Therefore the parts along PQ, QR will be unequally compressed, and the part which is compressed the less will be set in motion by that which is compressed the more.

Therefore there will not be rest; which is contrary to the hypothesis.

Hence the section of the surface will be the circumference of a circle whose centre is O; and so will all other sections by planes through O.

Therefore the surface is that of a sphere with centre O.

Proposition 3.

*Of solids those which, size for size, are of equal weight with
a fluid will, if let down into the fluid, be immersed so that they
do not project above the surface but do not sink lower.*

If possible, let a certain solid *EFHG* of equal weight,
volume for volume, with the fluid remain immersed in it so
that part of it, *EBCF*, projects above the surface.

Draw through *O*, the centre of the earth, and through the
solid a plane cutting the surface of the fluid in the circle
ABCD.

Conceive a pyramid with vertex *O* and base a parallelogram
at the surface of the fluid, such that it includes the immersed
portion of the solid. Let this pyramid be cut by the plane of
ABCD in *OL*, *OM*. Also let a sphere within the fluid and
below *GH* be described with centre *O*, and let the plane of
ABCD cut this sphere in *PQR*.

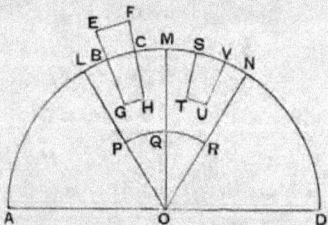

Conceive also another pyramid in the fluid with vertex *O*,
continuous with the former pyramid and equal and similar to
it. Let the pyramid so described be cut in *OM*, *ON* by the
plane of *ABCD*.

Lastly, let *STUV* be a part of the fluid within the second
pyramid equal and similar to the part *BGHC* of the solid, and
let *SV* be at the surface of the fluid.

Then the pressures on *PQ*, *QR* are unequal, that on *PQ*
being the greater. Hence the part at *QR* will be set in motion

by that at *PQ*, and the fluid will not be at rest; which is contrary to the hypothesis.

Therefore the solid will not stand out above the surface.

Nor will it sink further, because all the parts of the fluid will be under the same pressure.

Proposition 4.

A solid lighter than a fluid will, if immersed in it, not be completely submerged, but part of it will project above the surface.

In this case, after the manner of the previous proposition, we assume the solid, if possible, to be completely submerged and the fluid to be at rest in that position, and we conceive (1) a pyramid with its vertex at *O*, the centre of the earth, including the solid, (2) another pyramid continuous with the former and equal and similar to it, with the same vertex *O*, (3) a portion of the fluid within this latter pyramid equal to the immersed solid in the other pyramid, (4) a sphere with centre *O* whose surface is below the immersed solid and the part of the fluid in the second pyramid corresponding thereto. We suppose a plane to be drawn through the centre *O* cutting the surface of the fluid in the circle *ABC*, the solid in *S*, the first pyramid in *OA*, *OB*, the second pyramid in *OB*, *OC*, the portion of the fluid in the second pyramid in *K*, and the inner sphere in *PQR*.

Then the pressures on the parts of the fluid at *PQ*, *QR* are unequal, since *S* is lighter than *K*. Hence there will not be rest; which is contrary to the hypothesis.

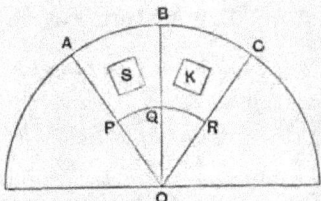

Therefore the solid *S* cannot, in a condition of rest, be completely submerged.

Proposition 5.

Any solid lighter than a fluid will, if placed in the fluid, be so far immersed that the weight of the solid will be equal to the weight of the fluid displaced.

For let the solid be *EGHF*, and let *BGHC* be the portion of it immersed when the fluid is at rest. As in Prop. 3, conceive a pyramid with vertex *O* including the solid, and another pyramid with the same vertex continuous with the former and equal and similar to it. Suppose a portion of the fluid *STUV* at the base of the second pyramid to be equal and similar to the immersed portion of the solid; and let the construction be the same as in Prop. 3.

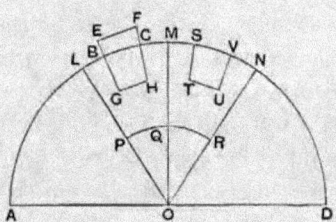

Then, since the pressure on the parts of the fluid at *PQ, QR* must be equal in order that the fluid may be at rest, it follows that the weight of the portion *STUV* of the fluid must be equal to the weight of the solid *EGHF*. And the former is equal to the weight of the fluid displaced by the immersed portion of the solid *BGHC*.

Proposition 6.

If a solid lighter than a fluid be forcibly immersed in it, the solid will be driven upwards by a force equal to the difference between its weight and the weight of the fluid displaced.

For let *A* be completely immersed in the fluid, and let *G* represent the weight of *A*, and (*G* + *H*) the weight of an equal volume of the fluid. Take a solid *D*, whose weight is *H* and add it to A. Then the weight of $(A + D)$ is less than that of an equal volume of the fluid; and, if $(A + D)$ is immersed in the fluid, it will project so that its weight will be equal to the weight of the fluid displaced. But its weight is $(G + H)$.

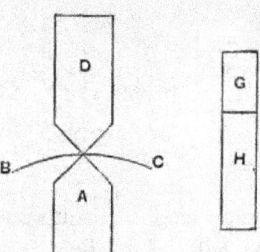

Therefore the weight of the fluid displaced is $(G + H)$, and hence the volume of the fluid displaced is the volume of the solid A. There will accordingly be rest with A immersed and D projecting.

Thus the weight of D balances the upward force exerted by the fluid on A, and therefore the latter force is equal to H, which is the difference between the weight of A and the weight of the fluid which A displaces.

Proposition 7.

A solid heavier than a fluid will, if placed in it, descend to the bottom of the fluid, and the solid will, when weighed in the fluid, be lighter than its true weight by the weight of the fluid displaced.

(1) The first part of the proposition is obvious, since the part of the fluid under the solid will be under greater pressure, and therefore the other parts will give way until the solid reaches the bottom.

(2) Let A be a solid heavier than the same volume of the fluid, and let $(G + H)$ represent its weight, while G represents the weight of the same volume of the fluid.

Take a solid B lighter than the same volume of the fluid, and such that the weight of B is G, while the weight of the same volume of the fluid is $(G+H)$.

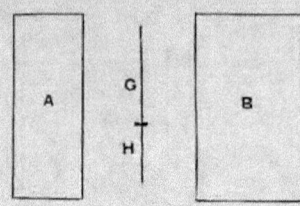

Let A and B be now combined into one solid and immersed. Then, since $(A+B)$ will be of the same weight as the same volume of fluid, both weights being equal to $(G+H)+G$, it follows that $(A+B)$ will remain stationary in the fluid.

Therefore the force which causes A by itself to sink must be equal to the upward force exerted by the fluid on B by itself. This latter is equal to the difference between $(G+H)$ and G [Prop. 6]. Hence A is depressed by a force equal to H, i.e. its weight in the fluid is H, or the difference between $(G+H)$ and G.

[This proposition may, I think, safely be regarded as decisive of the question how Archimedes determined the proportions of gold and silver contained in the famous crown (cf. Introduction, Chapter I.). The proposition suggests in fact the following method.

Let W represent the weight of the crown, w_1 and w_2 the weights of the gold and silver in it respectively, so that $W = w_1 + w_2$.

(1) Take a weight W of pure gold and weigh it in a fluid. The apparent loss of weight is then equal to the weight of the fluid displaced. If F_1 denote this weight, F_1 is thus known as the result of the operation of weighing.

It follows that the weight of fluid displaced by a weight w_1 of gold is $\dfrac{w_1}{W} \cdot F_1$.

(2) Take a weight W of pure silver and perform the same operation. If F_2 be the loss of weight when the silver is weighed in the fluid, we find in like manner that the weight of fluid displaced by w_2 is $\frac{w_2}{W} \cdot F_2$.

(3) Lastly, weigh the crown itself in the fluid, and let F be the loss of weight. Therefore the weight of fluid displaced by the crown is F.

It follows that

$$\frac{w_1}{W} \cdot F_1 + \frac{w_2}{W} \cdot F_2 = F,$$

or

$$w_1 F_1 + w_2 F_2 = (w_1 + w_2) F,$$

whence

$$\frac{w_1}{w_2} = \frac{F_2 - F}{F - F_1}.$$

This procedure corresponds pretty closely to that described in the poem *de ponderibus et mensuris* (written probably about 500 A.D.)[*] purporting to explain Archimedes' method. According to the author of this poem, we first take two equal weights of pure gold and pure silver respectively and weigh them against each other when both immersed in water; this gives the relation between their weights in water and therefore between their loss of weight in water. Next we take the mixture of gold and silver and an equal weight of pure silver and weigh them against each other in water in the same manner.

The other version of the method used by Archimedes is that given by Vitruvius[†], according to which he measured successively the *volumes* of fluid displaced by three equal weights, (1) the crown, (2) the same weight of gold, (3) the same weight of silver, respectively. Thus, if as before the weight of the crown is W, and it contains weights w_1 and w_2 of gold and silver respectively,

(1) the crown displaces a certain quantity of fluid, V say.

(2) the weight W of gold displaces a certain volume of

---

[*] Torelli's *Archimedes*, p. 364; Hultsch, *Metrol. Script.* II. 95 sq., and Prolegomena § 118.

[†] *De architect.* IX. 3.

fluid, V_1 say; therefore a weight w_1 of gold displaces a volume $\frac{w_1}{W} \cdot V_1$ of fluid.

(3) the weight W of silver displaces a certain volume of fluid, say V_2; therefore a weight w_2 of silver displaces a volume $\frac{w_2}{W} \cdot V_2$ of fluid.

It follows that $$V = \frac{w_1}{W} \cdot V_1 + \frac{w_2}{W} \cdot V_2,$$

whence, since $$W = w_1 + w_2,$$
$$\frac{w_1}{w_2} = \frac{V_2 - V}{V - V_1};$$

and this ratio is obviously equal to that before obtained, viz. $\frac{F_2 - F}{F - F_1}$.]

Postulate 2.

"Let it be granted that bodies which are forced upwards in a fluid are forced upwards along the perpendicular [to the surface] which passes through their centre of gravity."

Proposition 8.

If a solid in the form of a segment of a sphere, and of a substance lighter than a fluid, be immersed in it so that its base does not touch the surface, the solid will rest in such a position that its axis is perpendicular to the surface; and, if the solid be forced into such a position that its base touches the fluid on one side and be then set free, it will not remain in that position but will return to the symmetrical position.

[The proof of this proposition is wanting in the Latin version of Tartaglia. Commandinus supplied a proof of his own in his edition.]

Proposition 9.

If a solid in the form of a segment of a sphere, and of a substance lighter than a fluid, be immersed in it so that its base is completely below the surface, the solid will rest in such a position that its axis is perpendicular to the surface.

[The proof of this proposition has only survived in a
mutilated form. It deals moreover with only one case out of
three which are distinguished at the beginning, viz. that in
which the segment is greater than a hemisphere, while figures
only are given for the cases where the segment is equal to, or
less than, a hemisphere.]

Suppose, first, that the segment is greater than a hemisphere.
Let it be cut by a plane through its axis and the centre of the
earth; and, if possible, let it be at rest in the position shown
in the figure, where AB is the intersection of the plane with
the base of the segment, DE its axis, C the centre of the
sphere of which the segment is a part, O the centre of the
earth.

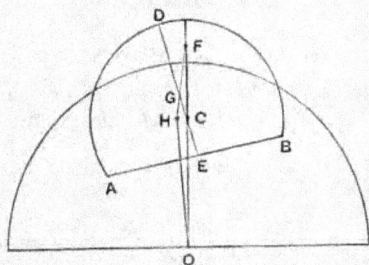

The centre of gravity of the portion of the segment outside
the fluid, as F, lies on OC produced, its axis passing through C.

Let G be the centre of gravity of the segment. Join FG,
and produce it to H so that

$FG : GH =$ (volume of immersed portion) : (rest of solid).
Join OH.

Then the weight of the portion of the solid outside the fluid
acts along FO, and the pressure of the fluid on the immersed
portion along OH, while the weight of the immersed portion
acts along HO and is by hypothesis less than the pressure of
the fluid acting along OH.

Hence there will not be equilibrium, but the part of the
segment towards A will ascend and the part towards B descend,
until DE assumes a position perpendicular to the surface of
the fluid.

ON FLOATING BODIES.

BOOK II.

___

Proposition 1.

If a solid lighter than a fluid be at rest in it, the weight of the solid will be to that of the same volume of the fluid as the immersed portion of the solid is to the whole.

Let $(A + B)$ be the solid, B the portion immersed in the fluid.

Let $(C + D)$ be an equal volume of the fluid, C being equal in volume to A and B to D.

Further suppose the line E to represent the weight of the solid $(A + B)$, $(F + G)$ to represent the weight of $(C + D)$, and G that of D.

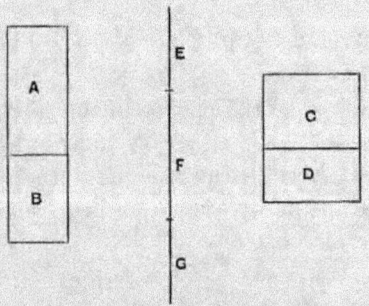

Then

weight of $(A + B)$: weight of $(C + D) = E : (F + G)$...(1).

And the weight of $(A + B)$ is equal to the weight of a volume B of the fluid [I. 5], i.e. to the weight of D.

That is to say, $E = G$.

Hence, by (1),

$$\text{weight of } (A + B) : \text{weight of } (C + D) = G : F + G$$
$$= D : C + D$$
$$= B : A + B.$$

Proposition 2.

If a right segment of a paraboloid of revolution whose axis is not greater than $\frac{3}{4} p$ (where p is the principal parameter of the generating parabola), and whose specific gravity is less than that of a fluid, be placed in the fluid with its axis inclined to the vertical at any angle, but so that the base of the segment does not touch the surface of the fluid, the segment of the paraboloid will not remain in that position but will return to the position in which its axis is vertical.

Let the axis of the segment of the paraboloid be AN, and through AN draw a plane perpendicular to the surface of the fluid. Let the plane intersect the paraboloid in the parabola BAB', the base of the segment of the paraboloid in BB', and the plane of the surface of the fluid in the chord QQ' of the parabola.

Then, since the axis AN is placed in a position not perpendicular to QQ', BB' will not be parallel to QQ'.

Draw the tangent PT to the parabola which is parallel to QQ', and let P be the point of contact*.

[From P draw PV parallel to AN meeting QQ' in V. Then PV will be a diameter of the parabola, and also the axis of the portion of the paraboloid immersed in the fluid.

---

* The rest of the proof is wanting in the version of Tartaglia, but is given in brackets as supplied by Commandinus.

Let C be the centre of gravity of the paraboloid BAB', and F that of the portion immersed in the fluid. Join FC and produce it to H so that H is the centre of gravity of the remaining portion of the paraboloid above the surface.

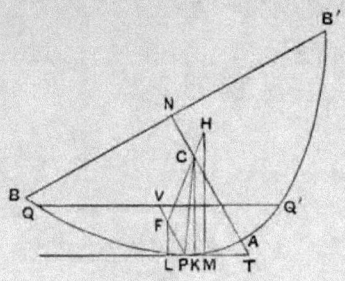

Then, since $$AN = \tfrac{3}{2}AC^*,$$

and $$AN \ngtr \tfrac{3}{4}p,$$

it follows that $$AC \ngtr \frac{p}{2}.$$

Therefore, if CP be joined, the angle CPT is acute†. Hence, if CK be drawn perpendicular to PT, K will fall between P and T. And, if FL, HM be drawn parallel to CK to meet PT, they will each be perpendicular to the surface of the fluid.

Now the force acting on the immersed portion of the segment of the paraboloid will act upwards along LF, while the weight of the portion outside the fluid will act downwards along HM.

Therefore there will not be equilibrium, but the segment

---

* As the determination of the centre of gravity of a segment of a paraboloid which is here assumed does not appear in any extant work of Archimedes, or in any known work by any other Greek mathematician, it appears probable that it was investigated by Archimedes himself in some treatise now lost.

† The truth of this statement is easily proved from the property of the sub-normal. For, if the normal at P meet the axis in G, AG is greater than $\frac{p}{2}$ except in the case where the normal is the normal at the vertex A itself. But the latter case is excluded here because, by hypothesis, AN is not placed vertically. Hence, P being a different point from A, AG is always greater than AC; and, since the angle TPG is right, the angle TPC must be acute.

will turn so that B will rise and B' will fall, until AN takes the vertical position.]

[For purposes of comparison the trigonometrical equivalent of this and other propositions will be appended.

Suppose that the angle NTP, at which in the above figure the axis AN is inclined to the surface of the fluid, is denoted by θ.

Then the coordinates of P referred to AN and the tangent at A as axes are

$$\frac{p}{4}\cot^2\theta, \quad \frac{p}{2}\cot\theta,$$

where p is the principal parameter.

Suppose that $AN = h, \quad PV = k$.

If now x' be the distance from T of the orthogonal projection of F on TP, and x the corresponding distance for the point C, we have

$$x' = \frac{p}{2}\cot^2\theta \cdot \cos\theta + \frac{p}{2}\cot\theta \cdot \sin\theta + \frac{2}{3}k\cos\theta,$$

$$x = \frac{p}{4}\cot^2\theta \cdot \cos\theta + \frac{2}{3}h\cos\theta,$$

whence $x' - x = \cos\theta\left\{\frac{p}{4}(\cot^2\theta + 2) - \frac{2}{3}(h-k)\right\}$.

In order that the segment of the paraboloid may turn in the direction of increasing the angle PTN, x' must be greater than x, or the expression just found must be positive.

This will always be the case, whatever be the value of θ, if

$$\frac{p}{2} \not< \frac{2h}{3},$$

or
$$h \not> \tfrac{3}{4}p.]$$

Proposition 3.

If a right segment of a paraboloid of revolution whose axis is not greater than $\frac{3}{4}p$ (where p is the parameter), and whose specific gravity is less than that of a fluid, be placed in the fluid with its axis inclined at any angle to the vertical, but so that its

base is entirely submerged, the solid will not remain in that posi-
tion but will return to the position in which the axis is vertical.

Let the axis of the paraboloid be AN, and through AN
draw a plane perpendicular to the surface of the fluid intersecting the paraboloid in the parabola BAB', the base of the
segment in BNB', and the plane of the surface of the fluid in
the chord QQ' of the parabola.

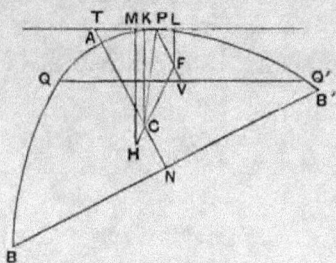

Then, since AN, as placed, is not perpendicular to the
surface of the fluid, QQ' and BB' will not be parallel.

Draw PT parallel to QQ' and touching the parabola at P.
Let PT meet NA produced in T. Draw the diameter PV
bisecting QQ' in V. PV is then the axis of the portion of the
paraboloid above the surface of the fluid.

Let C be the centre of gravity of the whole segment of the
paraboloid, F that of the portion above the surface. Join FC
and produce it to H so that H is the centre of gravity of
the immersed portion.

Then, since $AC \ngtr \dfrac{p}{2}$, the angle CPT is an acute angle, as in
the last proposition.

Hence, if CK be drawn perpendicular to PT, K will fall
between P and T. Also, if HM, FL be drawn parallel to CK,
they will be perpendicular to the surface of the fluid.

And the force acting on the submerged portion will act
upwards along HM, while the weight of the rest will act
downwards along LF produced.

Thus the paraboloid will turn until it takes the position
in which AN is vertical.

Proposition 4.

Given a right segment of a paraboloid of revolution whose axis AN is greater than $\frac{3}{4}p$ (where p is the parameter), and whose specific gravity is less than that of a fluid but bears to it a ratio not less than $(AN - \frac{3}{4}p)^2 : AN^2$, if the segment of the paraboloid be placed in the fluid with its axis at any inclination to the vertical, but so that its base does not touch the surface of the fluid, it will not remain in that position but will return to the position in which its axis is vertical.

Let the axis of the segment of the paraboloid be AN, and let a plane be drawn through AN perpendicular to the surface of the fluid and intersecting the segment in the parabola BAB', the base of the segment in BB', and the surface of the fluid in the chord QQ' of the parabola.

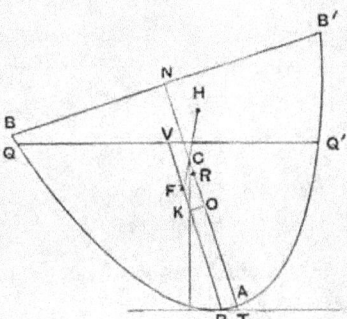

Then AN, as placed, will not be perpendicular to QQ'.

Draw PT parallel to QQ' and touching the parabola at P. Draw the diameter PV bisecting QQ' in V. Thus PV will be the axis of the submerged portion of the solid.

Let C be the centre of gravity of the whole solid, F that of the immersed portion. Join FC and produce it to H so that H is the centre of gravity of the remaining portion.

Now, since $AN = \frac{3}{2}AC,$

and $AN > \frac{3}{4}p,$

it follows that $AC > \dfrac{p}{2}.$

Measure CO along CA equal to $\frac{p}{2}$, and OR along OC equal to $\frac{1}{2}AO$.

Then, since $\qquad\qquad AN = \frac{3}{2}AC,$

and $\qquad\qquad\qquad AR = \frac{3}{2}AO,$

we have, by subtraction,

$$NR = \frac{3}{2}OC.$$

That is, $\qquad AN - AR = \frac{3}{2}OC$

$$= \frac{3}{4}p,$$

or $\qquad\qquad\qquad AR = (AN - \frac{3}{4}p).$

Thus $\qquad (AN - \frac{3}{4}p)^2 : AN^2 = AR^2 : AN^2,$

and therefore the ratio of the specific gravity of the solid to that of the fluid is, by the enunciation, not less than the ratio $AR^2 : AN^2$.

But, by Prop. 1, the former ratio is equal to the ratio of the immersed portion to the whole solid, i.e. to the ratio $PV^2 : AN^2$ [On Conoids and Spheroids, Prop. 24].

Hence $\qquad\qquad PV^2 : AN^2 \not< AR^2 : AN^2,$

or $\qquad\qquad\qquad PV \not< AR.$

It follows that

$$PF(= \tfrac{2}{3}PV) \not< \tfrac{2}{3}AR$$

$$\not< AO.$$

If, therefore, OK be drawn from O perpendicular to OA, it will meet PF between P and F.

Also, if CK be joined, the triangle KCO is equal and similar to the triangle formed by the normal, the subnormal and the ordinate at P (since $CO = \frac{1}{2}p$ or the subnormal, and KO is equal to the ordinate).

Therefore CK is parallel to the normal at P, and therefore perpendicular to the tangent at P and to the surface of the fluid.

Hence, if parallels to CK be drawn through F, H, they will be perpendicular to the surface of the fluid, and the force acting on the submerged portion of the solid will act upwards along the former, while the weight of the other portion will act downwards along the latter.

Therefore the solid will not remain in its position but will turn until AN assumes a vertical position.

[Using the same notation as before (note following Prop. 2), we have

$$x' - x = \cos\theta \left\{ \frac{p}{4}\left(\cot^2\theta + 2\right) - \frac{2}{3}(h - k) \right\},$$

and the *minimum* value of the expression within the bracket, for different values of θ, is

$$\frac{p}{2} - \frac{2}{3}(h - k),$$

corresponding to the position in which AM is vertical, or $\theta = \dfrac{\pi}{2}$.

Therefore there will be stable equilibrium in that position only, provided that

$$k \not< (h - \tfrac{3}{4}p),$$

or, if s be the ratio of the specific gravity of the solid to that of the fluid $(= k^2/h^2$ in this case),

$$s \not< (h - \tfrac{3}{4}p)^2/h^2.]$$

Proposition 5.

Given a right segment of a paraboloid of revolution such that its axis AN is greater than $\tfrac{3}{4}p$ (where p is the parameter), and its specific gravity is less than that of a fluid but in a ratio to it not greater than the ratio $\{AN^2 - (AN - \tfrac{3}{4}p)^2\} : AN^2$, if the segment be placed in the fluid with its axis inclined at any angle to the vertical, but so that its base is completely submerged, it will not remain in that position but will return to the position in which AN is vertical.

Let a plane be drawn through AN, as placed, perpendicular to the surface of the fluid and cutting the segment of the paraboloid in the parabola BAB', the base of the segment in BB', and the plane of the surface of the fluid in the chord QQ' of the parabola.

Draw the tangent PT parallel to QQ', and the diameter PV, bisecting QQ', will accordingly be the axis of the portion of the paraboloid above the surface of the fluid.

Let F be the centre of gravity of the portion above the surface, C that of the whole solid, and produce FC to H, the centre of gravity of the immersed portion.

As in the last proposition, $AC > \dfrac{p}{2}$, and we measure CO along CA equal to $\dfrac{p}{2}$, and OR along OC equal to $\frac{1}{2}AO$.

Then $AN = \frac{3}{2}AC$, and $AR = \frac{3}{2}AO$;

and we derive, as before,

$$AR = (AN - \tfrac{3}{4}p).$$

Now, by hypothesis,

(spec. gravity of solid) : (spec. gravity of fluid)

$$\ngtr \{AN^2 - (AN - \tfrac{3}{4}p)^2\} : AN^2$$
$$\ngtr (AN^2 - AR^2) : AN^2.$$

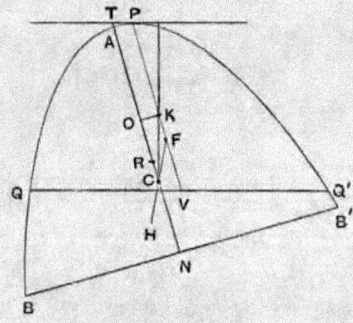

Therefore

(portion submerged) : (whole solid)

$$\ngtr (AN^2 - AR^2) : AN^2,$$

and   (whole solid) : (portion above surface)

$$\ngtr AN^2 : AR^2.$$

Thus $AN^2 : PV^2 \ngtr AN^2 : AR^2$,

whence $PV \nless AR,$

and $PF \nless \frac{2}{3}AR$

$$\nless AO.$$

Therefore, if a perpendicular to AC be drawn from O, it will meet PF in some point K between P and F.

And, since $CO = \frac{1}{2}p$, CK will be perpendicular to PT, as in the last proposition.

Now the force acting on the submerged portion of the solid will act upwards through H, and the weight of the other portion downwards through F, in directions parallel in both cases to CK; whence the proposition follows.

Proposition 6.

If a right segment of a paraboloid lighter than a fluid be such that its axis AM is greater than $\frac{3}{4}p$, but $AM : \frac{1}{2}p < 15 : 4$, and if the segment be placed in the fluid with its axis so inclined to the vertical that its base touches the fluid, it will never remain in such a position that the base touches the surface in one point only.

Suppose the segment of the paraboloid to be placed in the position described, and let the plane through the axis AM perpendicular to the surface of the fluid intersect the segment of the paraboloid in the parabolic segment BAB' and the plane of the surface of the fluid in BQ.

Take C on AM such that $AC = 2CM$ (or so that C is the centre of gravity of the segment of the paraboloid), and measure CK along CA such that

$$AM : CK = 15 : 4.$$

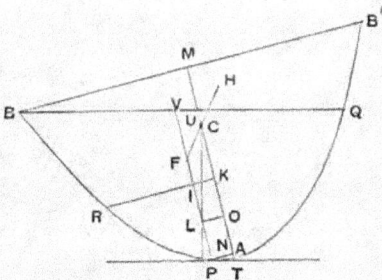

Thus $AM : CK > AM : \frac{1}{2}p$, by hypothesis; therefore $CK < \frac{1}{2}p$.

Measure CO along CA equal to $\frac{1}{2}p$. Also draw KR perpendicular to AC meeting the parabola in R.

Draw the tangent PT parallel to BQ, and through P draw the diameter PV bisecting BQ in V and meeting KR in I.

Then $$PV : PI \underset{\text{or}>}{=} KM : AK,$$
"*for this is proved.*"*

And $$CK = \tfrac{4}{15}AM = \tfrac{2}{5}AC;$$

whence $$AK = AC - CK = \tfrac{3}{5}AC = \tfrac{2}{5}AM.$$

Thus $$KM = \tfrac{2}{5}AM.$$

Therefore $$KM = \tfrac{3}{2}AK.$$

It follows that
$$PV \underset{\text{or}>}{=} \tfrac{3}{2}PI,$$

so that $$PI \underset{\text{or}<}{=} 2IV.$$

Let F be the centre of gravity of the immersed portion of the paraboloid, so that $PF = 2FV$. Produce FC to H, the centre of gravity of the portion above the surface.

Draw OL perpendicular to PV.

---

* We have no hint as to the work in which the proof of this proposition was contained. The following proof is shorter than Robertson's (in the Appendix to Torelli's edition).

Let BQ meet AM in U, and let PN be the ordinate from P to AM.

We have to prove that $PV . AK \underset{\text{or}>}{=} PI . KM$, or in other words that
$$(PV . AK - PI . KM) \text{ is } positive \text{ or } zero.$$

Now
$$PV . AK - PI . KM = AK . PV - (AK - AN)(AM - AK)$$
$$= AK^2 - AK(AM + AN - PV) + AM . AN$$
$$= AK^2 - AK . UM + AM . AN,$$

(since $AN = AT$).

Now $$UM : BM = NT : PN.$$

Therefore $$UM^2 : p . AM = 4AN^2 : p . AN,$$

whence $$UM^2 = 4AM . AN,$$

or $$AM . AN = \frac{UM^2}{4}.$$

Therefore
$$PV . AK - PI . KM = AK^2 - AK . UM + \frac{UM^2}{4}$$
$$= \left(AK - \frac{UM}{2}\right)^2,$$

and accordingly $(PV . AK - PI . KM)$ cannot be negative.

Then, since $CO = \frac{1}{2}p$, CL must be perpendicular to PT and therefore to the surface of the fluid.

And the forces acting on the immersed portion of the paraboloid and the portion above the surface act respectively upwards and downwards along lines through F and H parallel to CL.

Hence the paraboloid cannot remain in the position in which B just touches the surface, but must turn in the direction of increasing the angle PTM.

The proof is the same in the case where the point I is not on VP but on VP produced, as in the second figure*.

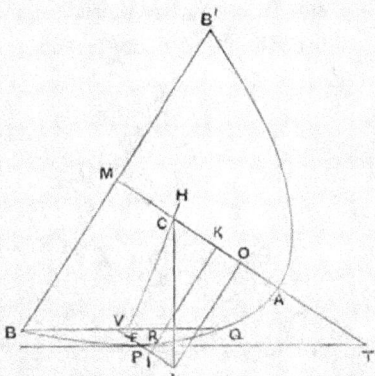

[With the notation used on p. 266, if the base BB' touch the surface of the fluid at B, we have

$$BM = BV \sin \theta + PN,$$

and, by the property of the parabola,

$$BV^2 = (p + 4AN)\,PV$$
$$= pk\,(1 + \cot^2 \theta).$$

Therefore $\sqrt{ph} = \sqrt{pk} + \frac{p}{2} \cot \theta.$

To obtain the result of the proposition, we have to eliminate k between this equation and

$$x' - x = \cos \theta \left\{ \frac{p}{4}(\cot^2 \theta + 2) - \frac{2}{3}(h - k) \right\}.$$

---

* It is curious that the figures given by Torelli, Nizze and Heiberg are all incorrect, as they all make the point which I have called I lie on BQ instead of VP produced.

We have, from the first equation,

$$k = h - \sqrt{ph} \cot \theta + \frac{p}{4} \cot^2 \theta,$$

or
$$h - k = \sqrt{ph} \cot \theta - \frac{p}{4} \cot^2 \theta.$$

Therefore

$$x' - x = \cos \theta \left\{ \frac{p}{4} (\cot^2 \theta + 2) - \frac{2}{3} \left(\sqrt{ph} \cot \theta - \frac{p}{4} \cot^2 \theta \right) \right\}$$

$$= \cos \theta \left\{ \frac{p}{4} (\tfrac{5}{3} \cot^2 \theta + 2) - \tfrac{2}{3} \sqrt{ph} \cot \theta \right\}.$$

If then the solid can never rest in the position described, but must turn in the direction of increasing the angle PTM, the expression within the bracket must be positive whatever be the value of θ.

Therefore $\qquad (\tfrac{2}{3})^2 ph < \tfrac{5}{6} p^2,$

or $\qquad\qquad h < \tfrac{15}{8} p.$]

Proposition 7.

Given a right segment of a paraboloid of revolution lighter than a fluid and such that its axis AM is greater than $\frac{3}{4} p$, but $AM : \frac{1}{2} p < 15 : 4$, if the segment be placed in the fluid so that its base is entirely submerged, it will never rest in such a position that the base touches the surface of the fluid at one point only.

Suppose the solid so placed that one point of the base only (B) touches the surface of the fluid. Let the plane through B and the axis AM cut the solid in the parabolic segment BAB' and the plane of the surface of the fluid in the chord BQ of the parabola.

Let C be the centre of gravity of the segment, so that $AC = 2CM$; and measure CK along CA such that

$$AM : CK = 15 : 4.$$

It follows that $\qquad CK < \tfrac{1}{2} p.$

Measure CO along CA equal to $\frac{1}{2} p$. Draw KR perpendicular to AM meeting the parabola in R.

18—2

Let PT, touching at P, be the tangent to the parabola which is parallel to BQ, and PV the diameter bisecting BQ, i.e. the axis of the portion of the paraboloid above the surface.

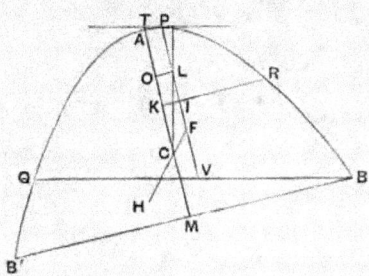

Then, as in the last proposition, we prove that

$$PV \underset{or >}{=} \tfrac{3}{2} PI,$$

and

$$PI \underset{or <}{=} 2IV.$$

Let F be the centre of gravity of the portion of the solid above the surface; join FC and produce it to H, the centre of gravity of the portion submerged.

Draw OL perpendicular to PV; and, as before, since $CO = \tfrac{1}{2} p$, CL is perpendicular to the tangent PT. And the lines through H, F parallel to CL are perpendicular to the surface of the fluid; thus the proposition is established as before.

The proof is the same if the point I is not on VP but on VP produced.

Proposition 8.

Given a solid in the form of a right segment of a paraboloid of revolution whose axis AM is greater than $\tfrac{3}{4} p$, but such that $AM : \tfrac{1}{2} p < 15 : 4$, and whose specific gravity bears to that of a fluid a ratio less than $(AM - \tfrac{3}{4} p)^2 : AM^2$, then, if the solid be placed in the fluid so that its base does not touch the fluid and its axis is inclined at an angle to the vertical, the solid will not return to the position in which its axis is vertical and will not *remain in any position except that in which its axis makes with the surface of the fluid a certain angle to be described.*

Let *am* be taken equal to the axis *AM*, and let *c* be a point on *am* such that $ac = 2cm$. Measure *co* along *ca* equal to $\frac{1}{2}p$, and *or* along *oc* equal to $\frac{1}{4}ao$.

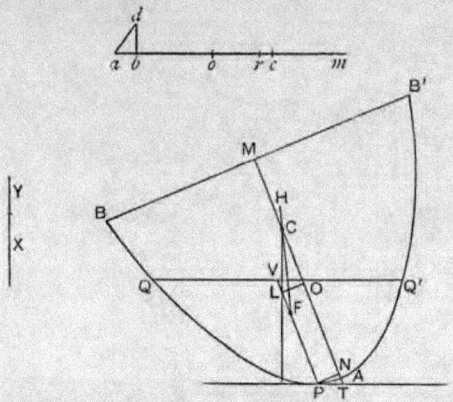

Let $X + Y$ be a straight line such that

$$(\text{spec. gr. of solid}) : (\text{spec. gr. of fluid}) = (X + Y)^2 : am^2 \ldots\ldots(\alpha),$$

and suppose $X = 2Y$.

Now
$$ar = \tfrac{3}{2} ao = \tfrac{3}{2}(\tfrac{2}{3} am - \tfrac{1}{2} p)$$
$$= am - \tfrac{3}{4} p$$
$$= AM - \tfrac{3}{4} p.$$

Therefore, by hypothesis,

$$(X + Y)^2 : am^2 < ar^2 : am^2,$$

whence $(X + Y) < ar$, and therefore $X < ao$.

Measure *ob* along *oa* equal to X, and draw *bd* perpendicular to *ab* and of such length that

$$bd^2 = \tfrac{1}{2} co \cdot ab \ldots\ldots\ldots\ldots\ldots\ldots(\beta).$$

Join *ad*.

Now let the solid be placed in the fluid with its axis *AM* inclined at an angle to the vertical. Through *AM* draw a plane perpendicular to the surface of the fluid, and let this plane cut the paraboloid in the parabola BAB' and the plane of the surface of the fluid in the chord QQ' of the parabola.

Draw the tangent PT parallel to QQ', touching at P, and let PV be the diameter bisecting QQ' in V (or the axis of the immersed portion of the solid), and PN the ordinate from P.

Measure AO along AM equal to ao, and OC along OM equal to oc, and draw OL perpendicular to PV.

I. Suppose the angle OTP greater than the angle dab.

Thus $PN^2 : NT^2 > db^2 : ba^2.$

But $PN^2 : NT^2 = p : 4AN$

$= co : NT,$

and $db^2 : ba^2 = \tfrac{1}{2}co : ab,$ by (β).

Therefore $NT < 2ab,$

or $AN < ab,$

whence $NO > bo$ (since $ao = AO$)

$> X.$

Now $(X + Y)^2 : am^2 = $ (spec. gr. of solid) : (spec. gr. of fluid)

$= $ (portion immersed) : (rest of solid)

$= PV^2 : AM^2,$

so that $X + Y = PV.$

But $PL \, (= NO) > X$

$> \tfrac{2}{3}(X + Y),$ since $X = 2Y,$

$> \tfrac{2}{3}PV,$

or $PV < \tfrac{3}{2}PL,$

and therefore $PL > 2LV.$

Take a point F on PV so that $PF = 2FV$, i.e. so that F is the centre of gravity of the immersed portion of the solid.

Also $AC = ac = \tfrac{2}{3}am = \tfrac{2}{3}AM$, and therefore C is the centre of gravity of the whole solid.

Join FC and produce it to H, the centre of gravity of the portion of the solid above the surface.

Now, since $CO = \frac{1}{2}p$, CL is perpendicular to the surface of the fluid; therefore so are the parallels to CL through F and H. But the force on the immersed portion acts upwards through F and that on the rest of the solid downwards through H.

Therefore the solid will not rest but turn in the direction of diminishing the angle MTP.

II. Suppose the angle OTP less than the angle dab. In this case, we shall have, instead of the above results, the following,

$$AN > ab,$$
$$NO < X.$$

Also $$PV > \tfrac{3}{2}PL,$$
and therefore $$PL < 2LV.$$

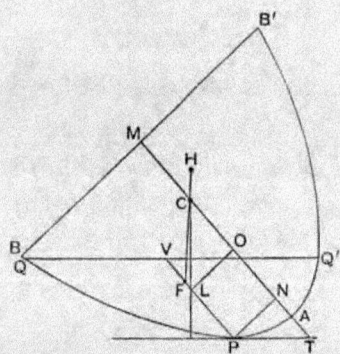

Make PF equal to $2FV$, so that F is the centre of gravity of the immersed portion.

And, proceeding as before, we prove in this case that the solid will turn in the direction of *increasing* the angle MTP.

III. When the angle MTP is equal to the angle dab, equalities replace inequalities in the results obtained, and L is itself the centre of gravity of the immersed portion. Thus all the forces act in one straight line, the perpendicular CL; therefore there is equilibrium, and the solid will rest in the position described.

[With the notation before used

$$x' - x = \cos \theta \left\{ \frac{p}{4} (\cot^2 \theta + 2) - \frac{2}{3} (h - k) \right\},$$

and a position of equilibrium is obtained by equating to zero the expression within the bracket. We have then

$$\frac{p}{4} \cot^2 \theta = \frac{2}{3} (h - k) - \frac{p}{2}.$$

It is easy to verify that the angle θ satisfying this equation is the identical angle determined by Archimedes. For, in the above proposition,

$$\frac{3X}{2} = PV = k,$$

whence
$$ab = \frac{2}{3} h - \frac{p}{2} - \frac{2}{3} k = \frac{2}{3} (h - k) - \frac{p}{2}.$$

Also
$$bd^2 = \frac{p}{4} . ab,$$

It follows that

$$\cot^2 dab = ab^2/bd^2 = \frac{4}{p} \left\{ \frac{2}{3} (h - k) - \frac{p}{2} \right\}.]$$

Proposition 9.

Given a solid in the form of a right segment of a paraboloid of revolution whose axis AM is greater than $\frac{3}{4} p$, but such that $AM : \frac{1}{2} p < 15 : 4$, and whose specific gravity bears to that of a fluid a ratio greater than $[AM^2 - (AM - \frac{3}{4} p)^2] : AM^2$, then, if the solid be placed in the fluid with its axis inclined at an angle to the vertical but so that its base is entirely below the surface, the solid will not return to the position in which its axis is vertical and will not remain in any position except that in which its axis makes with the surface of the fluid an angle equal to that described in the last proposition.

Take am equal to AM, and take c on am such that $ac = 2cm$. Measure co along ca equal to $\frac{1}{2} p$, and ar along ac such that $ar = \frac{3}{2} ao$.

Let $X + Y$ be such a line that

(spec. gr. of solid) : (spec. gr. of fluid) $= \{am^2 - (X + Y)^2\} : am^2$,

and suppose $X = 2Y$.

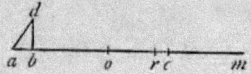

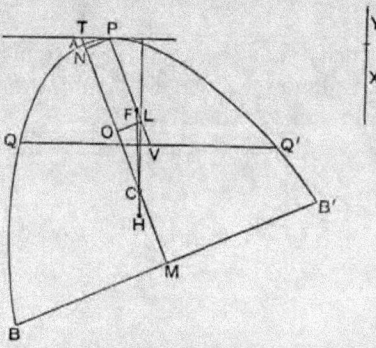

Now
$$ar = \tfrac{3}{2}ao$$
$$= \tfrac{3}{2}(\tfrac{2}{3}am - \tfrac{1}{2}p)$$
$$= AM - \tfrac{3}{4}p.$$

Therefore, by hypothesis,
$$am^2 - ar^2 : am^2 < \{am^2 - (X + Y)^2\} : am^2,$$

whence $\qquad\qquad\qquad X + Y < ar,$

and therefore $\qquad\qquad\quad X < ao.$

Make ob (measured along oa) equal to X, and draw bd perpendicular to ba and of such length that
$$bd^2 = \tfrac{1}{2}co \cdot ab.$$

Join ad.

Now suppose the solid placed as in the figure with its axis AM inclined to the vertical. Let the plane through AM perpendicular to the surface of the fluid cut the solid in the parabola BAB' and the surface of the fluid in QQ'.

Let PT be the tangent parallel to QQ', PV the diameter bisecting QQ' (or the axis of the portion of the paraboloid above the surface), PN the ordinate from P.

I. Suppose the angle MTP greater than the angle dab. Let AM be cut as before in C and O so that $AC = 2CM$, $OC = \frac{1}{2}p$, and accordingly AM, am are equally divided. Draw OL perpendicular to PV.

Then, we have, as in the last proposition,

$$PN^2 : NT^2 > db^2 : ba^2,$$

whence $co : NT > \frac{1}{2}co : ab,$

and therefore $AN < ab.$

It follows that $NO > bo$

$$> X.$$

Again, since the specific gravity of the solid is to that of the fluid as the immersed portion of the solid to the whole,

$$AM^2 - (X + Y)^2 : AM^2 = AM^2 - PV^2 : AM^2,$$

or $(X + Y)^2 : AM^2 = PV^2 : AM^2.$

That is, $X + Y = PV.$

And $PL \text{ (or } NO) > X$

$$> \tfrac{2}{3}PV,$$

so that $PL > 2LV.$

Take F on PV so that $PF = 2FV$. Then F is the centre of gravity of the portion of the solid above the surface.

Also C is the centre of gravity of the whole solid. Join FC and produce it to H, the centre of gravity of the immersed portion.

Then, since $CO = \frac{1}{2}p$, CL is perpendicular to PT and to the surface of the fluid; and the force acting on the immersed portion of the solid acts upwards along the parallel to CL through H, while the weight of the rest of the solid acts downwards along the parallel to CL through F.

Hence the solid will not rest but turn in the direction of diminishing the angle MTP.

II. Exactly as in the last proposition, we prove that, if the angle MTP be less than the angle dab, the solid will not remain in its position but will turn in the direction of increasing the angle MTP.

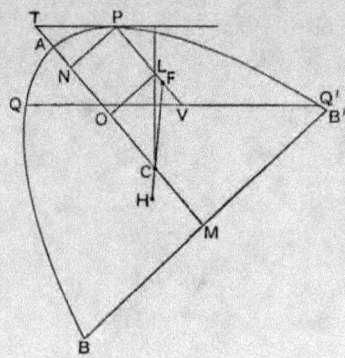

III. If the angle MTP is equal to the angle dab, the solid will rest in that position, because L and F will coincide, and all the forces will act along the one line CL.

Proposition 10.

Given a solid in the form of a right segment of a paraboloid of revolution in which the axis AM is of a length such that $AM : \frac{1}{2}p > 15 : 4$, and supposing the solid placed in a fluid of greater specific gravity so that its base is entirely above the surface of the fluid, to investigate the positions of rest.

(Preliminary.)

Suppose the segment of the paraboloid to be cut by a plane through its axis AM in the parabolic segment BAB_1 of which BB_1 is the base.

Divide AM at C so that $AC = 2CM$, and measure CK along CA so that

$$AM : CK = 15 : 4 \quad\ldots\ldots\ldots\ldots\ldots\ldots(\alpha),$$

whence, by the hypothesis, $CK > \frac{1}{2}p$.

Suppose CO measured along CA equal to $\frac{1}{2}p$, and take a point R on AM such that $MR = \frac{3}{2}CO$.

Thus

$$AR = AM - MR$$
$$= \tfrac{3}{2}(AC - CO)$$
$$= \tfrac{3}{2}AO.$$

Join BA, draw KA_2 perpendicular to AM meeting BA in A_2 bisect BA in A_3, and draw A_2M_2, A_3M_3 parallel to AM meeting BM in M_2, M_3 respectively.

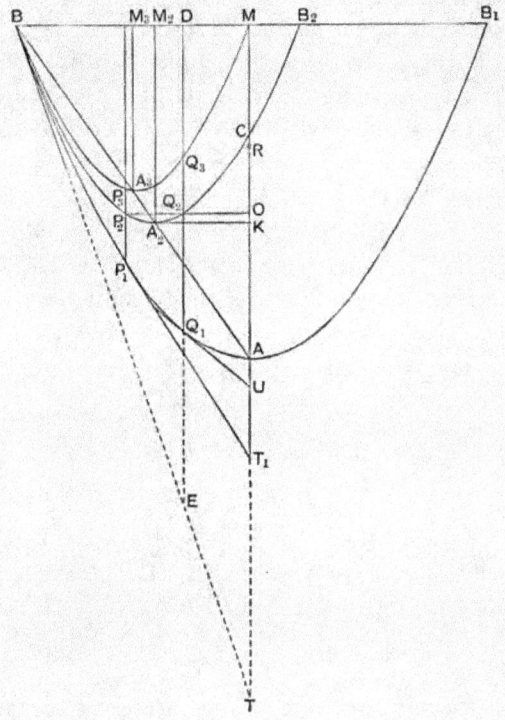

On A_2M_2, A_3M_3 as axes describe parabolic segments similar to the segment BAB_1. (It follows, by similar triangles, that BM will be the base of the segment whose axis is A_3M_3 and BB_2 the base of that whose axis is A_2M_2, where $BB_2 = 2BM_2$.)

The parabola BA_2B_2 will then pass through C.

[For $BM_2 : M_2M = BM_2 : A_2K$

$$= KM : AK$$

$$= CM + CK : AC - CK$$

$$= (\tfrac{1}{3} + \tfrac{4}{15}) AM : (\tfrac{2}{3} - \tfrac{4}{15}) AM$$

$$= 9 : 6\ldots\ldots\ldots\ldots\ldots\ldots\ldots\ldots(\beta)$$

$$= MA : AC.$$

Thus C is seen to be on the parabola BA_2B_2 by the converse of Prop. 4 of the *Quadrature of the Parabola*.]

Also, if a perpendicular to AM be drawn from O, it will meet the parabola BA_2B_2 in two points, as Q_2, P_2. Let $Q_1Q_2Q_3D$ be drawn through Q_2 parallel to AM meeting the parabolas BAB_1, BA_3M respectively in Q_1, Q_3 and BM in D; and let $P_1P_2P_3$ be the corresponding parallel to AM through P_2. Let the tangents to the outer parabola at P_1, Q_1 meet MA produced in T_1, U respectively.

Then, since the three parabolic segments are similar and similarly situated, with their bases in the same straight line and having one common extremity, and since $Q_1Q_2Q_3D$ is a diameter common to all three segments, it follows that

$$Q_1Q_2 : Q_2Q_3 = (B_2B_1 : B_1B).(BM : MB_2)^*.$$

Now $\qquad B_2B_1 : B_1B = MM_2 : BM \qquad$ (dividing by 2)

$$= 2 : 5, \qquad \text{by means of } (\beta) \text{ above.}$$

And $\qquad BM : MB_2 = BM : (2BM_2 - BM)$

$$= 5 : (6 - 5), \qquad \text{by means of } (\beta),$$

$$= 5 : 1.$$

* This result is assumed without proof, no doubt as being an easy deduction from Prop. 5 of the *Quadrature of the Parabola*. It may be established as follows.

First, since AA_2A_3B is a straight line, and $AN=AT$ with the ordinary notation (where PT is the tangent at P and PN the ordinate), it follows, by similar triangles, that the tangent at B to the outer parabola is a tangent to each of the other two parabolas at the same point B.

Now, by the proposition quoted, if $DQ_3Q_2Q_1$ produced meet the tangent BT in E,

$$EQ_3 : Q_3D = BD : DM,$$

whence $\qquad EQ_3 : ED = BD : BM.$ ⎫

Similarly $\qquad EQ_2 : ED = BD : BB_2,$ ⎬

and $\qquad EQ_1 : ED = BD : BB_1.$ ⎭

The first two proportions are equivalent to

$$EQ_3 : ED = BD.BB_2 : BM.BB_2,$$

and $\qquad EQ_2 : ED = BD.BM : BM.BB_2.$

By subtraction,

$$Q_2Q_3 : ED = BD.MB_2 : BM.BB_2.$$

Similarly $\qquad Q_1Q_3 : ED = BD.B_2B_1 : BB_2.BB_1.$

It follows that

$$Q_1Q_2 : Q_2Q_3 = (B_2B_1 : B_1B).(BM : MB_2).$$

It follows that

$$Q_1Q_2 : Q_2Q_3 = 2 : 1,$$

or

$$Q_1Q_2 = 2Q_2Q_3.\ \Big\}$$

Similarly

$$P_1P_2 = 2P_2P_3.\Big\}$$

Also, since

$$MR = \tfrac{3}{2}CO = \tfrac{3}{4}p,$$

$$AR = AM - MR$$

$$= AM - \tfrac{3}{4}p.$$

(Enunciation.)

If the segment of the paraboloid be placed in the fluid with its base entirely above the surface, then

(I.) *if*

(spec. gr. of solid) : (spec. gr. of fluid) $\not< AR^2 : AM^2$

$$[\not< (AM - \tfrac{3}{4}p)^2 : AM^2],$$

the solid will rest in the position in which its axis AM is vertical;

(II.) *if*

(spec. gr. of solid) : (spec. gr. of fluid) $< AR^2 : AM^2$

$$but > Q_1Q_3^2 : AM^2,$$

the solid will not rest with its base touching the surface of the fluid in one point only, but in such a position that its base does not touch the surface at any point and its axis makes with the surface an angle greater than U;

(III. a) *if*

(spec. gr. of solid) : (spec. gr. of fluid) $= Q_1Q_3^2 : AM^2$,

the solid will rest and remain in the position in which the base touches the surface of the fluid at one point only and the axis makes with the surface an angle equal to U;

(III. b) *if*

(spec. gr. of solid) : (spec. gr. of fluid) $= P_1P_3^2 : AM^2$,

the solid will rest with its base touching the surface of the fluid at one point only and with its axis inclined to the surface at an angle equal to T_1;

(IV.) *if*

(spec. gr. of solid) : (spec. gr. of fluid) $> P_1P_3^2 : AM^2$

$$but < Q_1Q_3^2 : AM^2,$$

the solid will rest and remain in a position with its base more submerged;

(V.) *if*

$$(spec.\ gr.\ of\ solid) : (spec.\ gr.\ of\ fluid) < P_1P_3^2 : AM^2,$$

the solid will rest in a position in which its axis is inclined to the surface of the fluid at an angle less than T_1, but so that the base does not even touch the surface at one point.

(Proof.)

(I.)   Since $AM > \frac{3}{4}p$, and

$$(spec.\ gr.\ of\ solid) : (spec.\ gr.\ of\ fluid) \nleq (AM - \tfrac{3}{4}p)^2 : AM^2,$$

it follows, by Prop. 4, that the solid will be in stable equilibrium with its axis vertical.

(II.)   In this case

$$(spec.\ gr.\ of\ solid) : (spec.\ gr.\ of\ fluid) < AR^2 : AM^2$$
$$but > Q_1Q_3^2 : AM^2.$$

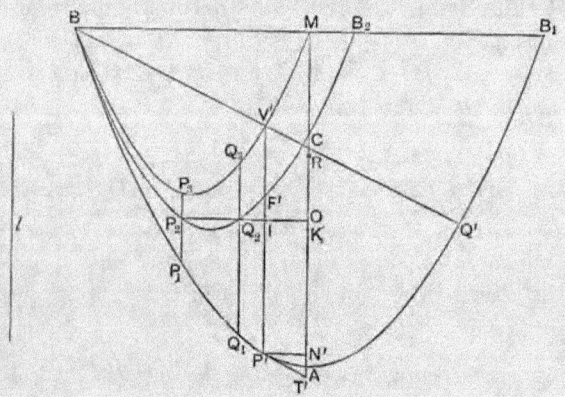

Suppose the ratio of the specific gravities to be equal to

$$l^2 : AM^2,$$

so that $l < AR$ but $> Q_1Q_3$.

Place $P'V'$ between the two parabolas BAB_1, BP_3Q_3M equal to l and parallel to AM^*; and let $P'V'$ meet the intermediate parabola in F'.

Then, by the same proof as before, we obtain

$$P'F' = 2F'V'.$$

Let $P'T'$, the tangent at P' to the outer parabola, meet MA in T', and let $P'N'$ be the ordinate at P'.

Join BV' and produce it to meet the outer parabola in Q'. Let OQ_2P_2 meet $P'V'$ in I.

Now, since, in two similar and similarly situated parabolic

---

* Archimedes does not give the solution of this problem, but it can be supplied as follows.

Let BR_1Q_1, BRQ_2 be two similar and similarly situated parabolic segments with their bases in the same straight line, and let BE be the common tangent at B.

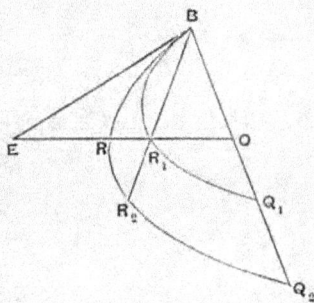

Suppose the problem solved, and let ERR_1O, parallel to the axes, meet the parabolas in R, R_1 and BQ_2 in O, making the intercept RR_1 equal to l.

Then, we have, as usual,

$$ER_1 : EO = BO : BQ_1$$
$$= BO . BQ_2 : BQ_1 . BQ_2,$$

and
$$ER : EO = BO : BQ_2$$
$$= BO . BQ_1 : BQ_1 . BQ_2.$$

By subtraction,
$$RR_1 : EO = BO . Q_1Q_2 : BQ_1 . BQ_2,$$

or
$$BO . OE = l . \frac{BQ_1 . BQ_2}{Q_1Q_2}, \text{ which is known.}$$

And the ratio $BO : OE$ is known. Therefore BO^2, or OE^2, can be found, and therefore O.

segments with bases BM, BB_1 in the same straight line, BV', BQ' are drawn making the same angle with the bases,

$$BV' : BQ' = BM : BB_1*$$
$$= 1 : 2,$$

so that $$BV' = V'Q'.$$

Suppose the segment of the paraboloid placed in the fluid, as described, with its axis inclined at an angle to the vertical, and with its base touching the surface at one point B only. Let the solid be cut by a plane through the axis and per-

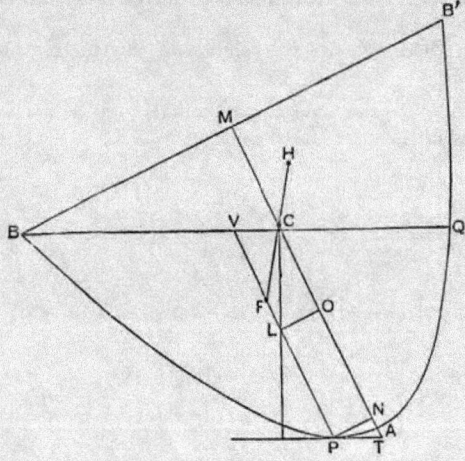

pendicular to the surface of the fluid, and let the plane intersect the solid in the parabolic segment BAB' and the plane of the surface of the fluid in BQ.

Take the points C, O on AM as before described. Draw

* To prove this, suppose that, in the figure on the opposite page, BR_1 is produced to meet the outer parabola in R_2.

We have, as before,
$$ER_1 : EO = BO : BQ_1,$$
$$ER : EO = BO : BQ_2,$$
whence $$ER_1 : ER = BQ_2 : BQ_1.$$
And, since R_1 is a point within the outer parabola,
$$ER : ER_1 = BR_1 : BR_2, \text{ in like manner.}$$
Hence $$BQ_1 : BQ_2 = BR_1 : BR_2.$$

H. A.   19

the tangent parallel to BQ touching the parabola in P and meeting AM in T; and let PV be the diameter bisecting BQ (i.e. the axis of the immersed portion of the solid).

Then

$$l^2 : AM^2 = \text{(spec. gr. of solid)} : \text{(spec. gr. of fluid)}$$
$$= \text{(portion immersed)} : \text{(whole solid)}$$
$$= PV^2 : AM^2,$$

whence $\qquad\qquad P'V' = l = PV.$

Thus the segments in the two figures, namely $BP'Q'$, BPQ, are equal and similar.

Therefore $\qquad \angle PTN = \angle P'T'N'.$

Also $\qquad AT = AT', AN = AN', PN = P'N'.$

Now, in the first figure, $P'I < 2IV'$.

Therefore, if OL be perpendicular to PV in the second figure,

$$PL < 2LV.$$

Take F on LV so that $PF = 2FV$, i.e. so that F is the centre of gravity of the immersed portion of the solid. And C is the centre of gravity of the whole solid. Join FC and produce it to H, the centre of gravity of the portion above the surface.

Now, since $CO = \frac{1}{2}p$, CL is perpendicular to the tangent at P and to the surface of the fluid. Thus, as before, we prove that the solid will not rest with B touching the surface, but will turn in the direction of increasing the angle PTN.

Hence, in the position of rest, the axis AM must make with the surface of the fluid an angle greater than the angle U which the tangent at Q_1 makes with AM.

(III. *a*) In this case

(spec. gr. of solid) : (spec. gr. of fluid) $= Q_1Q_3^2 : AM^2$.

Let the segment of the paraboloid be placed in the fluid so that its base nowhere touches the surface of the fluid, and its axis is inclined at an angle to the vertical.

Let the plane through AM perpendicular to the surface of the fluid cut the paraboloid in the parabola BAB' and the

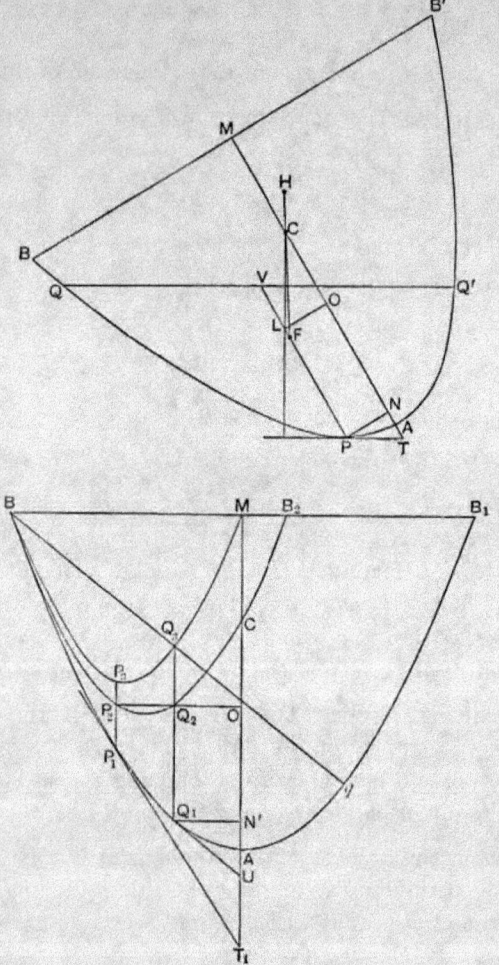

plane of the surface of the fluid in QQ'. Let PT be the tangent parallel to QQ', PV the diameter bisecting QQ', PN the ordinate at P.

Divide AM as before at C, O.

19—2

In the other figure let Q_1N' be the ordinate at Q_1. Join BQ_2 and produce it to meet the outer parabola in q. Then $BQ_3 = Q_3q$, and the tangent Q_1U is parallel to Bq. Now

$$Q_1Q_3^2 : AM^2 = (\text{spec. gr. of solid}) : (\text{spec. gr. of fluid})$$
$$= (\text{portion immersed}) : (\text{whole solid})$$
$$= PV^2 : AM^2.$$

Therefore $Q_1Q_3 = PV$; and the segments QPQ', BQ_1q of the paraboloid are equal in volume. And the base of one passes through B, while the base of the other passes through Q, a point nearer to A than B is.

It follows that the angle between QQ' and BB' is less than the angle B_1Bq.

Therefore $\angle U < \angle PTN,$

whence $AN' > AN,$

and therefore $N'O$ (or Q_1Q_2) $< PL,$

where OL is perpendicular to PV.

It follows, since $Q_1Q_2 = 2Q_2Q_3$, that

$$PL > 2LV.$$

Therefore F, the centre of gravity of the immersed portion of the solid, is between P and L, while, as before, CL is perpendicular to the surface of the fluid.

Producing FC to H, the centre of gravity of the portion of the solid above the surface, we see that the solid must turn in the direction of diminishing the angle PTN until one point B of the base just touches the surface of the fluid.

When this is the case, we shall have a segment BPQ equal and similar to the segment BQ_1q, the angle PTN will be equal to the angle U, and AN will be equal to AN'.

Hence in this case $PL = 2LV$, and F, L coincide, so that F, C, H are all in one vertical straight line.

Thus the paraboloid will remain in the position in which one point B of the base touches the surface of the fluid, and the axis makes with the surface an angle equal to U.

(III. *b*) In the case where

(spec. gr. of solid) : (spec. gr. of fluid) $= P_1P_3{}^2 : AM^2$,

we can prove in the same way that, if the solid be placed in the fluid so that its axis is inclined to the vertical and its base does not anywhere touch the surface of the fluid, the solid will take up and rest in the position in which one point only of the base touches the surface, and the axis is inclined to it at an angle equal to T_1 (in the figure on p. 284).

(IV.) In this case

(spec. gr. of solid) : (spec. gr. of fluid) $> P_1P_3{}^2 : AM^2$

but $< Q_1Q_3{}^2 : AM^2$.

Suppose the ratio to be equal to $l^2 : AM^2$, so that l is greater than P_1P_3 but less than Q_1Q_3.

Place $P'V'$ between the parabolas BP_1Q_1, BP_3Q_3 so that $P'V'$ is equal to l and parallel to AM, and let $P'V'$ meet the intermediate parabola in F' and OQ_2P_2 in I.

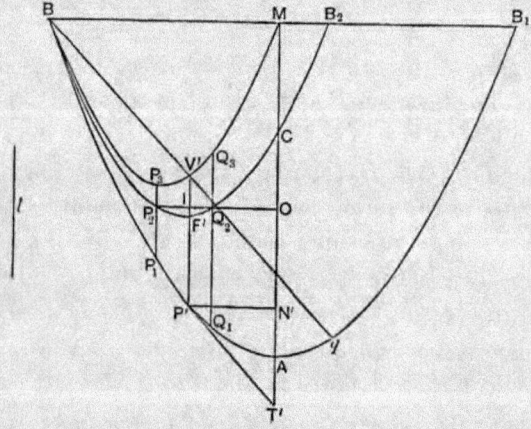

Join BV' and produce it to meet the outer parabola in q.

Then, as before, $BV' = V'q$, and accordingly the tangent $P'T'$ at P' is parallel to Bq. Let $P'N'$ be the ordinate of P'.

1. Now let the segment be placed in the fluid, *first*, with its axis so inclined to the vertical that its base does not anywhere touch the surface of the fluid.

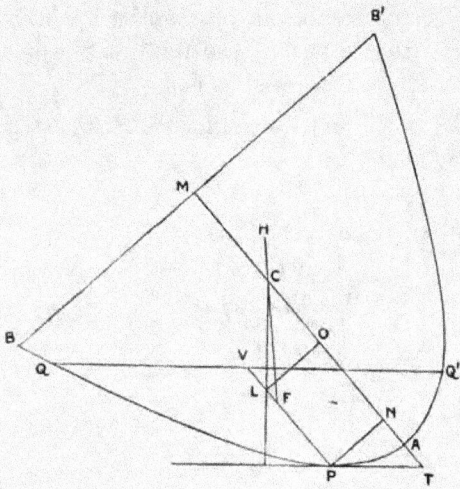

Let the plane through AM perpendicular to the surface of the fluid cut the paraboloid in the parabola BAB' and the plane of the surface of the fluid in QQ'. Let PT be the tangent parallel to QQ', PV the diameter bisecting QQ'. Divide AM at C, O as before, and draw OL perpendicular to PV.

Then, as before, we have $PV = l = P'V'$.

Thus the segments $BP'q$, QPQ' of the paraboloid are equal in volume; and it follows that the angle between QQ' and BB' is less than the angle B_1Bq.

Therefore $\angle P'T'N' < \angle PTN$,

and hence $AN' > AN$,

so that $NO > N'O$,

i.e. $PL > P'I$

$> P'F'$, *a fortiori*.

Thus $PL > 2LV$, so that F, the centre of gravity of the immersed portion of the solid, is between L and P, while CL is perpendicular to the surface of the fluid.

If then we produce FC to H, the centre of gravity of the portion of the solid above the surface, we prove that the solid will not rest but turn in the direction of diminishing the angle PTN.

2. Next let the paraboloid be so placed in the fluid that its base touches the surface of the fluid at one point B only, and let the construction proceed as before.

Then $PV = P'V'$, and the segments BPQ, $BP'q$ are equal and similar, so that

$$\angle PTN = \angle P'T'N'.$$

It follows that $\qquad AN = AN',\ NO = N'O,$

and therefore $\qquad P'I = PL,$

whence $\qquad PL > 2LV.$

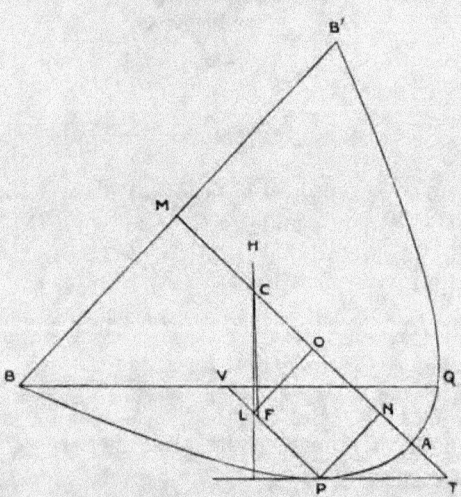

Thus F again lies between P and L, and, as before, the paraboloid will turn in the direction of diminishing the angle PTN, i.e. so that the base will be more submerged.

(V.) In this case

(spec. gr. of solid) : (spec. gr. of fluid) $< P_1P_3^2 : AM^2.$

If then the ratio is equal to $l^2 : AM^2$, $l < P_1P_3$. Place $P'V'$ between the parabolas BP_1Q_1 and BP_3Q_3 equal in length to l

and parallel to AM. Let $P'V'$ meet the intermediate parabola in F' and OP_2 in I.

Join BV' and produce it to meet the outer parabola in q. Then, as before, $BV' = V'q$, and the tangent $P'T'$ is parallel to Bq.

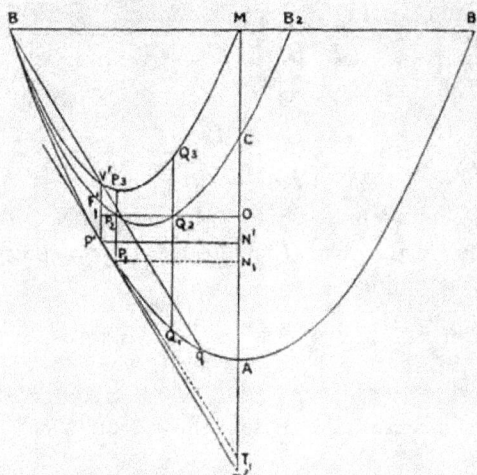

1. Let the paraboloid be so placed in the fluid that its base touches the surface at one point only.

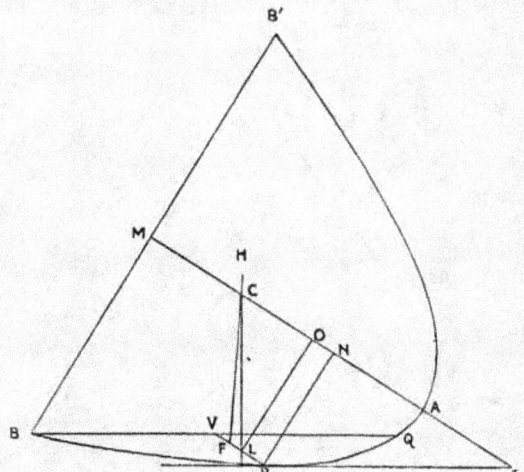

Let the plane through AM perpendicular to the surface of the fluid cut the paraboloid in the parabolic section BAB' and the plane of the surface of the fluid in BQ.

Making the usual construction, we find

$$PV = l = P'V',$$

and the segments BPQ, BP_1q are equal and similar.

Therefore $\angle PTN = \angle P'T'N'$,

and $AN = AN'$, $N'O = NO$.

Therefore $PL = P'I$,

whence it follows that $PL < 2LV$.

Thus F, the centre of gravity of the immersed portion of the solid, lies between L and V, while CL is perpendicular to the surface of the fluid.

Producing FC to H, the centre of gravity of the portion above the surface, we prove, as usual, that there will not be rest, but the solid will turn in the direction of increasing the angle PTN, so that the base will not anywhere touch the surface.

2.   The solid will however rest in a position where its axis makes with the surface of the fluid an angle less than T_1.

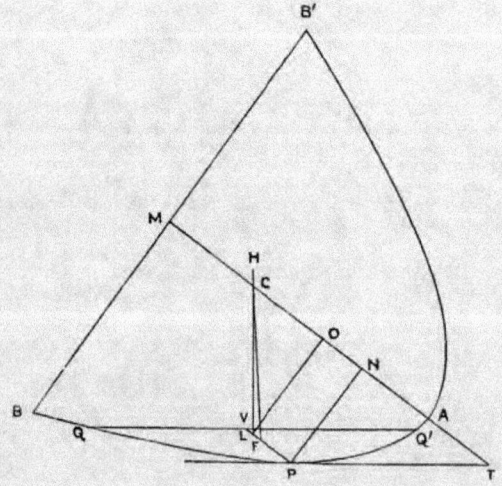

For let it be placed so that the angle PTN is not less than T_1.

Then, with the same construction as before, $PV = l = P'V'$.

And, since $\qquad\qquad \angle T \not< \angle T_1,$

$$AN \not> AN_1,$$

and therefore $NO \not< N_1O$, where P_1N_1 is the ordinate of P_1.

Hence $\qquad\qquad\qquad PL \not< P_1P_2.$

But $\qquad\qquad\qquad\quad P_1P_2 > P'F'.$

Therefore $\qquad\qquad\quad PL > \tfrac{2}{3}PV,$

so that F, the centre of gravity of the immersed portion of the solid, lies between P and L.

Thus the solid will turn in the direction of diminishing the angle PTN until that angle becomes less than T_1.

———

[As before, if x, x' be the distances from T of the orthogonal projections of C, F respectively on TP, we have

$$x' - x = \cos\theta \left\{ \frac{p}{4}(\cot^2\theta + 2) - \frac{2}{3}(h - k) \right\} \dots\dots(1),$$

where $h = AM, k = PV$.

Also, if the base BB' touch the surface of the fluid at one point B, we have further, as in the note following Prop. 6,

$$\sqrt{ph} = \sqrt{pk} + \frac{p}{2}\cot\theta \dots\dots\dots\dots\dots\dots(2),$$

and $\qquad\qquad h - k = \sqrt{ph}\cot\theta - \frac{p}{4}\cot^2\theta \dots\dots\dots\dots(3).$

Therefore, to find the relation between h and the angle θ at which the axis of the paraboloid is inclined to the surface of the fluid in a position of equilibrium with B just touching the surface, we eliminate k and equate the expression in (1) to zero; thus

$$\frac{p}{4}(\cot^2\theta + 2) - \frac{2}{3}\left(\sqrt{ph}\cot\theta - \frac{p}{4}\cot^2\theta\right) = 0,$$

or $\qquad\qquad 5p\cot^2\theta - 8\sqrt{ph}\cot\theta + 6p = 0 \dots\dots\dots\dots(4).$

The two values of θ are given by the equations

$$5\sqrt{p}\cot\theta = 4\sqrt{h} \pm \sqrt{16h - 30p} \quad \ldots\ldots\ldots\ldots \text{(5)}.$$

The lower sign corresponds to the angle U, and the upper sign to the angle T_1, in the proposition of Archimedes, as can be verified thus.

In the first figure of Archimedes (p. 284 above) we have

$$AK = \tfrac{2}{5}h,$$

$$M_2D^2 = \tfrac{3}{5}p \cdot OK = \tfrac{3}{5}p\left(\tfrac{2}{3}h - \tfrac{2}{5}h - \tfrac{1}{2}p\right)$$

$$= \frac{3p}{5}\left(\frac{4h}{15} - \frac{p}{2}\right).$$

If $P_1P_2P_3$ meet BM in D', it follows that

$$\left.\begin{matrix} M_3D \\ M_3D' \end{matrix}\right\} = M_2D \pm M_3M_2$$

$$= \sqrt{\frac{3p}{5}\left(\frac{4h}{15} - \frac{p}{2}\right)} \pm \frac{1}{10}\sqrt{ph},$$

and

$$\left.\begin{matrix} MD \\ MD' \end{matrix}\right\} = MM_2 \mp M_2D$$

$$= \frac{2}{5}\sqrt{ph} \mp \sqrt{\frac{3p}{5}\left(\frac{4h}{15} - \frac{p}{2}\right)}.$$

Now, from the property of the parabola,

$$\cot U = 2MD/p,$$

$$\cot T_1 = 2MD'/p,$$

so that

$$\frac{p}{2}\cot\left\{\begin{matrix} U \\ T_1 \end{matrix}\right\} = \frac{2}{5}\sqrt{ph} \mp \sqrt{\frac{3p}{5}\left(\frac{4h}{15} - \frac{p}{2}\right)},$$

or

$$5\sqrt{p}\cot\left\{\begin{matrix} U \\ T_1 \end{matrix}\right\} = 4\sqrt{h} \mp \sqrt{16h - 30p},$$

which agrees with the result (5) above.

To find the corresponding ratio of the specific gravities, or k^2/h^2, we have to use equations (2) and (5) and to express k in terms of h and p.

Equation (2) gives, on the substitution in it of the value of $\cot \theta$ contained in (5),

$$\sqrt{k} = \sqrt{h} - \tfrac{1}{10}(4\sqrt{h} \pm \sqrt{16h - 30p})$$

$$= \tfrac{3}{5}\sqrt{h} \mp \tfrac{1}{10}\sqrt{16h - 30p},$$

whence we obtain, by squaring,

$$k = \tfrac{13}{25}h - \tfrac{3}{10}p \mp \tfrac{3}{25}\sqrt{h(16h - 30p)} \ \ldots\ldots\ldots (6).$$

The lower sign corresponds to the angle U and the upper to the angle T_1, and, in order to verify the results of Archimedes, we have simply to show that the two values of k are equal to $Q_1 Q_3$, $P_1 P_3$ respectively.

Now it is easily seen that

$$Q_1 Q_3 = h/2 - MD^2/p + 2M_3 D^2/p,$$
$$P_1 P_3 = h/2 - MD'^2/p + 2M_3 D'^2/p.$$

Therefore, using the values of MD, MD', $M_3 D$, $M_3 D'$ above found, we have

$$\left.\begin{array}{c} Q_1 Q_3 \\ P_1 P_3 \end{array}\right\} = \frac{h}{2} + \frac{3}{5}\left(\frac{4h}{15} - \frac{p}{2}\right) - \frac{7h}{50} \pm \frac{6}{5}\sqrt{\frac{3h}{5}\left(\frac{4h}{15} - \frac{p}{2}\right)}$$

$$= \tfrac{13}{25}h - \tfrac{3}{10}p \pm \tfrac{3}{25}\sqrt{h(16h - 30p)},$$

which are the values of k given in (6) above.]

BOOK OF LEMMAS.

Proposition 1.

If two circles touch at A, and if BD, EF be parallel diameters in them, ADF is a straight line.

[The proof in the text only applies to the particular case where the diameters are perpendicular to the radius to the point of contact, but it is easily adapted to the more general case by one small change only.]

Let O, C be the centres of the circles, and let OC be joined and produced to A. Draw DH parallel to AO meeting OF in H.

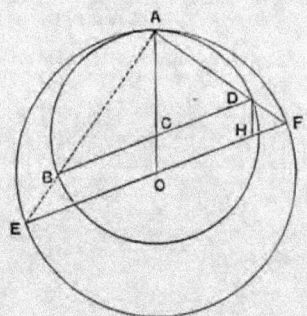

Then, since $$OH = CD = CA.$$
and $$OF = OA,$$
we have, by subtraction,
$$HF = CO = DH.$$
Therefore $$\angle HDF = \angle HFD.$$

Thus both the triangles CAD, HDF are isosceles, and the third angles ACD, DHF in each are equal. Therefore the equal angles in each are equal to one another, and

$$\angle ADC = \angle DFH.$$

Add to each the angle CDF, and it follows that

$$\angle ADC + \angle CDF = \angle CDF + \angle DFH$$

$$= (\text{two right angles}).$$

Hence ADF is a straight line.

The same proof applies if the circles touch externally*.

Proposition 2.

Let AB be the diameter of a semicircle, and let the tangents to it at B and at any other point D on it meet in T. If now DE be drawn perpendicular to AB, and if AT, DE meet in F,

$$DF = FE.$$

Produce AD to meet BT produced in H. Then the angle ADB in the semicircle is right; therefore the angle BDH is also right. And TB, TD are equal.

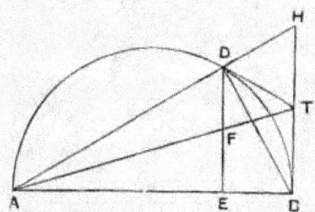

Therefore T is the centre of the semicircle on BH as diameter, which passes through D.

Hence $HT = TB$.

And, since DE, HB are parallel, it follows that $DF = FE$.

---

* Pappus assumes the result of this proposition in connexion with the ἄρβηλος (p. 214, ed. Hultsch), and he proves it for the case where the circles touch externally (p. 840).

Proposition 3.

Let P be any point on a segment of a circle whose base is AB, and let PN be perpendicular to AB. Take D on AB so that AN = ND. If now PQ be an arc equal to the arc PA, and BQ be joined,

BQ, BD shall be equal*.

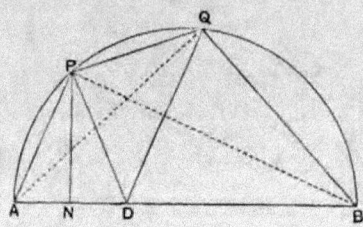

Join *PA, PQ, PD, DQ.*

* The segment in the figure of the ms. appears to have been a semicircle, though the proposition is equally true of any segment. But the case where the segment is a semicircle brings the proposition into close connexion with a proposition in Ptolemy's μεγάλη σύνταξις, I. 9 (p. 31, ed. Halma; cf. the reproduction in Cantor's *Gesch. d. Mathematik*, I. (1894), p. 389). Ptolemy's object is to connect by an equation the lengths of the chord of an arc and the chord of half the arc. Substantially his procedure is as follows. Suppose *AP, PQ* to be equal arcs, *AB* the diameter through *A*; and let *AP, PQ, AQ, PB, QB* be joined. Measure *BD* along *BA* equal to *BQ*. The perpendicular *PN* is now drawn, and it is proved that *PA = PD*, and *AN = ND*.

Then $AN = \frac{1}{2}(BA - BD) = \frac{1}{2}(BA - BQ) = \frac{1}{2}(BA - \sqrt{BA^2 - AQ^2})$.

And, by similar triangles, $AN : AP = AP : AB$.

Therefore
$$AP^2 = AB \cdot AN$$
$$= \frac{1}{2}(AB - \sqrt{AB^2 - AQ^2}) \cdot AB.$$

This gives *AP* in terms of *AQ* and the known diameter *AB*. If we divide by AB^2 throughout, it is seen at once that the proposition gives a geometrical proof of the formula

$$\sin^2 \frac{a}{2} = \frac{1}{2}(1 - \cos a).$$

The case where the segment is a semicircle recalls also the method used by Archimedes at the beginning of the second part of Prop. 3 of the *Measurement of a circle*. It is there proved that, in the figure above,

$$AB + BQ : AQ = BP : PA,$$

or, if we divide the first two terms of the proposition by *AB*,

$$(1 + \cos a)/\sin a = \cot \frac{a}{2}.$$

Then, since the arcs PA, PQ are equal,

$$PA = PQ.$$

But, since $AN = ND$, and the angles at N are right,

$$PA = PD.$$

Therefore $PQ = PD$,

and $\angle PQD = \angle PDQ.$

Now, since A, P, Q, B are concyclic,

$$\angle PAD + \angle PQB = \text{(two right angles)},$$

whence $\angle PDA + \angle PQB = \text{(two right angles)}$

$$= \angle PDA + \angle PDB.$$

Therefore $\angle PQB = \angle PDB$;

and, since the parts, the angles PQD, PDQ, are equal,

$$\angle BQD = \angle BDQ,$$

and $BQ = BD.$

Proposition 4.

If AB be the diameter of a semicircle and N any point on AB, and if semicircles be described within the first semicircle and having AN, BN as diameters respectively, the figure included between the circumferences of the three semicircles is "what Archimedes called an ἄρβηλος* "; and its area is equal to the circle on PN as diameter, where PN is perpendicular to AB and meets the original semicircle in P.

For $AB^2 = AN^2 + NB^2 + 2AN \cdot NB$

$$= AN^2 + NB^2 + 2PN^2.$$

But circles (or semicircles) are to one another as the squares of their radii (or diameters).

---

* ἄρβηλος is literally 'a shoemaker's knife.' Cf. note attached to the remarks on the *Liber Assumptorum* in the Introduction, Chapter II.

Hence

(semicircle on AB) = (sum of semicircles on AN, NB)

+ 2 (semicircle on PN).

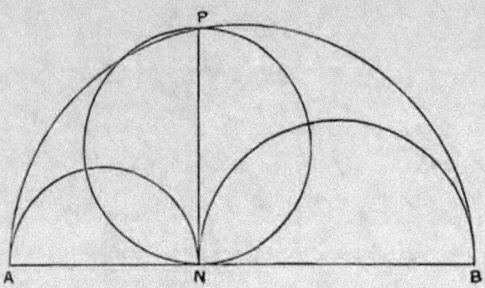

That is, the circle on PN as diameter is equal to the difference between the semicircle on AB and the sum of the semicircles on AN, NB, i.e. is equal to the area of the ἄρβηλος.

Proposition 5.

Let AB be the diameter of a semicircle, C any point on AB, and CD perpendicular to it, and let semicircles be described within the first semicircle and having AC, CB as diameters. Then, if two circles be drawn touching CD on different sides and each touching two of the semicircles, the circles so drawn will be equal.

Let one of the circles touch CD at E, the semicircle on AB in F, and the semicircle on AC in G.

Draw the diameter EH of the circle, which will accordingly be perpendicular to CD and therefore parallel to AB.

Join FH, HA, and FE, EB. Then, by Prop. 1, FHA, FEB are both straight lines, since EH, AB are parallel.

For the same reason AGE, CGH are straight lines.

Let AF produced meet CD in D, and let AE produced meet the outer semicircle in I. Join BI, ID.

Then, since the angles AFB, ACD are right, the straight lines AD, AB are such that the perpendiculars on each from the extremity of the other meet in the point E. Therefore, by the properties of triangles, AE is perpendicular to the line joining B to D.

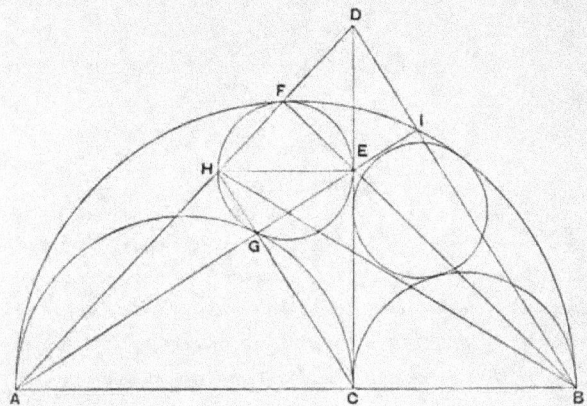

But AE is perpendicular to BI.

Therefore BID is a straight line.

Now, since the angles at G, I are right, CH is parallel to BD.

Therefore
$$AB : BC = AD : DH$$
$$= AC : HE,$$

so that
$$AC \cdot CB = AB \cdot HE.$$

In like manner, if d is the diameter of the other circle, we can prove that
$$AC \cdot CB = AB \cdot d.$$

Therefore $d = HE$, and the circles are equal*.

* The property upon which this result depends, viz. that
$$AB : BC = AC : HE,$$
appears as an intermediate step in a proposition of Pappus (p. 230, ed. Hultsch) which proves that, in the figure above,
$$AB : BC = CE^2 : HE^2.$$

The truth of the latter proposition is easily seen. For, since the angle CEH is a right angle, and EG is perpendicular to CH,
$$CE^2 : EH^2 = CG : GH$$
$$= AC : HE.$$

[As pointed out by an Arabian Scholiast Alkauhi, this proposition may be stated more generally. If, instead of one point C on AB, we have two points C, D, and semicircles be described on AC, BD as diameters, and if, instead of the perpendicular to AB through C, we take the radical axis of the two semicircles, then the circles described on different sides of the radical axis and each touching it as well as two of the semicircles are equal. The proof is similar and presents no difficulty.]

Proposition 6.

Let AB, the diameter of a semicircle, be divided at C so that $AC = \frac{3}{2} CB$ [or in any ratio]. Describe semicircles within the first semicircle and on AC, CB as diameters, and suppose a circle drawn touching all three semicircles. If GH be the diameter of this circle, to find the relation between GH and AB.

Let GH be that diameter of the circle which is parallel to AB, and let the circle touch the semicircles on AB, AC, CB in D, E, F respectively.

Join AG, GD and BH, HD. Then, by Prop. 1, AGD, BHD are straight lines.

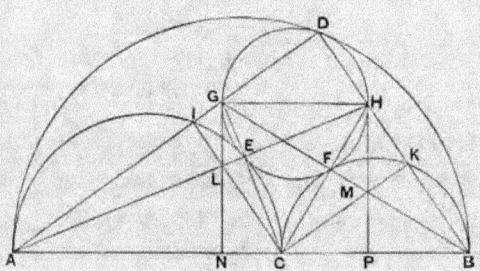

For a like reason AEH, BFG are straight lines, as also are CEG, CFH.

Let AD meet the semicircle on AC in I, and let BD meet the semicircle on CB in K. Join CI, CK meeting AE, BF

20—2

respectively in L, M, and let GL, HM produced meet AB in N, P respectively.

Now, in the triangle AGC, the perpendiculars from A, C on the opposite sides meet in L. Therefore, by the properties of triangles, GLN is perpendicular to AC.

Similarly HMP is perpendicular to CB.

Again, since the angles at I, K, D are right, CK is parallel to AD, and CI to BD.

Therefore $\qquad AC : CB = AL : LH$
$$= AN : NP,$$

and $\qquad\qquad BC : CA = BM : MG$
$$= BP : PN.$$

Hence $\qquad\qquad AN : NP = NP : PB,$

or AN, NP, PB are in continued proportion*.

Now, in the case where $AC = \frac{3}{2} CB$,
$$AN = \frac{3}{2} NP = \frac{9}{4} PB,$$
whence $\quad BP : PN : NA : AB = 4 : 6 : 9 : 19.$

Therefore $\qquad GH = NP = \frac{6}{19} AB.$

And similarly GH can be found when $AC : CB$ is equal to any other given ratio†.

* This same property appears incidentally in Pappus (p. 226) as an inter-mediate step in the proof of the "ancient proposition" alluded to below.

† In general, if $AC : CB = \lambda : 1$, we have
$$BP : PN : NA : AB = 1 : \lambda : \lambda^2 : (1 + \lambda + \lambda^2),$$
and $\qquad\qquad GH : AB = \lambda : (1 + \lambda + \lambda^2).$

It may be interesting to add the enunciation of the "ancient proposition" stated by Pappus (p. 208) and proved by him after several auxiliary lemmas.

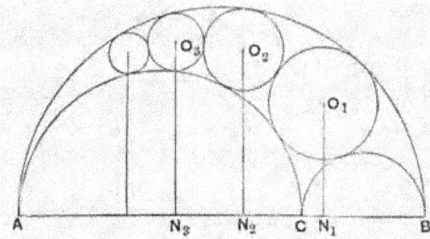

Proposition 7.

If circles be circumscribed about and inscribed in a square, the circumscribed circle is double of the inscribed circle.

For the ratio of the circumscribed to the inscribed circle is equal to that of the square on the diagonal to the square itself, i.e. to the ratio 2 : 1.

Proposition 8.

If AB be any chord of a circle whose centre is O, and if AB be produced to C so that BC is equal to the radius; if further CO meet the circle in D and be produced to meet the circle a second time in E, the arc AE will be equal to three times the arc BD.

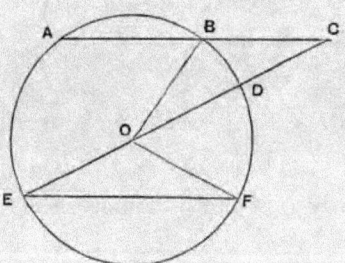

Draw the chord EF parallel to AB, and join OB, OF.

Let an ἄρβηλος be formed by three semicircles on AB, AC, CB as diameters, and let a series of circles be described, the first of which touches all three semicircles, while the second touches the first and two of the semicircles forming one end of the ἄρβηλος, the third touches the second and the same two semicircles, and so on. Let the diameters of the successive circles be d_1, d_2, d_3,... their centres O_1, O_2, O_3,... and O_1N_1, O_2N_2, O_3N_3,... the perpendiculars from the centres on AB. Then it is to be proved that

$$O_1N_1 = d_1,$$
$$O_2N_2 = 2d_2,$$
$$O_3N_3 = 3d_3,$$
$$\dotsc\dotsc\dotsc$$
$$O_nN_n = nd_n.$$

Then, since the angles OEF, OFE are equal,

$$\angle COF = 2 \angle OEF$$

$$= 2 \angle BCO, \text{ by parallels,}$$

$$= 2 \angle BOD, \text{ since } BC = BO.$$

Therefore

$$\angle BOF = 3 \angle BOD,$$

so that the arc BF is equal to three times the arc BD.

Hence the arc AE, which is equal to the arc BF, is equal to three times the arc BD*.

Proposition 9.

If in a circle two chords AB, CD which do not pass through the centre intersect at right angles, then

$$(\text{arc } AD) + (\text{arc } CB) = (\text{arc } AC) + (\text{arc } DB).$$

Let the chords intersect at O, and draw the diameter EF parallel to AB intersecting CD in H. EF will thus bisect CD at right angles in H, and

$$(\text{arc } ED) = (\text{arc } EC).$$

Also EDF, ECF are semicircles, while

$$(\text{arc } ED) = (\text{arc } EA) + (\text{arc } AD).$$

Therefore

(sum of arcs CF, EA, AD) = (arc of a semicircle).

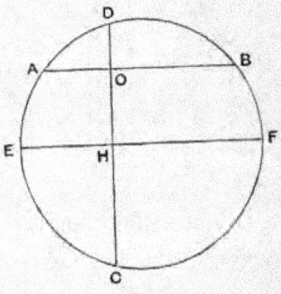

And the arcs AE, BF are equal.

Therefore

$$(\text{arc } CB) + (\text{arc } AD) = (\text{arc of a semicircle}).$$

---

* This proposition gives a method of reducing the trisection of any angle, i.e. of any circular arc, to a problem of the kind known as νεύσεις. Suppose that AE is the arc to be trisected, and that ED is the diameter through E of the circle of which AE is an arc. In order then to find an arc equal to one-third of AE, we have only *to draw through A a line ABC, meeting the circle again in B and ED produced in C, such that BC is equal to the radius of the circle.* For a discussion of this and other νεύσεις see the Introduction, Chapter V.

Hence the remainder of the circumference, the sum of the arcs AC, DB, is also equal to a semicircle; and the proposition is proved.

Proposition 10.

Suppose that TA, TB are two tangents to a circle, while TC cuts it. Let BD be the chord through B parallel to TC, and let AD meet TC in E. Then, if EH be drawn perpendicular to BD, it will bisect it in H.

Let AB meet TC in F, and join BE.

Now the angle TAB is equal to the angle in the alternate segment, i.e.

$$\angle TAB = \angle ADB$$
$$= \angle AET, \text{ by parallels.}$$

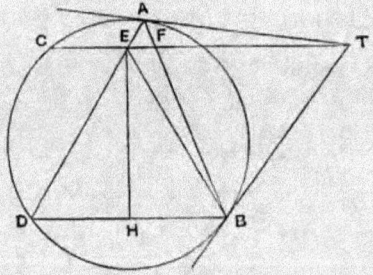

Hence the triangles EAT, AFT have one angle equal and another (at T) common. They are therefore similar, and

$$FT : AT = AT : ET.$$

Therefore

$$ET \cdot TF = TA^2$$
$$= TB^2.$$

It follows that the triangles EBT, BFT are similar.

Therefore
$$\angle TEB = \angle TBF$$
$$= \angle TAB.$$

But the angle TEB is equal to the angle EBD, and the angle TAB was proved equal to the angle EDB.

Therefore $\angle EDB = \angle EBD$.

And the angles at H are right angles.

It follows that $BH = HD*$.

Proposition 11.

If two chords AB, CD in a circle intersect at right angles in a point O, not being the centre, then

$$AO^2 + BO^2 + CO^2 + DO^2 = (diameter)^2.$$

Draw the diameter CE, and join AC, CB, AD, BE.

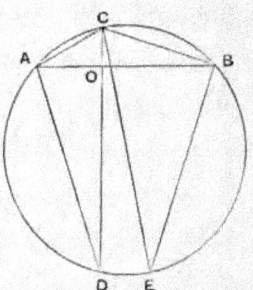

Then the angle CAO is equal to the angle CEB in the same segment, and the angles AOC, EBC are right; therefore the triangles AOC, EBC are similar, and

$$\angle ACO = \angle ECB.$$

It follows that the subtended arcs, and therefore the chords AD, BE, are equal.

---

* The figure of this proposition curiously recalls the figure of a problem given by Pappus (pp. 836–8) among his lemmas to the first Book of the treatise of Apollonius *On Contacts* (περὶ ἐπαφῶν). The problem is, *Given a circle and two points E, F* (neither of which is necessarily, as in this case, the middle point of the chord of the circle drawn through *E, F*), *to draw through E, F respectively two chords AD, AB having a common extremity A and such that DB is parallel to EF.* The analysis is as follows. Suppose the problem solved, *BD* being parallel to *FE*. Let *BT*, the tangent at *B*, meet *EF* produced in *T*. (*T* is not in general the pole of *AB*, so that *TA* is not generally the tangent at *A*.)

Then $\angle TBF = \angle BDA$, in the alternate segment,

$$= \angle AET, \text{ by parallels.}$$

Therefore A, E, B, T are concyclic, and

$$EF . FT = AF . FB.$$

But, the circle ADB and the point F being given, the rectangle $AF . FB$ is given. Also EF is given.

Hence FT is known.

Thus, to make the construction, we have only to find the length of FT from the data, produce EF to T so that FT has the ascertained length, draw the tangent TB, and then draw BD parallel to EF. DE, BF will then meet in A on the circle and will be the chords required.

Thus

$$(AO^2 + DO^2) + (BO^2 + CO^2) = AD^2 + BC^2$$
$$= BE^2 + BC^2$$
$$= CE^2.$$

Proposition 12.

If AB be the diameter of a semicircle, and TP, TQ the tangents to it from any point T, and if AQ, BP be joined meeting in R, then TR is perpendicular to AB.

Let TR produced meet AB in M, and join PA, QB.

Since the angle APB is right,

$$\angle PAB + \angle PBA = \text{(a right angle)}$$
$$= \angle AQB.$$

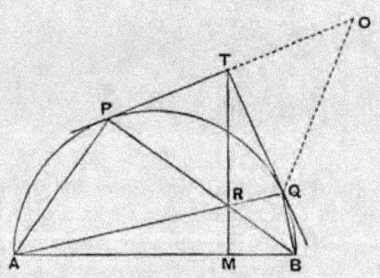

Add to each side the angle RBQ, and

$$\angle PAB + \angle QBA = \text{(exterior)} \angle PRQ.$$

But $\angle TPR = \angle PAB$, and $\angle TQR = \angle QBA$,

in the alternate segments;

therefore $\angle TPR + \angle TQR = \angle PRQ.$

It follows from this that $TP = TQ = TR$.

[For, if PT be produced to O so that $TO = TQ$, we have

$$\angle TOQ = \angle TQO.$$

And, by hypothesis, $\angle PRQ = \angle TPR + TQR.$

By addition, $\angle POQ + \angle PRQ = \angle TPR + OQR.$

It follows that, in the quadrilateral $OPRQ$, the opposite angles are together equal to two right angles. Therefore a circle will go round $OPQR$, and T is its centre, because $TP = TO = TQ$.   Therefore $TR = TP$.]

Thus $\angle TRP = \angle TPR = \angle PAM$.

Adding to each the angle PRM,

$$\angle PAM + \angle PRM = \angle TRP + \angle PRM$$

$$= \text{(two right angles)}.$$

Therefore $\angle APR + \angle AMR = \text{(two right angles)}$,

whence $\angle AMR = \text{(a right angle)}$*.

Proposition 13.

If a diameter AB of a circle meet any chord CD, not a diameter, in E, and if AM, BN be drawn perpendicular to CD, then

$$CN = DM\dagger.$$

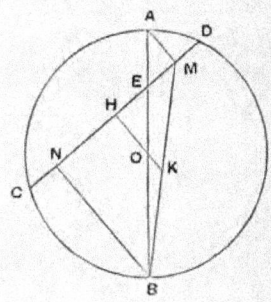

Let O be the centre of the circle, and OH perpendicular to CD. Join BM, and produce HO to meet BM in K.

Then $CH = HD$.

And, by parallels,

since $BO = OA$,

$$BK = KM.$$

Therefore $NH = HM$.

Accordingly $CN = DM$.

---

* TM is of course the polar of the intersection of PQ, AB, as it is the line joining the poles of PQ, AB respectively.

† This proposition is of course true whether M, N lie on CD or on CD produced each way. Pappus proves it for the latter case in his first lemma (p. 788) to the second Book of Apollonius' νεύσεις.

Proposition 14.

Let ACB be a semicircle on AB as diameter, and let AD, BE be equal lengths measured along AB from A, B respectively. On AD, BE as diameters describe semicircles on the side towards C, and on DE as diameter a semicircle on the opposite side. Let the perpendicular to AB through O, the centre of the first semicircle, meet the opposite semicircles in C, F respectively.

Then shall the area of the figure bounded by the circumferences of all the semicircles ("which Archimedes calls 'Salinon'"*) be equal to the area of the circle on CF as diameter†.

By Eucl. II. 10, since ED is bisected at O and produced to A,
$$EA^2 + AD^2 = 2(EO^2 + OA^2),$$
and $$CF = OA + OE = EA.$$

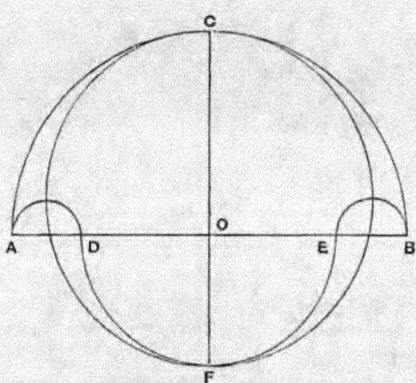

* For the explanation of this name see note attached to the remarks on the *Liber Assumptorum* in the Introduction, Chapter II. On the grounds there given at length I believe σάλινον to be simply a Graecised form of the Latin word *salinum*, 'salt-cellar.'

† Cantor (*Gesch. d. Mathematik*, 1. p. 285) compares this proposition with Hippocrates' attempt to square the circle by means of lunes, but points out that the object of Archimedes may have been the converse of that of Hippocrates. For, whereas Hippocrates wished to find the area of a circle from that of other figures of the same sort, Archimedes' intention was possibly to equate the area of figures bounded by different curves to that of a circle regarded as already known.

Therefore

$$AB^2 + DE^2 = 4\left(EO^2 + OA^2\right) = 2\left(CF^2 + AD^2\right).$$

But circles (and therefore semicircles) are to one another as the squares on their radii (or diameters).

Therefore

(sum of semicircles on AB, DE)

$= $ (circle on CF) $+$ (sum of semicircles on AD, BE).

Therefore

(area of 'salinon') $=$ (area of circle on CF as diam.).

Proposition 15.

Let AB be the diameter of a circle, AC a side of an inscribed regular pentagon, D the middle point of the arc AC. Join CD and produce it to meet BA produced in E; join AC, DB meeting in F, and draw FM perpendicular to AB. Then

$$EM = (radius\ of\ circle)^*.$$

Let O be the centre of the circle, and join DA, DM, DO, CB.

Now　　　　　　　　　$\angle ABC = \frac{2}{5}$ (right angle),

and　　　　$\angle ABD = \angle DBC = \frac{1}{5}$ (right angle),

whence　　　　　　　$\angle AOD = \frac{2}{5}$ (right angle).

---

* Pappus gives (p. 418) a proposition almost identical with this among the lemmas required for the comparison of the five regular polyhedra. His enunciation is substantially as follows. If DH be half the side of a pentagon inscribed in a circle, while DH is perpendicular to the radius OHA, and if HM be made equal to AH, then OA is divided at M in extreme and mean ratio, OM being the greater segment.

In the course of the proof it is first shown that AD, DM, MO are all equal, as in the proposition above.

Then, the triangles ODA, DAM being similar,

$$OA : AD = AD : AM,$$

or (since $AD = OM$)　　　$OA : OM = OM : MA.$

Further, the triangles FCB, FMB are equal in all respects.

Therefore, in the triangles DCB, DMB, the sides CB, MB being equal and BD common, while the angles CBD, MBD are equal,

$$\angle BCD = \angle BMD = \tfrac{6}{5}\,(\text{right angle}).$$

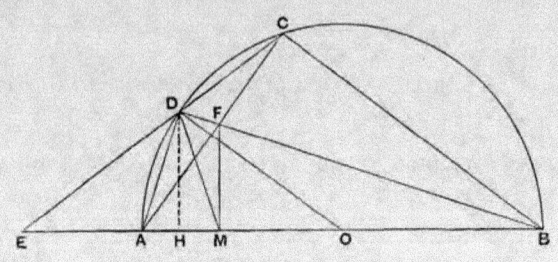

But $\angle BCD + \angle BAD = (\text{two right angles})$

$$= \angle BAD + \angle DAE$$

$$= \angle BMD + \angle DMA,$$

so that $\angle DAE = \angle BCD,$

and $\angle BAD = \angle AMD.$

Therefore $AD = MD.$

Now, in the triangle DMO,

$$\angle MOD = \tfrac{2}{5}\,(\text{right angle}),$$

$$\angle DMO = \tfrac{6}{5}\,(\text{right angle}).$$

Therefore $\angle ODM = \tfrac{2}{5}\,(\text{right angle}) = AOD;$

whence $OM = MD.$

Again $\angle EDA = (\text{supplement of } ADC)$

$$= \angle CBA$$

$$= \tfrac{2}{5}\,(\text{right angle})$$

$$= \angle ODM.$$

Therefore, in the triangles EDA, ODM,

$$\angle EDA = \angle ODM,$$

$$\angle EAD = \angle OMD,$$

and the sides AD, MD are equal.

Hence the triangles are equal in all respects, and

$$EA = MO.$$

Therefore $EM = AO.$

Moreover $DE = DO$; and it follows that, since DE is equal to the side of an inscribed hexagon, and DC is the side of an inscribed decagon, EC is divided at D in extreme and mean ratio [i.e. $EC : ED = ED : DC$]; "and this is proved in the book of the Elements." [Eucl. XIII. 9, "If the side of the hexagon and the side of the decagon inscribed in the same circle be put together, the whole straight line is divided in extreme and mean ratio, and the greater segment is the side of the hexagon."]

THE CATTLE-PROBLEM.

It is required to find the number of bulls and cows of each of four colours, or to find 8 unknown quantities. The first part of the problem connects the unknowns by seven simple equations; and the second part adds two more conditions to which the unknowns must be subject.

Let W, w be the numbers of white bulls and cows respectively,

X, x	„	„	black	„	„	„
Y, y	„	„	yellow	„	„	„
Z, z	„	„	dappled	„	„	„

First part.

(I)
$$W = (\tfrac{1}{2} + \tfrac{1}{3}) X + Y \dots\dots\dots\dots(\alpha),$$
$$X = (\tfrac{1}{4} + \tfrac{1}{5}) Z + Y \dots\dots\dots\dots(\beta),$$
$$Z = (\tfrac{1}{6} + \tfrac{1}{7}) W + Y \dots\dots\dots\dots(\gamma),$$

(II)
$$w = (\tfrac{1}{3} + \tfrac{1}{4})(X + x) \dots\dots\dots\dots(\delta),$$
$$x = (\tfrac{1}{4} + \tfrac{1}{5})(Z + z) \dots\dots\dots\dots(\epsilon),$$
$$z = (\tfrac{1}{5} + \tfrac{1}{6})(Y + y) \dots\dots\dots\dots(\zeta),$$
$$y = (\tfrac{1}{6} + \tfrac{1}{7})(W + w) \dots\dots\dots\dots(\eta).$$

Second part.
$$W + X = \text{a square} \dots\dots\dots\dots\dots(\theta),$$
$$Y + Z = \text{a triangular number} \dots\dots\dots(\iota).$$

[There is an ambiguity in the language which expresses the condition (θ). Literally the lines mean "When the white bulls joined in number with the black, they stood firm ($\check{\epsilon}\mu\pi\epsilon\delta\text{o}\nu$) with depth and breadth of equal measurement ($\iota\sigma\acute{o}\mu\epsilon\tau\rho\text{o}\iota$ $\epsilon\grave{\iota}\varsigma$ $\beta\acute{a}\theta\text{o}\varsigma$ $\epsilon\grave{\iota}\varsigma$ $\epsilon\mathring{\nu}\rho\acute{o}\varsigma$ $\tau\epsilon$); and the plains of Thrinakia, far-stretching all ways, were filled with their multitude" (reading, with Krumbiegel, $\pi\lambda\acute{\eta}\theta\text{o}\nu\varsigma$ instead of $\pi\lambda\acute{\iota}\nu\theta\text{o}\nu$). Considering that, if the bulls were packed together so as to form a square *figure*, the number of them need not be a square *number*, since a bull is longer than it is broad, it is clear that one possible interpretation would be to take the 'square' to be a square *figure*, and to understand condition (θ) to be simply

$W + X =$ a rectangle (i.e. a product of two factors).

The problem may therefore be stated in two forms:

(1) the simpler one in which, for the condition (θ), there is substituted the mere requirement that

$W + X =$ a product of two whole numbers;

(2) the complete problem in which all the conditions have to be satisfied including the requirement (θ) that

$W + X =$ a square number.

The simpler problem was solved by Jul. Fr. Wurm and may be called

Wurm's Problem.

The solution of this is given (together with a discussion of the complete problem) by Amthor in the *Zeitschrift für Math. u. Physik* (*Hist. litt. Abtheilung*), XXV. (1880), p. 156 sqq.

Multiply (α) by 336, (β) by 280, (γ) by 126, and add; thus

$$297\,W = 742\,Y, \text{ or } 3^3 . 11\,W = 2 . 7 . 53\,Y \ldots\ldots(\alpha').$$

Then from (γ) and (β) we obtain

$$891\,Z = 1580\,Y, \text{ or } 3^4 . 11\,Z = 2^2 . 5 . 79\,Y \ldots\ldots(\beta'),$$

and $\qquad 99\,X = 178\,Y, \text{ or } 3^2 . 11\,X = 2 . 89\,Y \ldots\ldots\ldots(\gamma').$

Again, if we multiply (δ) by 4800, (ϵ) by 2800, (ζ) by 1260, (η) by 462, and add, we obtain

$$4657\,w = 2800\,X + 1260\,Z + 462\,Y + 143\,W;$$

and, by means of the values in (α'), (β'), (γ'), we derive

$$297 \cdot 4657w = 2402120 \, Y,$$

or $$3^3 \cdot 11 \cdot 4657w = 2^3 \cdot 5 \cdot 7 \cdot 23 \cdot 373 \, Y \quad \ldots\ldots(\delta').$$

Hence, by means of (η), (ζ), (ϵ), we have

$$3^2 \cdot 11 \cdot 4657y = 13 \cdot 46489 \, Y \ldots\ldots\ldots\ldots\ldots(\epsilon'),$$

$$3^3 \cdot 4657z = 2^2 \cdot 5 \cdot 7 \cdot 761 \, Y \ldots\ldots\ldots\ldots(\zeta'),$$

and $$3^2 \cdot 11 \cdot 4657x = 2 \cdot 17 \cdot 15991 \, Y \ldots\ldots\ldots\ldots(\eta').$$

And, since all the unknowns must be whole numbers, we see from the equations (α'), (β'), ... (η') that Y must be divisible by $3^4 \cdot 11 \cdot 4657$, i.e. we may put

$$Y = 3^4 \cdot 11 \cdot 4657n = 4149387n.$$

Therefore the equations (α'), (β'),...(η') give the following values for all the unknowns in terms of n, viz.

$$\left.\begin{array}{llll}
W = 2 \cdot 3 \cdot 7 \cdot 53 \cdot 4657n & = 10366482n \\
X = 2 \cdot 3^2 \cdot 89 \cdot 4657n & = 7460514n \\
Y = 3^4 \cdot 11 \cdot 4657n & = 4149387n \\
Z = 2^2 \cdot 5 \cdot 79 \cdot 4657n & = 7358060n \\
w = 2^3 \cdot 3 \cdot 5 \cdot 7 \cdot 23 \cdot 373n = & 7206360n \\
x = 2 \cdot 3^2 \cdot 17 \cdot 15991n & = 4893246n \\
y = 3^2 \cdot 13 \cdot 46489n & = 5439213n \\
z = 2^2 \cdot 3 \cdot 5 \cdot 7 \cdot 11 \cdot 761n = & 3515820n
\end{array}\right\} \quad \ldots\ldots\ldots(A).$$

If now $n = 1$, the numbers are the smallest which will satisfy the seven equations (α), (β),...(η); and we have next to find such an integral value for n that the equation (ι) will be satisfied also. [The modified equation (θ) requiring that $W + X$ must be a product of two factors is then simultaneously satisfied.]

Equation (ι) requires that

$$Y + Z = \frac{q(q+1)}{2},$$

where q is some positive integer.

Putting for Y, Z their values as above ascertained, we have

$$\frac{q(q+1)}{2} = (3^{4}.11 + 2^{2}.5.79).4657n$$

$$= 2471.4657n$$

$$= 7.353.4657n.$$

Now q is either even or odd, so that either $q = 2s$, or $q = 2s - 1$, and the equation becomes

$$s(2s \pm 1) = 7.353.4657n.$$

As n need not be a prime number, we suppose $n = u . v$, where u is the factor in n which divides s without a remainder and v the factor which divides $2s \pm 1$ without a remainder; we then have the following sixteen alternative pairs of simultaneous equations:

(1)	$s =$	u,	$2s \pm 1 = 7.353.4657v$,	
(2)	$s =$	$7u$,	$2s \pm 1 =$	$353.4657v$,
(3)	$s =$	$353u$,	$2s \pm 1 =$	$7.4657v$,
(4)	$s =$	$4657u$,	$2s \pm 1 =$	$7.353v$,
(5)	$s =$	$7.353u$,	$2s \pm 1 =$	$4657v$,
(6)	$s =$	$7.4657u$,	$2s \pm 1 =$	$353v$,
(7)	$s =$	$353.4657u$,	$2s \pm 1 =$	$7v$,
(8)	$s = 7.353.4657u$,		$2s \pm 1 =$	v.

In order to find the least value of n which satisfies all the conditions of the problem, we have to choose from the various positive integral solutions of these pairs of equations that particular one which gives the smallest value for the product uv or n.

If we solve the various pairs and compare the results, we find that it is the pair of equations

$$s = 7u, \quad 2s - 1 = 353.4657v,$$

which leads to the solution we want; this solution is then

$$u = 117423, \quad v = 1,$$

so that $\qquad n = uv = 117423 = 3^{3}.4349,$

whence it follows that

$$s = 7u = 821961,$$

and $\qquad q = 2s - 1 = 1643921.$

Thus $\qquad Y + Z = 2471.4657u$

$$= 2471.4657.117423$$

$$= 1351238949081$$

$$= \frac{1643921.1643922}{2},$$

which is a triangular number, as required.

The number in equation (θ) which has to be the product of two integers is now

$$W + X = 2.3.(7.53 + 3.89).4657u$$

$$= 2^2.3.11.29.4657u$$

$$= 2^2.3.11.29.4657.117423$$

$$= 2^2.3^4.11.29.4657.4349$$

$$= (2^2.3^4.4349).(11.29.4657)$$

$$= 1409076.1485583,$$

which is a rectangular number with nearly equal factors.

The solution is then as follows (substituting for n its value 117423):

$$W = 1217263415886$$
$$X = 876035935422$$
$$Y = 487233469701$$
$$Z = 864005479380$$
$$w = 846192410280$$
$$x = 574579625058$$
$$y = 638688708099$$
$$z = 412838131860$$
$$\text{and the sum} = 5916837175686$$

The complete problem.

In this case the seven original equations (α), (β),...(η) have to be satisfied, and the following further conditions must hold,

$$W + X = \text{a square number} = p^2, \text{ say,}$$

$$Y + Z = \text{a triangular number} = \frac{q(q+1)}{2}, \text{ say.}$$

Using the values found above (A), we have in the first place

$$p^2 = 2.3.(7.53 + 3.89).4657n$$

$$= 2^2.3.11.29.4657n,$$

and this equation will be satisfied if

$$n = 3.11.29.4657\xi^2 = 4456749\xi^2,$$

where ξ is any integer.

Thus the first 8 equations (α), (β),...(η), (θ) are satisfied by the following values:

$W = 2.3^2.7.11.29.53.4657^2.\xi^2$ $= 46200808287018.\xi^2$

$X = 2.3^3.11.29.89.4657^2.\xi^2$ $= 33249638308986.\xi^2$

$Y = 3^5.11^2.29.4657^2.\xi^2$ $= 18492776362863.\xi^2$

$Z = 2^2.3.5.11.29.79.4657^2.\xi^2$ $= 32793026546940.\xi^2$

$w = 2^3.3^2.5.7.11.23.29.373.4657.\xi^2 = 32116937723640.\xi^2$

$x = 2.3^3.11.17.29.15991.4657.\xi^2$ $= 21807969217254.\xi^2$

$y = 3^3.11.13.29.46489.4657.\xi^2$ $= 24241207098537.\xi^2$

$z = 2^2.3^2.5.7.11^2.29.761.4657.\xi^2$ $= 15669127269180.\xi^2$

It remains to determine ξ so that equation (ι) may be satisfied, i.e. so that

$$Y + Z = \frac{q(q+1)}{2}.$$

Substituting the ascertained values of Y, Z, we have

$$\frac{q(q+1)}{2} = 51285802909803.\xi^2$$

$$= 3.7.11.29.353.4657^2.\xi^2.$$

Multiply by 8, and put

$$2q + 1 = t, \qquad 2 . 4657 . \xi = u,$$

and we have the " Pellian " equation

$$t^2 - 1 = 2 . 3 . 7 . 11 . 29 . 353 . u^2,$$

that is, $t^2 - 4729494 \, u^2 = 1.$

Of the solutions of this equation the smallest has to be chosen for which u is divisible by $2 . 4657$.

When this is done,

$$\xi = \frac{u}{2 . 4657} \text{ and is a whole number;}$$

whence, by substitution of the value of ξ so found in the last system of equations, we should arrive at the solution of the complete problem.

It would require too much space to enter on the solution of the " Pellian " equation

$$t^2 - 4729494 \, u^2 = 1,$$

and the curious reader is referred to Amthor's paper itself. Suffice it to say that he develops $\sqrt{4729494}$ in the form of a continued fraction as far as the period which occurs after 91 convergents, and, after an arduous piece of work, arrives at the conclusion that

$$W = 1598 \, \langle 206541 \rangle,$$

where $\langle 206541 \rangle$ represents the fact that there are 206541 more digits to follow, and that, with the same notation,

the whole number of cattle $= 7766 \, \langle 206541 \rangle.$

One may well be excused for doubting whether Archimedes solved the complete problem, having regard to the enormous size of the numbers and the great difficulties inherent in the work. By way of giving an idea of the space which would be required for merely writing down the results when obtained, Amthor remarks that the large seven-figured logarithmic tables contain on one page 50 lines with 50 figures or so in each, say altogether 2500 figures; therefore *one* of the eight unknown quantities would, when found, occupy $82\frac{1}{2}$ such pages, and to write down all the eight numbers would require a volume of 660 pages!]

CAMBRIDGE: PRINTED BY J. AND C. F. CLAY, AT THE UNIVERSITY PRESS.